Applied Mathematical Sciences

EDITORS

Fritz John
Courant Institute of
Mathematical Sciences
New York University
New York, N. Y. 10003

Joseph P. LaSalle
Division of
Applied Mathematics
Brown University
Providence, R. I. 02912

Lawrence Sirovich
Division of
Applied Mathematics
Brown University
Providence, R. I. 02912

EDITORIAL STATEMENT

The mathematization of all sciences, the fading of traditional scientific boundaries, the impact of computer technology, the growing importance of mathematical-computer modelling and the necessity of scientific planning all create the need both in education and research for books that are introductory to and abreast of these developments.

The purpose of this series is to provide such books, suitable for the user of mathematics, the mathematician interested in applications, and the student scientist. In particular, this series will provide an outlet for material less formally presented and more anticipatory of needs than finished texts or monographs, yet of immediate interest because of the novelty of its treatment of an application or of mathematics being applied or lying close to applications.

The aim of the series is, through rapid publication in an attractive but inexpensive format, to make material of current interest widely accessible. This implies the absence of excessive generality and abstraction, and unrealistic idealization, but with quality of exposition as a goal.

Many of the books will originate out of and will stimulate the development of new undergraduate and graduate courses in the applications of mathematics. Some of the books will present introductions to new areas of research, new applications and act as signposts for new directions in the mathematical sciences. This series will often serve as an intermediate stage of the publication of material which, through exposure here, will be further developed and refined and appear later in the Mathematics in Science Series of books in applied mathematics also published by Springer-Verlag and in the same spirit as this series.

MANUSCRIPTS

The Editors welcome all inquiries regarding the submission of manuscripts for the series. Final preparation of all manuscripts will take place in the editorial offices of the series in the Division of Applied Mathematics, Brown University, Providence, Rhode Island.

Published by SPRINGER-VERLAG NEW YORK INC., 175 Fifth Avenue, New York, N.Y. 10010.

Printed in U.S.A.

Applied Mathematical Sciences | Volume 8

Applied Mathematical Sciences

F. John
PARTIAL DIFFERENTIAL EQUATIONS
ISBN 0-387-90021-7

L. Sirovich
TECHNIQUES OF ASYMPTOTIC ANALYSIS
ISBN 0-387-90022-5

J. Hale
FUNCTIONAL DIFFERENTIAL EQUATIONS
ISBN 0-387-90023-3

J. K. Percus
COMBINATORIAL METHODS
ISBN 0-387-90027-6

R. von Mises and K. O. Friedrichs
FLUID DYNAMICS
ISBN 0-387-90028-4

W. Freiberger and U. Grenander
A SHORT COURSE IN COMPUTATIONAL PROBABILITY AND STATISTICS
ISBN 0-387-90029-2

A. C. Pipkin
LECTURES ON VISCOELASTICITY THEORY
ISBN 0-387-90030-6

G. E. O. Giacaglia

Perturbation Methods in Non-Linear Systems

Springer-Verlag New York · Heidelberg · Berlin
1972

Georgio Eugenio Oscare Giacaglia
University of Sao Paulo
and
University of Texas at Austin

Library of Congress Card Catalog Number 72-87714.
Softcover reprint of the hardcover 1st Edition 1972

ISBN-13: 978-0-387-90054-4 e-ISBN-13: 978-1-4612-6400-2
DOI: 10.1007/978-1-4612-6400-2

PREFACE

This volume is intended to provide a comprehensive treatment of recent developments in methods of perturbation for nonlinear systems of ordinary differential equations. In this respect, it appears to be a unique work.

The main goal is to describe perturbation techniques, discuss their advantages and limitations and give some examples. The approach is founded on analytical and numerical methods of nonlinear mechanics.

Attention has been given to the extension of methods to high orders of approximation, required now by the increased accuracy of measurements in all fields of science and technology.

The main theorems relevant to each perturbation technique are outlined, but they only provide a foundation and are not the objective of these notes.

Each chapter concludes with a detailed survey of the pertinent literature, supplemental information and more examples to complement the text, when necessary, for better comprehension.

The references are intended to provide a guide for background information and for the reader who wishes to analyze any particular point in more detail. The main sources referenced are in the fields of differential equations, nonlinear oscillations and celestial mechanics.

Thanks are due to Katherine MacDougall and Sandra Spinacci for their patience and competence in typing these notes.

Partial support from the Mathematics Program of the Office of Naval Research is gratefully acknowledged.

April, 1972

G. E. O. Giacaglia
Austin, Texas

TABLE OF CONTENTS

Perturbation Methods in Non-Linear Systems

INTRODUCTION

In what follows we are going to describe in short the basic problem of perturbations the way it is to be developed in these notes. We shall here make free and simple statements without entering mathematical details on the functions involved. The necessary hypotheses will be made in the subsequent chapters. Historically we consider Lindstedt's (1882) problem of obtaining a series solution, free from secular and/or mixed secular terms, of the equation

$$\ddot{x} + \omega_0^2 x = \epsilon f(x, \dot{x}, t)$$

where $0 < \epsilon < 1$ is a parameter. The possibility of obtaining a solution

$$x = x_0(t) + \epsilon x_1(t) + \epsilon^2 x_2(t) + \dots$$
$$\dot{x} = \dot{x}_0(t) + \epsilon \dot{x}_1(t) + \epsilon^2 \dot{x}_2(t) + \dots$$

of the above equation, with $x_j(t)$, $\dot{x}_j(t)$ bounded functions for all $t \in R$ was found to depend essentially on the nature of f and its derivatives up to some order. The reference solution introduced by Lindstedt, that is, $(x_0(t), \dot{x}_0(t))$ was given by

$$x_0 = a \cos(\omega t + \sigma)$$
$$\dot{x}_0 = -a\omega \sin(\omega t + \sigma)$$

where ω is a priori unknown but, by assumption, developable in a power series

$$\omega = \omega_0 + \epsilon\omega_1 + \epsilon^2\omega_2 + \epsilon^3\omega_3 + \dots$$

where $\omega_1, \omega_2, \dots$ are constants depending on ω_0, a and f. Strictly speaking the very first attempt of dealing with perturbed oscillatory systems had been made by Euler (1772) in his researches on the motion of the Moon. Delaunay was the second in line to recognize that the major difficulty in the avoidance of unbounded terms in the series solution of such systems was the choice of a reference frequency, a fact which lead him to produce perhaps the first systematic series process of determining what are today called Floquet's characteristic exponents (Delaunay, 1860). The transformation of Delaunay's method of successive canonical transformation to a method utilizing a generating function was first foreseen by Tisserand (1868). After some time the work of Lindstedt was published (1882) and, right after, reduced to a systematic averaging procedure by Poincaré (1886) for Hamiltonian, but not necessarily conservative, systems. In essence, the whole second volume of his "Mécanique Céleste" is devoted to this method and related questions, among the most important, the problem of resonance, in the nonlinear sense. He accomplished a great deal of unification of all previous works including the milestone works of Bohlin and Gyldèn. In chronological order it is again in Celestial Mechanics that new efforts were made on the problem by von Zeipel (1911), by generalizing Poincaré's ideas. We shall not endeavor into details along these works and refer to several surveys on the subject (Cesari, 1959; Giacaglia, 1965; Kyner, 1967). It was only at least a decade later that similar problems and questions arose in nonlinear circuit theory leading to the averaging methods of Krylov and Bogoliubov (1942) made available to the western mathematicians by the efforts of Lefschetz. The work by Brown (1931) on nonlinear resonance came well after Poincaré's dealing with the problem and it is actually based on the examples he produced to illustrate Bohlin's method. Modern literature on perturbation methods and averaging procedures becomes highly dense after about 1950 and specific reference on these will be done along the work, at the proper moment. So far for purely analytic works which aimed the quantitive approach, typical of the classical analysis of last century and beginning of this, of obtaining an explicit time solution for a system of differential equations.

Along different lines, it was Poincaré (1912) who tried to understand, for the first time, the geometry of a differential system. His conjecture on the existence of fixed points for area preserving mapping, associated to the solution of a conservative system, was proved to be right by Birkhoff (1915) whose work is to be considered as one of the most deep changes ever introduced in the concept of solution of a differential system. He was certainly the forerunner of Topology and introduced important concepts like invariant sets, wandering points, etc., all related to the geometric behavior of the integral curves of a system. Along these lines the basic approach is probably best explained by Moser's celebrated work on the area preserving mapping of a circle into itself (1962), by Hale's work on integral manifolds of perturbed systems (1961) and by the work of Krylov and Bogoliubov (1934). Again we shall refer more specifically to the current literature when dealing with perturbations of invariant sets.

The classical and perhaps the oldest methods of perturbations are of the Euler-Lagrange type, generalized by Poisson. Their conservative analogues are condensed in Jacobi's Theorem on the variation of canonical variables. Since Poisson's method is the most general, it is worth mentioning here, but it will be done in an heuristic manner. We consider a differential system

$$\dot{x} = f(x,t) \tag{1}$$

where x, f are n-vectors. For simplicity we assume f to be analytic in a certain domain D of the vector space x and for $t \in R$. Let, in D,

$$\sigma = \sigma(x,t) \tag{2}$$

be a first uniform integral of (1). It follows that, along any solution of (1) in D, we have

$$\dot{\sigma} = \frac{\partial\sigma}{\partial x}\dot{x} + \frac{\partial\sigma}{\partial t} = 0$$

where σ is an m-vector $(m \leq n)$, so that $\partial\sigma/\partial x$ is a rectangular Jacobian matrix $(m \times n)$. We have, for every $x \in D$, the identity

$$\frac{\partial\sigma}{\partial x} f(x,t) + \frac{\partial\sigma}{\partial t} = 0. \tag{3}$$

Consider now the perturbed system

$$\dot{x} = f(x,t) + g(x,t) \tag{4}$$

where, again, $g(x,t)$ is supposed analytic in $D \times R$. We consider the variation of (2) along system (4), that is,

$$\dot{\sigma} = \frac{\partial\sigma}{\partial x}[\, f(x,t) + g(x,t)] + \frac{\partial\sigma}{\partial t}$$

or, in view of (3),

$$\dot{\sigma} = \frac{\partial\sigma}{\partial x} g(x,t). \tag{5}$$

Equation (5) is generally credited to Poisson (1956) and contains as particular examples Lagrange's Equations for the variation of arbitrary constants and Jacobi's theorem. In the particular case of a dynamical system

$$\ddot{x} = f(x,\dot{x},t) + g(x,\dot{x},t)$$

Poisson's equation becomes

$$\dot{\sigma} = \frac{\partial\sigma}{\partial\dot{x}} g(x,\dot{x},t) \tag{6}$$

where σ is an integral for $g \equiv 0$. Interestingly enough, all basic theorems of Classical Mechanics are immediately derivable from (6). In fact, if σ is the Energy Integral

$$E = \frac{1}{2}\dot{x}^2 + V(x,t)$$

it follows that, $E_{\dot{x}} = \dot{x}$ and

$$\dot{E} = \dot{x}^T g(x,\dot{x},t)$$

which is the basic law of energy and work. If σ is the Angular Momentum Integral

$$L = x \times \dot{x}$$

it follows that $L_{\dot{x}} = x \times$ and

$$\dot{L} = x \times g(x,\dot{x},t)$$

which is the basic law of angular momentum and torque.

If (1) is a Hamiltonian system (x is a $2n$-vector), that is,

$$\dot{x} = MH_x^T \tag{7}$$

where $H = H(x,t)$ and M is the canonical matrix $2n \times 2n$,

$$M = \begin{pmatrix} 0_n & I_n \\ -I_n & 0_n \end{pmatrix}$$

and we let

$$H = H_o + H_1,$$

if σ is a first integral of (7), in involution with H_o, for $H = H_o$, it follows that

$$\dot{\sigma} = \frac{\partial\sigma}{\partial x} M\left(\frac{\partial H_1}{\partial x}\right)^T . \tag{8}$$

If, furthermore, the Jacobian matrix $J = \frac{\partial\sigma}{\partial x}$ is symplectic (that is, the transformation $x \to \sigma$ is canonical) it follows that

$$\dot{\sigma} = \frac{\partial\sigma}{\partial x} M\left(\frac{\partial\sigma}{\partial x}\right)^T \left(\frac{\partial H_1}{\partial\sigma}\right)^T = M\left(\frac{\partial H_1}{\partial\sigma}\right)^T \tag{9}$$

which is Jacobi's theorem. We observe that if σ is a $2n$-vector, (8) are Lagrange's equations for the variation of arbitrary constants in case of conservative forces.

The classical approach to (9) is to assume for σ a power series in a small parameter and reduce the problem to a method of successive approximations. In most cases this procedure leads to secular and mixed secular terms and therefore the series cannot converge for all time. If we limit the time, convergence can eventually be obtained and the earliest reference to this question is probably the work by MacMillan (1910). We refer to this work since it is simple yet quite rigorous.

In the more sophisticated methods of averaging it is generally assumed (Hamiltonian system) that the Hamiltonian function is 2π periodic in every angular variable $y_1, y_2, \ldots, y_n$ and representable in a convergent Fourier series

$$H = \sum_j A_j(x) \exp(j \cdot y) \tag{10}$$

where $j = (j_1, j_2, \ldots, j_n)$ is an "integer" vector.

The equations generated by (10) are

$$\dot{x} = -\left(\frac{\partial H}{\partial y}\right)^T$$

$$\dot{y} = +\left(\frac{\partial H}{\partial x}\right)^T \qquad (11)$$

If we consider only the part of H corresponding to $j = 0$,

$$H_o = A_o(x)$$

system (11) is obviously integrable and

$$x = x_o$$

$$y = \omega(x_o)t + y_o \qquad (12)$$

where

$$\omega_j(x_o) = \partial H_o/\partial x_j\big|_{x=x_o}$$

If in a certain region, the $A_j(x)$ for $j \neq 0$, are such that their derivatives are small (in some sense) with respect to the $\omega_j(x)$, then we can treat $H\text{-}H_o$ as a perturbation. Classically it was assumed that if this situation occurs, than the solution of (11) never departs too much from the solution (12). Such supposition is evidently false and seldom verified, even considering "orbital proximity" regardless of the time. It is actually the "time proximity" of corresponding points which is the most affected by the perturbation, and such phenomenon is well known as the "in-track error". The analogy with the concept of stability is that it is easier to have orbital than Lyapunov stability.

In any event, using (12) as a reference solution with modified frequency vector $\nu(x_o)$ and iterating, we obtain formal series

$$x = x_o + \sum_j \frac{C_j(x_o)}{j\cdot\nu(x_o)} \exp[i(j\cdot\nu)t]$$

$$y = \nu(x_o)t + y_o + \sum_j \frac{D_j(x_o)}{j\cdot\nu(x_o)} \exp[i(j\cdot\nu)t] \qquad (13)$$

where $\nu = \omega_o + \epsilon\omega_1 + \epsilon^2\omega_2 + \dots$. It is evident that the products $j\cdot\nu =$

$j_1\nu_1 + j_2\nu_2 + \ldots + j_n\nu_n$ present in the denominators may become arbitrarily small for $j_1, j_2, \ldots, j_n$ covering the all set of integers. In this form, Poincaré concluded that such series were therefore divergent for a set of frequency everywhere dense, which is in fact the case. Nevertheless, as Kolmogorov (1954) suggested, there exists a set of frequencies, of non-zero measure (density as close to one as ϵ is close to zero), where the series converge. This is basically due to the fact that it is possible, for all integers $j_1, j_2, \ldots, j_n$ to set a lower bound on the numbers $j_1\nu_1 + \ldots + j_n\nu_n$, as shown by the diophantive approximation. The way one can arrive at the series will be shown, in a pure formal fashion, in Chapter II, while the subsequent chapters will be devoted to the problem of convergence of the methods introduced. Chapter I is devoted to the introduction of a basic background and terminology to be used throughout these notes. The last chapter will be devoted to the question of nonlinear resonance.

CHAPTER I

CANONICAL TRANSFORMATION THEORY AND GENERALIZATIONS

1. Introduction.

In this chapter we deal with the terminology and basic well known results, which are necessary to the development of the subsequent chapters. It is not the scope of this chapter to describe Hamiltonian Systems and their general properties. They are found in several books and monographs, among which we wish to mention the classics of Birkhoff (1927), Siegel (1956), Wintner (1947), Abraham (1966), Moser (1968). We avoid any and every sophistication in arriving at intrinsic representations and definitions of Hamiltonian systems on manifolds, not because they are not important, but because they are of no essential necessity in what has to follow.

Initially, we remember the definitions of Lagrange's and Poisson's matrices. They arise naturally from the method of variation of arbitrary constants. We consider the transformation $(y,x) \to (\eta,\xi)$ to be C^2 and invertible in some domain of a 2n-dimensional space. The vectors y, x are n-dimensional as well as the vectors η, ξ. Also, let $z = \mathrm{col}(y,x)$ and $\zeta = \mathrm{col}(\eta,\xi)$ be 2n-dimensional vectors. The Lagrange Matrix $\mathcal{L}(\zeta)$ is defined as

$$\mathcal{L}(\zeta) = J^T M J \tag{1.1.1}$$

where M is the $2n \times 2n$ canonical matrix

$$M = \begin{pmatrix} 0 & I \\ -I & 0 \end{pmatrix}$$

and J the Jacobian matrix of the transformation $z \to \zeta$, that is

$$J = \frac{\partial z}{\partial \zeta} . \tag{1.1.2}$$

It is easily verified that

$$\mathcal{L}(\zeta) = \left(\frac{\partial y}{\partial \zeta}\right)^T \left(\frac{\partial x}{\partial \zeta}\right) - \left(\frac{\partial x}{\partial \zeta}\right)^T \left(\frac{\partial y}{\partial \zeta}\right) \tag{1.1.3}$$

and, therefore,

$$\mathcal{L}_{ij}(\zeta) = [\zeta_i, \zeta_j] = \sum_{k=1}^{n} \left(\frac{\partial y_k}{\partial \zeta_i} \frac{\partial x_k}{\partial \zeta_j} - \frac{\partial x_k}{\partial \zeta_i} \frac{\partial y_k}{\partial \zeta_j} \right) \quad . \tag{1.1.4}$$

The following properties are obvious

$$\mathcal{L}^T = J^T M^T J = -J^T M J = -\mathcal{L},$$

$$|\mathcal{L}| = |J|^2,$$

where $|A| \overset{\Delta}{=} \det A$, for any square matrix A.

The Poisson matrix $P(z)$ is defined by

$$P(z) = JMJ^T \tag{1.1.5}$$

and one verifies that

$$P(z) = \left(\frac{\partial z}{\partial \eta}\right)\left(\frac{\partial z}{\partial \xi}\right)^T - \left(\frac{\partial z}{\partial \xi}\right)^T\left(\frac{\partial z}{\partial \eta}\right) \tag{1.1.6}$$

so that

$$P_{ij}(z) = (z_i, z_j) = \sum_{k=1}^{n} \left(\frac{\partial z_i}{\partial \eta_k} \frac{\partial z_j}{\partial \xi_k} - \frac{\partial z_i}{\partial \xi_k} \frac{\partial z_j}{\partial \eta_k} \right) \quad . \tag{1.1.7}$$

Also

$$P^T = -P,$$

$$|P| = |J^{-1}|^2 = 1/|J|^2$$

$$\mathcal{L}^{-1}(\zeta) = J^{-1}M^{-1}(J^T)^{-1} = J^{-1}M(J^{-1})^T = -P(\zeta).$$

The expressions (1.1.4) and (1.1.7) are called Lagrange's Brackets and Poisson's Parentheses, respectively.

If one considers the system of n second order ordinary differential equations

$$\ddot{y} = f(y, \dot{y}, t) \tag{1.1.8}$$

and a solution

$$y = y^{o}(t;\alpha,\beta)$$

$$\dot{y} = x^{o}(t;\alpha,\beta) = \frac{\partial y^{o}}{\partial t} \tag{1.1.9}$$

corresponding to the initial conditions

$$y^{o}(0;\alpha,\beta) = y_{o}$$

$$x^{o}(0;\alpha,\beta) = \dot{y}_{o} \; , \tag{1.1.10}$$

one verifies

$$\frac{\partial x^{o}}{\partial t} = f(y,\dot{y},t).$$

For a perturbed system (1.1.8) one has

$$\ddot{y} = f(y,\dot{y},t) + g(y,\dot{y},t) \tag{1.1.11}$$

and assumes the solution to be of the same form as (1.1.9), where, of course, α, β are now in general variable. It follows that

$$\frac{dy}{dt} = \frac{\partial y^{o}}{\partial t} + \frac{\partial y^{o}}{\partial \alpha}\dot{\alpha} + \frac{\partial y^{o}}{\partial \beta}\dot{\beta} = x^{o}(t,\alpha,\beta)$$

and, therefore,

$$\frac{\partial y^{o}}{\partial \alpha}\dot{\alpha} + \frac{\partial y^{o}}{\partial \beta}\dot{\beta} = 0 \tag{1.1.12}$$

where α, β are, evidently, n-vectors. Moreover

$$\frac{d\dot{y}}{dt} = \frac{\partial x^{o}}{\partial t} + \frac{\partial x^{o}}{\partial \alpha}\dot{\alpha} + \frac{\partial x^{o}}{\partial \beta}\dot{\beta} = f(y,\dot{y},t) + g(y,\dot{y},t)$$

and, therefore,

$$\frac{\partial x^{o}}{\partial \alpha}\dot{\alpha} + \frac{\partial x^{o}}{\partial \beta}\dot{\beta} = g(y^{o}(t;\alpha,\beta),\; x^{o}(t;\alpha,\beta),t). \tag{1.1.13}$$

The system of 2n first order ordinary differential equations (1.1.12) and (1.1.13) are Lagrange's equations for the variation of arbitrary constants. They can be written in terms of a unique system using, for example, Lagrange's matrix $\mathcal{L}(\gamma)$ where γ = column (α,β). The result is

$$\mathcal{L}(\gamma)\dot{\gamma} = (\frac{\partial x^o}{\partial \gamma})^T \; g(y^o(t;\gamma),\; x^o(t;\gamma),t). \tag{1.1.14}$$

Evidently, equation (1.1.14) defines γ under the standard condition

$$|\mathcal{L}(\gamma)| \neq 0$$

that is,

$$|\frac{\partial(y^o,x^o)}{\partial(\alpha,\beta)}| \neq 0$$

which is met by the fact that we assumed (y^o,x^o) to be the general solution of (1.1.8) under arbitrary initial conditions (y_o,x_o) or constants of integration (α,β). Moreover, we require that

$$P(\gamma)(\frac{\partial y^o}{\partial \gamma})^T g(y^o,x^o,t)$$

is Lipschitzian in some domain of the γ-space. Strictly speaking all of the above statements have a local character, but it is important, as far as applications are concerned, that they extend to some domain of the variables. Also, the functions we are dealing with are assumed to be continuously differentiable in t, generally for any real t.

Lagrange's and Poisson's matrices satisfy an ordinary differential equation with some remarkable properties. In fact, consider the system of $2n$ differential equations

$$\dot{z} = \phi(z;t)$$

and a solution $z(\gamma;t) \in C^2$ in the $2n$ integration constants γ, and t, in some domain of the γ space and for all $|t| < T$. Let $J = \partial z/\partial\gamma$ be the non-singular Jacobian matrix of the transformation $\gamma \rightarrow z$, which, by hypothesis, is C^2. Thus

$$\dot{J} = \frac{d}{dt}\frac{\partial z}{\partial \gamma} = \frac{\partial}{\partial t}\frac{\partial z}{\partial \gamma}(\gamma;t) = \frac{\partial}{\partial \gamma}\dot{z}(\gamma;t)$$

$$= \frac{\partial}{\partial \gamma}\phi(z(\gamma;t);t) = \frac{\partial \phi}{\partial z} J$$

or

$$\dot{J} = GJ \tag{1.1.15}$$

where $G = \partial\phi/\partial z$ is a $2n \times 2n$ non-singular matrix. Let us now consider

$$\mathcal{L}(\gamma;t) = J^T M J$$

so that, making use of (1.1.15), one finds

$$\dot{\mathcal{L}} = J^T(G^T M + MG)J. \tag{1.1.16}$$

<u>Lemma</u>. <u>The Lagrange matrix $\mathcal{L}(\gamma;t)$ of the transformation $\gamma \to z$ is constant if, and only if, the matrix MG is symmetric.</u>

In fact, suppose MG is symmetric, that is

$$MG = (MG)^T = -G^T M.$$

Then $G^T M + MG = 0$ and $\dot{\mathcal{L}} = 0$. Reciprocally let $\dot{\mathcal{L}} = 0$. Under the foregoing hypotheses, it follows that

$$G^T M + MG = 0$$

or

$$G^T M = -MG = M^T G = (G^T M)^T$$

which completes the proof. From (1.1.16) and the above Lemma it follows that <u>the flow of a Hamiltonian system is conservative</u>. (<u>Liouville's Theorem</u>). In fact, in this case, if $H = H(z)$ is the Hamiltonian, one has

$$\dot{z} = MH_z^T$$

so that

$$G = \frac{\partial}{\partial z}(MH_z^T) = MH_{zz}$$

and

$$MG = -H_{zz}$$

is therefore symmetric. It follows that $\dot{\mathcal{L}} = 0$ or

$$\frac{d}{dt}(J^T M J) = 0$$

or $J^T M J = \text{constant}$. If γ is the vector of initial conditions z_o, $J_o = I$ (the identity matrix), and therefore

$$J^T M J = M \tag{1.1.17}$$

and also, in particular,

$$|J| = \text{const.} = 1$$

which proves the theorem. (The case $|J| = -1$ is discarded for reasons of continuity.) If the 2n-vector z is composed by the n-vectors y and x (coordinates and momenta), one can, more precisely, write

$$J = \begin{pmatrix} \frac{\partial y}{\partial y_o} & \frac{\partial y}{\partial x_o} \\ \frac{\partial x}{\partial y_o} & \frac{\partial x}{\partial x_o} \end{pmatrix}$$

and at $t = 0$,

$$J_o = \begin{pmatrix} I_n & 0 \\ 0 & I_n \end{pmatrix} = I_{2n} .$$

It follows that the mapping $z_o \to z$ can be represented by

$$\begin{aligned} y &= y_o + \tilde{Y}(x_o, y_o; t) \\ x &= x_o + \tilde{X}(x_o, y_o; t) \end{aligned} \tag{1.1.18}$$

where $Y(x_o, y_o; 0) = X(x_o, y_o; 0) = 0$, so that, for t sufficiently small,

$$\tilde{Y}(x_o, y_o; t) = t\, Y(x_o, y_o; t)$$

and

$$\tilde{X}(x_o,y_o;t) = t\,X(x_o,y_o;t).$$

The situation can also be viewed from another point. Since at $t = 0$ the mapping $z_o \to z$ is the identity, there exists a generating function

$$S = x_o \cdot y + tW(x_o;y;t) \tag{1.1.19}$$

such that

$$x = S_y^T = x_o + t\,W_y^T$$

and

$$y_o = S_{x_o}^T = y + t\,W_{x_o}^T \tag{1.1.20}$$

which should be equivalent to (1.1.18).

2. Canonical Transformations.

A transformation $z \to \zeta$, non-singular and C^2 is canonical if it transforms every Hamiltonian system $\dot{z} = MH_z^T$ into a Hamiltonian system $\dot{\zeta} = MK_\zeta^T$. The property is purely local, but, again, the usefulness of such definition and what follows relies on the possible global extension into some domain of the phase space. We consider $z = \text{col}(y;x)$, $\zeta = \text{col}(\eta;\xi)$ to be $2n$-dimensional vectors. The invariance of the Hamiltonian Form implies that the transformation is canonical if and only if the form

$$\Phi(H) = \sum_{k=1}^{n} (\dot{\eta}_k \delta\xi_k - \dot{\xi}_k \delta\eta_k) \tag{1.2.1}$$

is an exact differential, for all H.

From (1.2.1) we shall derive the necessary and sufficient condition for the transformation to be canonical (Breves, 1972). We observe that (1.2.1) can be written

$$\Phi(H) = \dot{\zeta}^T M\delta\zeta. \tag{1.2.2}$$

Moreover, given the transformation

$$\zeta = \zeta(z;t) \tag{1.2.3}$$

we have

$$\dot{\zeta} = J\dot{z} + \zeta_t \tag{1.2.4}$$

where J is the Jacobian matrix

$$J = \frac{\partial\zeta}{\partial z} .$$

It follows that

$$\dot{\zeta} = J\ M\ H_z^T + \zeta_t$$

and, from (1.2.2),

$$\Phi(H) = (-H_z M J^T + \zeta_t^T) M\ \delta\zeta,$$

or, with $\delta\zeta = J\delta z$,

$$\Phi(H) = -H_z M\ \mathcal{L}(z)\,\delta z + \zeta_t^T MJ\ \delta z$$

or

$$\Phi(H) = -H_z M\ \mathcal{L}(z)\,\delta z + \mathcal{L}^*(t,z)\,\delta z \tag{1.2.5}$$

where

$$\mathcal{L}(z) = \left(\frac{\partial\zeta}{\partial z}\right)^T M \left(\frac{\partial\zeta}{\partial z}\right)$$

and

$$\mathcal{L}^*(t,z) = \left(\frac{\partial\zeta}{\partial t}\right)^T M \left(\frac{\partial\zeta}{\partial z}\right).$$

The quantity $\mathcal{L}^*(t,z)$ is, evidently, a row vector, whose elements are the Lagrange brackets $[t, z_k]$.

The conditions of integrability of $\Phi(H)$, for all H, can be translated into conditions of integrability for

$$\Phi(0) = \mathcal{L}^*(t,z)\,\delta z = \sum_k [t, z_k]\,\delta z_k,$$

$$\Phi(y_k) = -\sum_{\ell} [x_k, z_\ell]\delta z_\ell + \Phi(0) ,$$

$$\Phi(x_k) = \sum_{\ell} [y_k, z_\ell]\delta z_\ell + \Phi(0)$$

$$\Phi(y_k x_j) = \sum_{\ell} \{[y_j, z_\ell]y_k - [x_k, z_\ell]x_j\}\delta z_\ell + \Phi(0) .$$

It follows that

$$\frac{\partial}{\partial z_j}[t, z_k] = \frac{\partial}{\partial z_k}[t, z_j] ,$$

$$\frac{\partial}{\partial z_j}[x_k, z_\ell] = \frac{\partial}{\partial z_\ell}[x_k, z_j] ,$$

$$\frac{\partial}{\partial z_j}[y_k, z_\ell] = \frac{\partial}{\partial z_\ell}[y_k, z_j] ,$$

and

$$[y_j, z_\ell] = 0 \quad \text{for} \quad z_\ell \neq x_j ,$$

$$[x_k, z_\ell] = 0 \quad \text{for} \quad z_\ell \neq y_k ,$$

and (1.2.6)

$$[y_k, x_k] = -[x_\ell, y_\ell] = \text{const.} = \lambda.$$

The last relation is obtained in view of the first three from where we conclude, using Jacobi's identity, that

$$\frac{\partial}{\partial t}[z_k, z_j] = 0 , \quad \text{and}$$

$$\frac{\partial}{\partial z_\ell}[z_k, z_j] = 0 .$$

In matrix notation, conditions (1.2.6) can be written as

$$\mathcal{L}(z) = J^T M J = \lambda M \tag{1.2.7}$$

and since, by hypothesis, $|J| \neq 0$, the constant λ cannot be zero. Equation (1.2.7) is the necessary and sufficient condition for a transformation to be canonical. On the other hand, since $P(z) = -\mathcal{L}(z)$, such condition can also be expressed in terms of Poisson's Matrix

$$P(z) = JMJ^T = \lambda M. \tag{1.2.8}$$

That the condition is sufficient follows immediately from the substitution of (1.2.7) into (1.2.5) which gives

$$\Phi(H) = \lambda H_z \delta z + \mathcal{L}^*(t,z)\,\delta z = \delta(\lambda H + W) \tag{1.2.9}$$

where $W(z;t)$ is a function such that

$$W_z \delta z = \mathcal{L}^*(t,z)\,\delta z = \Phi(0), \tag{1.2.10}$$

an exact differential form. Under the circumstance, one can easily conclude the following result.

Theorem (Jacobi-Poincaré). "A necessary and sufficient condition that a transformation C^2 and non-singular $z \to \zeta$ be canonical and the new Hamiltonian be

$$K = \lambda H + W \tag{1.2.11}$$

is that the form

$$\psi = \lambda x^T dy - \xi^T d\eta + W dt \tag{1.2.12}$$

be an exact differential."

In fact,

$$\psi = (\lambda x^T - \xi^T \frac{\partial \eta}{\partial y})\,dy - \xi^T \frac{\partial \eta}{\partial x} dx + (W - \xi^T \frac{\partial \eta}{\partial t})\,dt$$

and the integrability conditions for ψ are

$$\frac{\partial}{\partial x}(\lambda x^T - \xi^T \frac{\partial \eta}{\partial y}) = \frac{\partial}{\partial y}(-\xi^T \frac{\partial \eta}{\partial x})$$

$$\frac{\partial}{\partial t}(\lambda x^T - \xi^T \frac{\partial \eta}{\partial y}) = \frac{\partial}{\partial y}(W - \xi^T \frac{\partial \eta}{\partial t})$$

$$\frac{\partial}{\partial t}(-\xi^T \frac{\partial \eta}{\partial x}) = \frac{\partial}{\partial x}(W - \xi^T \frac{\partial \eta}{\partial t})$$

or, in component form,

$$[z_k, z_\ell] = 0 \qquad (z_k \neq x_\ell,\ z_\ell \neq x_k),$$

$$[y_k, x_k] = \lambda,$$

$$[t, z_k] = \frac{\partial W}{\partial z_k},$$

which completes the proof. We finally arrive at the Jacobi-Poincaré relation. From (1.2.12),

$$\psi = \lambda x^T dy - \xi^T d\eta + (K - \lambda H)\,dt$$

and, therefore, "the necessary and sufficient condition for a transformation to be canonical can be expressed by the fact that ψ has to be an exact differential, that is,

$$\lambda x^T dy - \xi^T d\eta + (K - \lambda H)\,dt = dF \tag{1.2.13}$$

when expressed in terms of the variables η, ξ."

The set of all matrices A satisfying the condition

$$A^T M A = M$$

constitutes a group (with respect to matrix multiplication), which is called the Symplectic Group. The case $\lambda \neq 1$ is generally excluded from the definition. Canonical (and therefore, Symplectic) transformations with $\lambda \neq 1$ are also usually excluded since they are the product of a canonical transformation $(\lambda = 1)$ and the trivial canonical transformation $(\lambda \neq 1)$ given by

$$\xi = -\lambda x$$

$$\eta = y$$

for, in this case,

$$J_o = \frac{\partial\,(\eta, \xi)}{\partial\,(y, x)} = \begin{pmatrix} I & 0 \\ 0 & -\lambda I \end{pmatrix}$$

and it is easily seen that

$$J_o^T M J_o = \lambda M,$$

as discussed by Siegel (1956).

<u>Excluded such case, the necessary and sufficient condition for a canonical transformation is</u>

$$\mathcal{L}(z) = J^T M J = M,$$

or (1.2.14)

$$P(z) = J M J^T = M,$$

where

$$J = \frac{\partial \zeta(z;t)}{\partial z} \quad .$$

The Jacobi-Poincaré condition is reduced to

$$x^T dy - \xi^T d\eta + (K-H)dt = dF \tag{1.2.15}$$

and if the transformation does not depend explicitly of t is called <u>completely canonical</u> and if $dF = 0$, <u>homogeneous</u>.

From the results obtained in Section 1, we also conclude that <u>the transformation defined by the solution of a Hamiltonian system, mapping the phase space into itself, is canonical.</u> The volume preserving property was already established. In more precise form:

<u>"Let $\dot{z} = MH_z^T$ be a Hamiltonian system of differential equations and let there exist a unique solution $z = z(\zeta,t)$ going through the point $z = \zeta$ at $t = t_o$, and assume $z(\zeta,t)$ to be C^2 with respect to the $2n+1$ variables $(z;t)$ in a neighborhood of $z = \zeta$ and for $|t-t_o|$ sufficiently small. Then the mapping $\zeta \to z$ defined by $z = z(\zeta,t)$ is volume preserving and canonical."</u>

3. Hamilton - Jacobi Equation. Generalizations.

Consider the non-singular C^2 transformation

$$y = y(\eta;\xi;t)$$

(1.3.1)

$$x = x(\eta;\xi;t)$$

and suppose the particular situation

$$\left|\frac{\partial y}{\partial \eta}\right| \neq 0, \quad \|\eta-\eta_0\| < \delta, \tag{1.3.2}$$

so that, locally, one can solve the first system for η,

$$\eta = \eta(y;\xi;t)$$

and, therefore

$$x = x(y;\xi;t).$$

<u>If there exists a function $S(y;\xi;t)$ such that</u>

$$\left|\frac{\partial^2 S}{\partial y \partial \xi}\right| \neq 0,$$

<u>and S is C^2, the transformation defined by</u>

$$x = S_y^{\,T}$$

$$\eta = S_\xi^{\,T}$$

<u>is canonical, and the new Hamiltonian is given by</u>

$$K(\eta;\xi;t) = H(y(\eta;\xi;t);\ x(\eta;\xi;t);t)$$

$$+ \frac{\partial S}{\partial t}(y(\eta;\xi;t);\xi;t).$$

In fact, let us write, in (1.2.15),

$$\xi^T d\eta = d(\xi^T\eta) - \eta^T d\xi$$

and we have

$$x^T dy + \eta^T d\xi + (K-H)dt = d(F-\xi^T\eta). \tag{1.3.3}$$

If we let

$$S = F - \xi^T\eta = S(y;\xi;t)$$

then

$$dS = \frac{\partial S}{\partial y} dy + \frac{\partial S}{\partial \xi} d\xi + \frac{\partial S}{\partial t} dt,$$

and from (1.3.3),

$$\begin{aligned} x^T &= \frac{\partial S}{\partial y} = x^T(y;\xi;t) \quad , \\ \eta^T &= \frac{\partial S}{\partial \xi} = \eta^T(y;\xi;t) \quad , \\ K &= H + S_t \quad . \end{aligned} \tag{1.3.4}$$

For the transformation to be written in explicit form we require that

$$\left|\frac{\partial^2 S}{\partial y \partial \xi}\right| \neq 0 ,$$

in which case one obtains

$$\xi = \xi(y;x;t)$$

and therefore

$$\eta = \eta(y;x;t),$$

with the evident condition that $|\partial\eta/\partial y| \neq 0$. Since S is supposed to be C^2, this implies $|\partial y/\partial\eta| \neq 0$ and therefore, through (1.3.2), the recovery of (1.3.1).

The important result, to our purposes, is the last of equations (1.3.4), which we write explicitly,

$$\begin{aligned} &K(\eta(y;\xi;t);\xi;t) \\ &\quad = H(y;x(y;\xi;t);t) + \frac{\partial S}{\partial t}(y;\xi;t). \end{aligned} \tag{1.3.5}$$

If the transformation is time independent, that is, $S_t = 0$, the new Hamiltonian is simply the image of the old one through the mapping $z \to \zeta$.

The basic problem of Hamilton-Jacobi is whether there exists a transformation, generated by S, and such that the new Hamiltonian reduces to an absolute constant, or, which is equivalent, to a function identically zero. In other words, we seek the solution of the partial differential equation

$$H(y;S_y;t) + S_t = 0 \tag{1.3.6}$$

with $S = S(y;\xi;t)$. As is well known, Jacobi has shown that a general solution is not needed but only a complete solution, in the sense of a function $S(y;\xi;t)$ depending on n arbitrary constants ξ and such that $\|\partial S/\partial\xi\| \neq 0$. In such case the new variables $(\eta;\xi)$ are constants and the relations

$$\eta = \eta(y;x;t) \quad ,$$

$$\xi = \xi(y;x;t) \quad ,$$

which are obtained from (1.3.4) are $2n$ integrals of motion. Obviously, if the original Hamiltonian system is integrable in the sense of existence and uniqueness of solution of the equations

$$z = MH_z^T,$$

a generating function $S(y;\xi;t)$ must exist (which might not be expressible in terms of elementary functions). In fact, since the solution defines a canonical mapping $z = z(\zeta,t)$ where ζ is the vector of initial conditions, and since for $t = t_o$, $\partial y/\partial\eta = I$ (the identity matrix), then for $|t-t_o|$ sufficiently small $|\partial y/\partial\eta| \neq 0$, and therefore

$$S = \xi^T y + (t-t_o)F(y;\xi;t) \tag{1.3.7}$$

for $t-t_o$ sufficiently small, in agreement with (1.1.19).

The problem of Hamilton-Jacobi can be generalized by relaxing the condition that the new Hamiltonian be an absolute constant. As far as canonical perturbation methods are concerned the following generalized problem is of great relevance.

We ask if there exists a canonical transformation generated by $S(y;\xi;t)$, such that the new Hamiltonian has fewer degrees of freedom than the old one. One of the ways to translate this, is to produce a Hamiltonian

$$K(\eta;\xi;t) = H(y;x;t) + S_t(y;\xi;t)$$

such that

$$\frac{\partial K}{\partial \eta_k} = 0 \tag{1.3.8}$$

for $k = 1,2,\ldots,p \leq n$. The resulting system is obviously reduced to quadratures in the cases $p = n$ or $p = n-1$. This is the least one requires from the transformation, but still it is a much weaker requirement than that proposed by Jacobi. One may also require that the new Hamiltonian does not depend on time explicitly. This process of elimination is generally called an averaging method (Burstein and Solovev, 1961) and is usually applied when H is a periodic function of t. One can also easily generalize the concept for the case of almost period functions of t. If H depends on a small parameter say ϵ, and admits a Taylor series about $\epsilon = 0$, it can be shown that there is a formal series in ϵ which solves S up to any desired power. The convergence properties of such series are not known in general. The problem of existence of such series and its convergence is strictly related to the theory of periodic surfaces (Diliberto, 1961; Diliberto et. al., 1961) and to the theory of Moser (1962) on invariant curves of area preserving mapping. This last subject will be dealt with in some detail in Chapter IV. A qualitative description of these problems are described by Kyner (1964) in relation to the motion of a satellite in the oblate field of a planet. We shall not deal with Diliberto's theory. Such approach is indeed relevant to the subject, but it is dealt in details elsewhere (e.g. Diliberto, 1961; Hale, 1961).

A new approach to canonical transformations can be viewed by introducing a theory formulated by Lie (1888). Lie Series in problems of dynamics have been used in several occasions and a good reference to the subject, as a general background, is the work by Leimanis (1965). Quite recently they have been introduced as a mean to perturbation methods in non-linear Hamiltonian systems and also have been extended to systems of ordinary differential equations with few restrictions and no requirement for such systems to have a Hamiltonian form. Such applications will be discussed in Chapters II and V. Here, we wish to describe whatever is necessary for the understanding of such applications. The motivation for such series is the simple fact that given a system depending on a parameter, one usually knows the solution when that parameter is set equal to zero. A series solution is then constructed as a power series of the parameter, or, in conservative systems, it can be generated by a canonical transformation which, again, is given by power series on the parameter. Generally speaking, little is known about the convergence of such series, but in many applications they have proved invaluable. Such applicability has been actually checked against precise numerical integrations or observations of the system. At this moment, it is perhaps appropriate to repeat some of the words of Professor Siegel (1941), about the normalization of Hamiltonian functions. "On account of the small divisors appearing in the coefficients of the transformation, it seemed to be probable that the series would diverge in general, but no single example had hitherto been found. From Poincaré's well known theorem on the analytic integrals of canonical differential equations we can only infer that those series do not uniformly converge... whereas this theorem cannot be applied to a fixed function H." Later, about a specific problem he says "In particular, it would be interesting to decide, whether H is regular or singular (i.e., reducible or not to normal form by convergent series) in the special case ... But this seems to be beyond the power of the known methods of analysis." Moser (1955) analyzed similar questions but could not, in essence, prove any general new theorem on denseness of regular Hamiltonians, beside the results of Siegel in 1954 (see Chap. IV, Notes).

4. Lie Series and Lie Transforms.

The subject to be dealt with in this section is related to the following fact (to be proved in the text).

Let $S(y;x;\epsilon)$ and $f(y;x;\epsilon)$ be functions of the n-vectors y (coordinates) and x (conjugate momenta), and let ϵ be a dimensionless parameter. We assume S and f to be real analytic functions of the $2n+1$ arguments. Let us define an operator

$$\Delta_W f = (f,W) + \frac{\partial f}{\partial \epsilon} \qquad (1.4.1)$$

where (f,W) is Poisson parenthesis. Finally, consider the operator

$$E_W f = \sum_{n\geq 0}^{\infty} \frac{\epsilon^n}{n!} (\Delta_W^n f)_{\epsilon=0} \tag{1.4.2}$$

where

$$\Delta_W^0 f = f$$

$$\Delta_W^1 f = \Delta_W f$$

$$\Delta_W^n f = \Delta_W \Delta_W^{n-1} f \quad (n = 2,3,\ldots).$$

The main result is that, under the foregoing conditions, if the series (1.4.2) converges, the transformation

$$\begin{aligned} \eta_k &= E_W y_k \\ \xi_k &= E_W x_k \end{aligned} \tag{1.4.3}$$

is completely canonical. Moreover, any function $g(y;x)$ real analytic is expressed in the new variables $(\eta;\xi)$ by

$$g(y(\eta;\xi;\epsilon),\ x(\eta;\xi;\epsilon)) = E_W g(\eta;\xi). \tag{1.4.4}$$

Lie's Theorem (1888). The original application of Lie's series to perturbations methods was introduced by Hori (1966). He considered the operator $D_S^n f$ defined by

$$\begin{aligned} D_S^0 f &= f \\ D_S^1 f &= (f,S) \\ D_S^n f &= D_S^1 D_S^{n-1} f \end{aligned} \tag{1.4.5}$$

where f, S are real analytic functions of $2n$ variables $(\eta;\xi)$, $\eta = (\eta_1,\ldots,\eta_n)$, $\xi = (\xi_1,\ldots,\xi_n)$, canonically conjugate, and wrote Lie's theorem as follows: "A set of $2n$ variables $(y;x)$ defined by the equation

$$f(y;x) = \sum_{n=0}^{\infty} \frac{\epsilon^n}{n!} D_S^n f(\eta;\xi) \tag{1.4.6}$$

is canonical if the series converges for ϵ sufficiently small and independent of $(\eta;\xi)$." The proof of such theorem is quite elementary. One introduces the canonical system of differential equations $(j = 1,2,\ldots,n)$:

$$\frac{d\eta_j}{d\tau} = \frac{\partial S}{\partial \xi_j}, \quad \frac{d\xi_j}{d\tau} = -\frac{\partial S}{\partial \eta_j} \tag{1.4.7}$$

where τ is any parameter, and let $\eta_j(\tau)$, $\xi_j(\tau)$ be the solution of the system which is unique in the region where S is real analytic. It follows that, from (1.4.6)

$$f(y;x) = \sum_{n=0}^{\infty} \frac{\epsilon^n}{n!} \left.\frac{d^n f}{d\tau^n}\right|_{\epsilon=0} = f(\eta(\tau+\epsilon);\ \xi(\tau+\epsilon))$$

or, since $f(y;x)$ is analytic

$$y_j = \eta_j(\tau+\epsilon), \quad x_j = \xi_j(\tau+\epsilon) \tag{1.4.8}$$

for $j = 1,2,\ldots,n$ and ϵ sufficiently small. Since (1.4.8) are solutions of the Hamiltonian system (1.4.7), it follows that $(y;x)$ are canonical, because the mapping (1.4.8) is canonical.

If the "generator" S is given, the transformation has the explicit form

$$\begin{aligned} y_j &= \eta_j + \sum_{n=1}^{\infty} \frac{\epsilon^n}{n!} D_S^{n-1} \frac{\partial S}{\partial \xi_j} \\ x_j &= \xi_j - \sum_{n=1}^{\infty} \frac{\epsilon^n}{n!} D_S^{n-1} \frac{\partial S}{\partial \eta_j} \end{aligned} \tag{1.4.9}$$

which follow from (1.4.6). The apparent incongruence in the application of such theory to a perturbation method is that the functions f and S are to be considered power series in ϵ and such dependence is not taken care in the formulation. A modified approach to the question was introduced by Deprit (1969) and later was shown to be equivalent to Hori's formulation by several authors (e.g., Mersman, 1970). The equivalence of the generalized Hamilton-Jacobi transformation theory and Lie's transformations as used by Poincaré, Hori and Deprit, will be dealt with at the end of Chapter II. Here, we limit the presentation to the basic theorems involved in Lie's series transformation in the case when f and/or S

are functions of ϵ. The main purpose is to establish (1.4.3) and (1.4.4). The exposition follows the lines of Deprit's (1969) work.

Consider f and S to be real analytic functions of $2n$ canonically conjugate variables $(y;x)$. The Poisson's parenthesis (f,S) may be written

$$(f,S) = \frac{\partial f}{\partial y}\left(\frac{\partial S}{\partial x}\right)^T - \frac{\partial f}{\partial x}\left(\frac{\partial S}{\partial y}\right)^T \tag{1.4.10}$$

where, as usual, the derivative of a scalar function with respect to a vector is supposed to be a row matrix. One can define the $2n$-vector $z = (y;x)$ and the 2-vector (f,S) and write the Poisson's 2×2 matrix

$$P_z(f,S) = J_z M J_z^T \tag{1.4.11}$$

where $J_z = \frac{\partial(f,S)}{\partial z}$ is a $2 \times 2n$ matrix and M is the $2n \times 2n$ canonical matrix. Then

$$P_z(f,S) = (f,S)_z \begin{pmatrix} 0 & 1 \\ -1 & 0 \end{pmatrix}.$$

For a nontrivial canonical transformation $z = \mathbf{z}(\zeta)$ one has

$$J^T M J = M$$

where $J = \partial \mathbf{z}/\partial\zeta$. Then

$$J_\zeta = J_z J\ . \tag{1.4.12}$$

Now one has

$$\begin{aligned} P_\zeta(f,S) &= J_\zeta M J_\zeta^T = J_z J M J^T J_z^T \\ &= J_z M J_z^T = P_z(f,S). \end{aligned}$$

which shows the invariance of P with respect to a canonical transformation.

The <u>Lie Derivative</u> of f generated by S is simply

$$L_S f = (f,S), \tag{1.4.13}$$

and the following properties follow from the fact that $L_S f$ is a bilinear form in f, S (α,β are constants):

$$\begin{aligned} &a. \quad L_S(\alpha f + \beta g) = \alpha L_S f + \beta L_S g \\ &b. \quad L_S(f\cdot g) = f\cdot L_S g + g\cdot L_S f \\ &c. \quad L_S(f,g) = (f,L_S g) + (L_S f,g) \\ &d. \quad L_S L_{S'} f = L_{S'} L_S f + L_{(S,S')} f. \end{aligned} \tag{1.4.14}$$

Defining $L_S^0 f = f$, the n iterate of the Lie Derivative is

$$L_S^n f = L_S L_S^{n-1} f \ .$$

For this n iterate, the following properties are easily verified:

$$\begin{aligned} &a. \quad L_S^n(\alpha f + \beta g) = \alpha L_S^n f + \beta L_S^n g \\ &b. \quad L_S^n(f\cdot g) = \sum_{m=0}^{n} \binom{n}{m} L_S^m f\cdot L_S^{n-m} g \\ &c. \quad L_S^n(f,g) = \sum_{m=0}^{n} \binom{n}{m} (L_S^m f, L_S^{n-m} g) \ . \end{aligned} \tag{1.4.15}$$

If the function S is real analytic one may choose ϵ sufficiently small so that the series

$$\sum_{n=0}^{\infty} \frac{\epsilon^n}{n!} L_S^n f = \exp(\epsilon\, L_S) f \tag{1.4.16}$$

converges when applied to an analytic function f. Again, one can easily verify that

$$\begin{aligned} &a. \quad \exp(\epsilon L_S)\,(\alpha f + \beta g) = \alpha \exp(\epsilon L_S) f + \beta \exp(\epsilon L_S) g \\ &b. \quad \exp(\epsilon L_S)\,(f\cdot g) = \exp(\epsilon L_S) f\cdot \exp(\epsilon L_S) g \\ &c. \quad \exp(\epsilon L_S)\,(f,g) = (\exp(\epsilon L_S) f, \exp(\epsilon L_S) g) \end{aligned} \tag{1.4.17}$$

From the last of the above relations one concludes the Theorem: "Let ϵ be a constant parameter and consider the transformation $z = z(\zeta)$ from the $2n$-vector

$z = (y;x)$ where y, x are canonically conjugate, to the $2n$-vector $\zeta = (\eta;\xi)$. If there exists a real analytic function $S(z)$ such that the series

$$\zeta = \exp(\epsilon L_S)z \tag{1.4.18}$$

converges in some domain of the z-space, the transformation is canonical."

Note that this is essentially Lie's Theorem as stated before. The proof, under the present approach, follows immediately by considering

$$\zeta_i = \exp(\epsilon L_S)z_i$$

$$\zeta_j = \exp(\epsilon L_S)z_j$$

and from (1.4.17)c,

$$(\zeta_i,\zeta_j) = (\exp(\epsilon L_S)z_i, \exp(\epsilon L_S)z_j)$$

$$= \exp(\epsilon L_S)(z_i,z_j)$$

or

$$P(\zeta) = \exp(\epsilon L_S)P(z).$$

From the fact the z is a canonical set, $P(z) = M$ and, therefore

$$P(\zeta) = M$$

so that ζ is canonical.

Another important result gives the transformation law for any function of z into a function of ζ. Theorem: "The image of every real analytic function $f(z)$ under the transformation

$$z = \exp(\epsilon L_S)\zeta \tag{1.4.19}$$

is

$$\tilde{f}(\zeta;\epsilon) = f(\exp(\epsilon L_S)\zeta) = \exp(\epsilon L_S)f(\zeta)". \tag{1.4.20}$$

In fact,

$$L_S\tilde{f}(\zeta;\epsilon) = \frac{\partial f}{\partial z} L_S z \tag{1.4.21}$$

where $\partial f/\partial z$ is the row matrix $\{\partial f/\partial z_k\}$ and $L_S z$ is the column matrix $\{(z_k, S)\}$.

Differentiating (1.4.19) with respect to ϵ,

$$\frac{\partial z}{\partial \epsilon} = \sum_{m=0}^{\infty} \frac{\epsilon^m}{m!} L_S^{m+1}\zeta = L_S z,$$

so that

$$L_S \tilde{f}(\zeta;\epsilon) = \frac{\partial f}{\partial z}\frac{\partial z}{\partial \epsilon} = \frac{\partial \tilde{f}}{\partial \epsilon} .$$

The n-th iterate of such an operation gives

$$L_S^n \tilde{f} = \frac{\partial^n \tilde{f}}{\partial \epsilon^n}$$

or

$$\left.\frac{\partial^n \tilde{f}}{\partial \epsilon^n}\right|_{\epsilon=0} = L_S^n \tilde{f}(\zeta;0) = f(\zeta) \qquad (1.4.22)$$

from (1.4.20). Hence, the Taylor's expansion of $\tilde{f}(\zeta;\epsilon)$ is given by

$$\tilde{f}(\zeta;\epsilon) = \sum_{n=0}^{\infty} \frac{\epsilon^n}{n!} \left.\frac{\partial^n \tilde{f}}{\partial \epsilon^n}\right|_{\epsilon=0}$$

$$= \sum_{n=0}^{\infty} \frac{\epsilon^n}{n!} L_S^n f(\zeta) = \exp(\epsilon L_S) f(\zeta)$$

which completes the proof.

From this last theorem we conclude a corollary which, ultimately, will establish the validity of Hori's approach who considered S an explicit function of ϵ. Corollary: "If the function $f(z,\epsilon)$ admits a Taylor series in the neighborhood of $\epsilon = 0$, that is,

$$f(z;\epsilon) = \sum_{n=0}^{\infty} \frac{\epsilon^n}{n!} f_n(z) \qquad (1.4.23)$$

then, under the canonical mapping (1.4.19),

$$f(z(\zeta;\epsilon);\epsilon) = \sum_{n=0}^{\infty} \frac{\epsilon^n}{n!} \sum_{m=0}^{\infty} \binom{n}{m} L_S^m f_{n-m}(\zeta)." \tag{1.4.24}$$

In fact, from (1.4.20),

$$f_n(z(\zeta;\epsilon)) = \sum^{\infty} \frac{1}{m!} \epsilon^m L_S^m f_n(\zeta)$$

which substituted into (1.4.23) gives the desired result, upon collection of like powers of ϵ.

Finally we prove the following theorem about the inverse of a canonical transformation defined by Lie's Series:

Theorem: "The inverse of the canonical transformation

$$z = \exp(\epsilon L_S)\zeta$$

is given by

$$\zeta = \exp(\epsilon L_{-S}) z." \tag{1.4.25}$$

In fact

$$\zeta = \exp(\epsilon L_{S'}) z = \exp(\epsilon L_{S'})(\exp(\epsilon L_S)\zeta)$$

$$= \exp(\epsilon(L_{S'} + L_S))\zeta .$$

The operator

$$\exp(\epsilon(L_{S'} + L_S))$$

must reduce to the identity transformation, that is, $L_{S'} + L_S = 0$, and, therefore, $S' = -S$, necessarily.

5. Lie Transform Depending on a Parameter.

As was stated earlier, canonical transformations associated with perturbation methods are necessarily functions of a parameter, generally small, for the solution is known when such parameter is set equal to zero (or any fixed numerical value). In terms of the Lie Transformation Theory presented in the previous section, this means that one should allow for the Generator to depend explicitly on the parameter ϵ. This can be accomplished by defining (Deprit, 1969) the operator

$$\Delta_S = L_S + \frac{\partial}{\partial\epsilon} \qquad (1.5.1)$$

with the obvious properties:

$$\begin{aligned} &\text{a.}\quad \Delta_S(\alpha f + \beta g) = \alpha\Delta_S f + \beta\Delta_S g \\ &\text{b.}\quad \Delta_S(f\cdot g) = f\cdot\Delta_S g + g\cdot\Delta_S f \\ &\text{c.}\quad \Delta_S(f,g) = (\Delta_S f,g) + (f,\Delta_S g) \\ &\text{d.}\quad \Delta_S\Delta_{S'}f = \Delta_{S'}\Delta_S f + L_{(S',S)}f + L_{S'_\epsilon - S_\epsilon} \end{aligned} \qquad (1.5.2)$$

where

$$S = S(z;\epsilon),$$

$$S_\epsilon = \frac{\partial S}{\partial\epsilon}.$$

It is also legitimate to define the n iterate of $\Delta_S f$ by

$$\Delta_S^n f = \Delta_S(\Delta_S^{n-1}f)$$

$$\Delta_S^0 f = f$$

and easily obtain the relations corresponding to (1.5.2).

We also define

$$f_n(\zeta;0) = [\Delta^n_{S(\zeta;\epsilon)} f(\zeta;\epsilon)]_{\epsilon=0} \qquad (1.5.3)$$

and the new operator

$$E_S f = \sum_{n=0}^{\infty} \frac{\epsilon^n}{n!} f_n(\zeta;0). \qquad (1.5.4)$$

Evidently, if there exist a finite quantity A such that

$$f_n(\zeta,0) < A^n$$

for ζ in some neighborhood of a point ζ_o, the series (1.5.4) certainly converges.

The following relations are easily verified:

$$\begin{aligned}
&\text{a.}\quad E_S(\alpha f + \beta g) = \alpha\, E_S f + \beta\, E_S g \\
&\text{b.}\quad E_S(f\cdot g) = E_S f\cdot E_S g \qquad\qquad (1.5.5)\\
&\text{c.}\quad E_S(f,g) = (E_S f,\, E_S g).
\end{aligned}$$

As done previously with the operator L_S, one shows that the transformation $(\zeta;\epsilon) \to z$ defined by

$$z = E_S(\zeta) = \sum_{n=0}^{\infty} \frac{\epsilon^n}{n!}\, z_n(\zeta;0). \qquad (1.5.6)$$

is canonical provided the series converges. In order to establish a Lie Generator for the above transformation, we prove the following theorem.

Theorem. "The transformation $z = E_S(\zeta)$ is the solution of the Hamiltonian system

$$\frac{dz}{d\epsilon} = M\left(\frac{\partial S}{\partial z}\right)^T \qquad (1.5.7)$$

corresponding to the initial conditions $z = \zeta$ at $\epsilon = 0$ and where $S(z;\epsilon)$ is related to $E_S(\zeta)$ through (1.5.4) and (1.5.3)."

In fact, considering (1.5.1),

$$\begin{aligned}
\Delta_S z(\zeta;\epsilon) &= L_S z(\zeta;\epsilon) + \frac{\partial z}{\partial \epsilon} \\
&= \frac{\partial z}{\partial y}\left(\frac{\partial S}{\partial x}\right)^T - \frac{\partial z}{\partial x}\left(\frac{\partial S}{\partial y}\right)^T + \frac{\partial z}{\partial \epsilon},
\end{aligned}$$

where $z = \text{col}(y,x)$. From (1.5.7), with $S = S(\zeta;\epsilon)$,

$$\left(\frac{\partial S}{\partial x}\right)^T = \frac{dy}{d\epsilon}$$

and

$$\left(\frac{\partial S}{\partial y}\right)^T = -\frac{dx}{d\epsilon},$$

so that

$$\Delta_S z(\zeta;\epsilon) = \frac{\partial z}{\partial y}\frac{dy}{d\epsilon} + \frac{\partial z}{\partial x}\frac{dx}{d\epsilon} + \frac{\partial z}{\partial \epsilon} = \frac{dz}{d\epsilon}\ . \tag{1.5.8}$$

The transformation $z(\zeta;\epsilon)$ being supposed real analytic, we obtain

$$\Delta_S^n\ z(\zeta;\epsilon) = \frac{d^n z}{d\epsilon^n}\ , \tag{1.5.9}$$

and for $\epsilon = 0$ there results

$$\Delta_S^n\ z(\zeta;\epsilon)\Big|_{\epsilon=0} = \frac{d^n z}{d\epsilon^n}\Big|_{\epsilon=0} = z_n(\zeta;0) \tag{1.5.10}$$

so that, using (1.5.6)

$$z = E_S(\zeta) = \sum_{n=0}^{\infty} \frac{\epsilon^n}{n!}\frac{d^n z}{d\epsilon^n}\Big|_{\epsilon=0} = z(\zeta;\epsilon).$$

which completes the proof.

The transformation of a real analytic function $f(z;\epsilon)$ under the canonical mapping $z = z(\zeta;\epsilon) = E_S(\zeta)$ defined by (1.5.6), is simply obtained as

$$f(E_S(\zeta);\epsilon) = E_S f(\zeta;\epsilon). \tag{1.5.11}$$

In fact, along the solution $z = z(\zeta;\epsilon)$, going through $z = \zeta$ at $\epsilon = 0$, of system (1.5.7),

$$f(z(\zeta;\epsilon);\epsilon)$$

$$= \sum_{n=0}^{\infty} \frac{\epsilon^n}{n!}\left(\frac{d^n f}{d\epsilon^n}\right)_{\epsilon=0} = \sum_{n=0}^{\infty} \frac{\epsilon^n}{n!}(\Delta_S^n f)_{\epsilon=0}$$

as shown by (1.5.10). Therefore, by definition of E_S,

$$f(z(\zeta;\epsilon);\epsilon) = E_S f(\zeta;\epsilon),$$

which is (1.5.11).

A particular case of interest for the transformation rule (1.5.11) is when both $S(\zeta;\epsilon)$ and $f(\zeta;\epsilon)$ are power series in ϵ, that is,

$$S(\zeta;\epsilon) = \sum_{n=0}^{\infty} \frac{\epsilon^n}{n!} S_{n+1}(\zeta) \tag{1.5.12}$$

and

$$f(\zeta;\epsilon) = \sum_{n=0}^{\infty} \frac{\epsilon^n}{n!} f_n(\zeta). \tag{1.5.13}$$

In this case, let us define

$$L_{S_p} = L_p \quad (p \geq 1)$$

so that, from the results of the previous section, one finds

$$\frac{\partial}{\partial\epsilon} f(\zeta;\epsilon) = \sum_{n=0}^{\infty} \frac{\epsilon^n}{n!} f_{n+1}(\zeta)$$

and

$$L_S f(\zeta;\epsilon) = \sum_{n=0}^{\infty} \frac{\epsilon^n}{n!} \sum_{m=0}^{n} \binom{n}{m} L_{m+1} f_{n-m}(\zeta).$$

Thus, representing $\Delta_S f$ by the series

$$\Delta_S f = \sum_{n=0}^{\infty} \frac{\epsilon^n}{n!} f_n^{(1)}(\zeta),$$

one finds

$$f_n^{(1)}(\zeta) = f_{n+1}(\zeta) + \sum_{m=0}^{n} \binom{n}{m} L_{m+1} f_{n-m}(\zeta)$$

and therefore

$$f_o^{(1)}(\zeta) = f_1 + L_1 f_o = f_1 + (f_o, S_1).$$

In the same manner, introducing the series

$$\Delta_S^2 f = \sum_{n=0}^{\infty} \frac{\epsilon^n}{n!} f_n^{(2)}(\zeta)$$

we find

$$f_n^{(2)}(\zeta) = f_{n+1}^{(1)}(\zeta) + \sum_{m=0}^{n} \binom{n}{m} L_{m+1} f_{n-m}^{(1)}(\zeta)$$

and therefore

$$f_o^{(2)}(\zeta) = f_1^{(1)} + L_1 f_o^{(1)},$$

or, using the expression for $f_o^{(1)}$, $f_1^{(1)}$, it follows that

$$f_o^{(2)}(\zeta) = f_2 + 2(f_1,S_1) + (f_o,S_2) + ((f_o,S_1),S_1).$$

A general recurrence algorithm is thus obtained for the transformation of $f(z;\epsilon)$ under a Lie Series Transform generated by $S(z;\epsilon)$ when both these functions are real analytic in all variables and for ϵ in the neighborhood of $\epsilon = 0$:

$$f_n^{(k)}(\zeta) = f_{n+1}^{(k-1)} + \sum_{m=0}^{n} \binom{n}{m} L_{m+1} f_{n-m}^{(k-1)} \quad . \tag{1.5.14}$$

This is represented in the following triangular map:

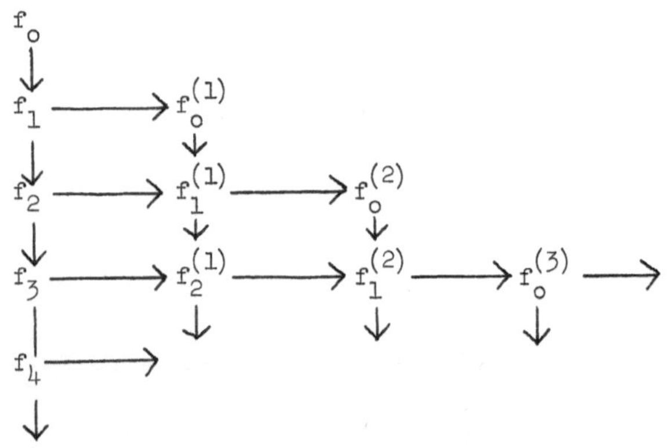

A particular case of interest is the transformation of the vector $\mathbf{z} = \text{col}(y;x)$. The canonical transformation is

$$y = E_S(\eta) = \sum_{n=0}^{\infty} \frac{\epsilon^n}{n!} \eta_o^{(n)}(\zeta;0)$$

$$x = E_S(\xi) = \sum_{n=0}^{\infty} \frac{\epsilon^n}{n!} \xi_o^{(n)}(\zeta;0) \tag{1.5.15}$$

and the recurrence procedure above described gives the coefficients $\eta_o^{(n)} = \eta_o^n(\zeta;0)$ and $\xi_o^{(n)} = \xi_o^{(n)}(\zeta;0)$. In (1.5.15) it is worth noting that, obviously, $\eta_o^{(o)} = \eta$ and $\xi_o^{(o)} = \xi$. The all procedure can be extended to the case in which the canonical transformation depends explicitly on time. One way to produce the corresponding result is by simply taking the time as an additional canonical coordinate, the conjugate momentum being the Hamiltonian itself. This leads directly to the algorithm described in detail by Deprit (1969).

6. Equivalence Relations.

In previous sections we have described ways of producing canonical transformations as power series of a parameter ϵ. Such transformations are written

$$y = y(\eta;\xi;\epsilon)$$

$$x = x(\eta;\xi;\epsilon) \tag{1.6.1}$$

or

$$z = z(\zeta;\epsilon), \tag{1.6.2}$$

where x, y, η, ξ are n-vectors and z, ζ are $2n$-vectors. In terms of a generator satisfying Hamilton-Jacobi's equation, that is, the one required to develop Poincaré's method of perturbations, the transformation (1.6.1) is produced by

$$y = \eta + \left(\frac{\partial W}{\partial x}\right)^T = y(\eta;x;\epsilon)$$

$$\xi = x + \left(\frac{\partial W}{\partial \eta}\right)^T = \xi(\eta;x;\epsilon) \tag{1.6.3}$$

where $W = W(\eta;x;\epsilon)$. The condition

$$W(\eta;x;0) = 0 \tag{1.6.4}$$

indicates the fact that the transformation (1.6.1) is "near" the identity for ϵ sufficiently small.

A transformation of the same character is produced, as was seen, by a generator $S = S(y;x;\epsilon)$, through the solution of the Hamiltonian system

$$\frac{dy}{d\epsilon} = \left(\frac{\partial S}{\partial x}\right)^T$$
$$\frac{dx}{d\epsilon} = -\left(\frac{\partial S}{\partial y}\right)^T \tag{1.6.5}$$

with the initial conditions $y = \eta$, $x = \xi$ at $\epsilon = 0$. We have the following basic equivalence statement: Theorem (Shniad, 1970): "The generators W and S, satisfying the foregoing conditions, satisfy the relation

$$S(y;x;\epsilon) = \frac{\partial W}{\partial \epsilon}(\eta;x;\epsilon) \tag{1.6.6}$$

where

$$y = \eta + \left(\frac{\partial W}{\partial x}\right)^T = y(\eta;x;\epsilon) \text{ .''} \tag{1.6.7}$$

In fact, applying the canonical transformation (1.6.3) to the system (1.6.5), the new Hamiltonian $S'(\eta;\xi;\epsilon)$ is given, according to Hamilton-Jacobi theory, by

$$S'(\eta;\xi(\eta;x;\epsilon);\epsilon) = S(y(\eta;x;\epsilon);x;\epsilon) - \frac{\partial W}{\partial \epsilon}(\eta;x;\epsilon). \tag{1.6.8}$$

On the other hand, by definition, η and ξ are constants and, therefore, the Hamiltonian $S'(\eta;\xi;\epsilon)$ must be identically zero, which proves the theorem.

Now, both W and S are generally defined as power series in ϵ and (1.6.6) provides the relations among the coefficients of these two series. In fact, since S' is identically zero, the corresponding relation

$$S\left(\eta + \left(\frac{\partial W}{\partial x}\right)^T;x;\epsilon\right) - \frac{\partial W}{\partial \epsilon}(\eta;x;\epsilon) = 0 \tag{1.6.9}$$

must be identically satisfied as a function of the $2n+1$ independent variables $(\eta;x;\epsilon)$.

Let us assume for S and W the series

$$S(y;x;\epsilon) = \sum_{n=0}^{\infty} S_{n+1}(y;x)\,\epsilon^n$$
$$W(\eta;x;\epsilon) = \sum_{n=1}^{\infty} W_n(\eta;x)\,\epsilon^n \qquad (1.6.10)$$

where y is defined by (1.6.7).

Substitution of (1.6.10) and (1.6.7) into (1.6.9) leads to the recurrence relations

$$W_1 = S_1$$
$$2W_2 = S_2 + \left(\frac{\partial S_1}{\partial \eta}\right)\left(\frac{\partial W_1}{\partial x}\right)^T$$
$$3W_3 = S_3 + \left(\frac{\partial S_1}{\partial \eta}\right)\left(\frac{\partial W_2}{\partial x}\right)^T + \left(\frac{\partial S_2}{\partial \eta}\right)\left(\frac{\partial W_1}{\partial x}\right)^T + \frac{1}{2}\frac{\partial W_1}{\partial x}\frac{\partial^2 S_1}{\partial \eta \partial \eta}\left(\frac{\partial W_1}{\partial x}\right)^T$$

where $\frac{\partial S_n}{\partial \eta}$ and higher derivatives stand for $\left.\frac{\partial S_n}{\partial y}\right|_{y=\eta}$. In general, Mersman (1971) finds that

$$W_1 = S_1$$
$$(n+1)W_{n+1} = S_{n+1} + \sum_{k=1}^{n} \frac{1}{k!} \sum_{p} \frac{\partial^k S_{p_0}}{\partial \eta_{i_1} \partial \eta_{i_2} \cdots \partial \eta_{i_k}} \cdot \frac{\partial W_{p_1}}{\partial x_{i_1}} \frac{\partial W_{p_2}}{\partial x_{i_2}} \cdots \frac{\partial W_{p_k}}{\partial x_{i_k}} \qquad (1.6.11)$$

where the second summation is over all sets of $k+1$ positive integers $(p_0, p_1, p_2, \ldots, p_k)$ such that $p_0+p_1+p_2+ \ldots + p_k = n+1$. Relation (1.6.11) is totally equivalent to the one originally obtained by Giacaglia (1964) in the development of explicit relations for the von Zeipel (Poincaré) method. The recurrence formula (1.6.11) can now be used to establish explicit relations among the generators defined in Poincaré's method and those given by Hori and Deprit by means of Lie Series. These relations are given in detail by Mersman (1971). The equivalence of Hori's and Deprit's formulations establishes, indirectly, a justification of the fact that in Hori's original approach the generator S could be considered a function of ϵ, although, apparently the proof of Lie's Theorem falls short in such case. A discussion over the above question was originally presented by Campbell and Jefferys (1970) with respect to some negative remarks by Deprit (1969) about Hori's Theory. Their argument is essentially the one of assuming the generator S imbedded on a one parameter family (parameter ϵ_o), constructing the transformation for a fixed value of the parameter and showing the validity for any value ϵ of ϵ_o. An analogous reasoning was quite successfully applied by Poincaré (1892) in a problem where the same parameter is fictitiously labeled by two different names, an expansion in one name parameter is carried on and finally the two names are identified again.

As an example of Poincaré's remark, consider

$$f(\epsilon) = \sin \frac{\epsilon}{1-\epsilon}$$

and the Taylor series of $f(\epsilon)$ about $\epsilon = 0$. One can produce such series as follows. Let

$$\sin \frac{\epsilon}{1-\epsilon} = \sin \frac{\epsilon}{1-\mu}$$

and the Taylor series is

$$\sin \frac{\epsilon}{1-\mu} = \sum_{n=0}^{\infty} \frac{\epsilon^n}{2n!} \frac{1}{(1-\mu)^n} [1 - (-1)^n]$$

$$= \sum_{n=0}^{\infty} \frac{\epsilon^n}{2n!} [1 - (-1)^n] \sum_{m=0}^{\infty} \binom{-n}{m} (-1)^m \mu^m.$$

Identification of μ and ϵ gives

$$\sin \frac{\epsilon}{1-\epsilon} = \sum_{p=0}^{\infty} \{ \sum_{n=0}^{p} \frac{1}{2n!} \binom{-n}{p-n} (-1)^{p-n} [1 - (-1)^n]\} \epsilon^p$$

which in fact is the correct Taylor series of $f(\epsilon)$ as it is readily verified.

In the case under question, recalling the operator

$$\exp(\epsilon L_S) = \sum_{n=0}^{\infty} \frac{\epsilon^n}{n!} L_S^n ,$$

the property

$$\exp(\epsilon L_S)(f,g) = ((\exp \epsilon L_S)f, (\exp \epsilon L_S)g)$$

does not depend on the fact that S is dependent or independent of ϵ. Therefore, since the above relation is basically the proof of the transformation

$$z = \exp(\epsilon L_S)\zeta$$

to be canonical, it can likewise be applied to Hori's development, as a proof independent of the Hamiltonian system of differential equations generated by S.

7. General Transformations induced by Lie Series.

Consider an n-dimensional vector space and a non-singular real analytic transformation from a point x to a point y of this space, defined by

$$y = x + \sum_{m=1}^{\infty} \frac{\epsilon^m}{m!} y_m(x) \tag{1.7.1}$$

where y_m are n-vectors, and ϵ a parameter independent of x. For $\epsilon = 0$, (1.7.1) reduces to the identity transformation and for ϵ small (1.7.1) is "near" the identity if the series converges. We shall however consider (1.7.1) as a formal series and apply the rules of operations with convergent series (e.g. Cartan, 1963).

One of the goals of the following discussion is to construct a simple algorithm for the transformation, under (1.7.1), of a vector function $F(y;\epsilon)$. We wish the result to be a power series in ϵ, that is, we wish to find the coefficients $F_n(x)$ in the expansion

$$F(y(x,\epsilon);\epsilon) = \sum_{n=0}^{\infty} \frac{\epsilon^n}{n!} F_n(x). \tag{1.7.2}$$

Obviously, the vector function F should be real analytic in ϵ at $\epsilon = 0$ so that the series (1.7.2) exists. We shall also assume it is real analytic in y.

Two different algorithms were developed independently by Hori (1970) and Kamel (1970), having as major goal the solution by formal series of problems in non-linear oscillations. The description of such applications will be given in the next chapter. Here, we limit ourselves to the description of the formal expansion discussed above.

By hypothesis, one can expand $F(x;\epsilon)$ as

$$F(x;\epsilon) = \sum_{n=0}^{\infty} \frac{\epsilon^n}{n!} \left(\frac{\partial^n F(x;\epsilon)}{\partial \epsilon^n}\right)_{\epsilon=0} = \sum_{n=0}^{\infty} \frac{\epsilon^n}{n!} F_n(x) \qquad (1.7.3)$$

and also,

$$F(x(y;\epsilon);\epsilon) = \sum_{n=0}^{\infty} \frac{\epsilon^n}{n!} F^{(n)}(y) \qquad (1.7.4)$$

where

$$F^{(n)} = \frac{d^n}{d\epsilon^n} F(x;\epsilon)\Big|_{\epsilon=0}$$

$$= \left(\frac{\partial}{\partial \epsilon} + \frac{\partial x}{\partial \epsilon}\frac{\partial}{\partial x}\right)^n F(x(y;\epsilon);\epsilon)\Big|_{\epsilon=0} ,$$

and

$$x = x(y;\epsilon)$$

is the inverse of (1.7.1) which we suppose exists.

We also have, writing the inverse of (1.7.1) as

$$x = y + \sum_{n=1}^{\infty} \frac{\epsilon^n}{n!} X^{(n)}(y), \qquad (1.7.5)$$

that

$$\frac{\partial x}{\partial \epsilon} = \sum_{n=0}^{\infty} \frac{\epsilon^n}{n!} X^{(n+1)}(y). \qquad (1.7.6)$$

The expression $\partial x/\partial \epsilon$ clearly indicates that y is kept fixed. From (1.7.6) we can write

$$\frac{\partial x}{\partial \epsilon} = T(x;\epsilon) \qquad (1.7.7)$$

where

$$T(x;\epsilon) = \sum_{n=0}^{\infty} \frac{\epsilon^n}{n!} X^{(n+1)}(y)$$
$$= \sum_{n=0} \frac{\epsilon^n}{n!} T_{n+1}(x). \tag{1.7.8}$$

We have that

$$\frac{d}{d\epsilon} = \frac{\partial}{\partial\epsilon} + \frac{\partial x}{\partial\epsilon}\frac{\partial}{\partial x} = \frac{\partial}{\partial\epsilon} + T(x;\epsilon)\frac{\partial}{\partial x} = \frac{\partial}{\partial\epsilon} + L_T \tag{1.7.9}$$

where the operator L_T is defined by

$$L_T = T(x;\epsilon)\frac{\partial}{\partial x} \tag{1.7.10}$$

acting on a real analytic function $f(x,\epsilon)$. In the above relations we have assumed $T(x;\epsilon)$ to be an n-dimensional row vector and $\frac{\partial}{\partial x}$ an n-dimensional column vector. Now we have

$$\frac{d}{d\epsilon} F(x;\epsilon) = \frac{d}{d\epsilon}\sum_{n=0}^{\infty}\frac{\epsilon^n}{n!}F_n(x)$$
$$= \sum_{n=0}^{\infty}\frac{\epsilon^n}{n!}F_{n+1}(x) + \sum_{n=0}^{\infty}\frac{\epsilon^n}{n!}\sum_{m=0}^{\infty}\frac{\epsilon^m}{m!}T_{n+1}\frac{\partial F_n(x)}{\partial x} \tag{1.7.11}$$
$$= \sum_{n=0}\frac{\epsilon^n}{n!}F_n^{(1)}(x)$$

where

$$F_n^{(1)}(x) = F_{n+1}(x) + \sum_{m=0}^{n}\binom{n}{m}T_{n-m+1}(x)\frac{\partial F_m}{\partial x}$$
$$= F_{n+1}(x) + \sum_{p=0}^{n}\binom{n}{p}T_{p+1}(x)\frac{\partial F_{n-p}}{\partial x}. \tag{1.7.12}$$

In general, we obtain

$$\frac{d^k}{d\epsilon^k}F(x;\epsilon) = \sum_{n=0}^{\infty}\frac{\epsilon^n}{n!}F_n^{(k)}(x), \tag{1.7.13}$$

where

$$F_n^{(k)}(x) = F_{n+1}^{(k-1)}(x) + \sum_{m=0}^{n} \binom{n}{m} T_{m+1}(x) \frac{\partial F_{n-m}^{(k-1)}}{\partial x} \qquad (1.7.14)$$

for $k \geq 1$ and $n \geq 0$, where

$$F_n^{(o)}(x) = F_n(x), \quad F_o^{(k)}(x) = F^{(k)}(x) = F^{(k)}(y)\Big|_{y=x} \qquad (1.7.15)$$

The equation (1.7.14) is a recursive algorithm to construct the coefficients $F^{(n)}(x)$ from $F_n(x)$ of the series (1.7.4) and (1.7.3). The variable's name is, obviously, dummy. The corresponding formula to construct the coefficients $F_n(x)$ from $F^{(n)}(x)$ is

$$F_n^{(k)} = F_{n-1}^{(k+1)} - \sum_{m=0}^{n-1} \binom{n-1}{m} T_{m+1}(x) \frac{\partial F_{n-m-1}^{(k)}}{\partial x} \qquad (1.7.16)$$

for $n \geq 1$ and $k \geq 0$.

Successive substitution of (1.7.16) into itself from $n = 1$ up, gives

$$F_n^{(k)} = \sum_{j=0}^{n} \binom{n}{j} N_j(F^{(k+n-j)}) \qquad (1.7.17)$$

where $n \geq 1$, $k \geq 0$ and N_j $(j \geq 0)$ is a linear operator given by

$$N_o = 1$$

$$N_j = -\sum_{m=1}^{j} \binom{j-1}{m-1} N_{j-m}\{T_m(x) \frac{\partial}{\partial x}\} = \qquad (1.7.18)$$

$$= -\sum_{m=1}^{j} \binom{j-1}{m-1} N_{j-m} L_m$$

for $j \geq 1$, and where

$$L_m = T_m(x) \frac{\partial}{\partial x} \quad . \qquad (1.7.19)$$

For instance, the first few operators N_j are

$$N_o = 1$$

$$N_1 = -L_1$$

$$N_2 = -N_1L_1 - L_2$$

$$N_3 = -N_2L_1 - 2N_1L_2 - L_3$$

In particular, for $k = 0$, Eq. (1.7.17) yields

$$F_n = \sum_{j=0}^{n} \binom{n}{j} N_j (F^{(n-j)})$$

which may be written as

$$F_n = \sum_{j=0}^{n} \binom{n}{j} F_{j,n-j} \qquad (1.7.20)$$

where

$$F_{j,k} = -\sum_{m=1}^{j} \binom{j-1}{m-1} L_m F_{j-m,k} \qquad (1.7.21)$$

and, by definition

$$F_{o,k} = F^{(k)} .$$

Formula (1.7.20) gives the $F^{(n)}$ recursively in terms of the F_n or the F_n recursively in terms of the $F^{(n)}$. This is the simplest possible form, as derived by Kamel.

Vector Transformation.

The coefficients $y_n(x)$ in (1.7.1) are easily obtained now from (1.7.16) for the special case of (1.7.3) when one takes

$$F^{(o)} = F = y$$

$$F^{(k)} = 0, \quad k > 0$$

$$F_o = y_o(x) = x$$

$$F_n^{(o)} = F_n = y_n(x) .$$

In fact, (1.7.16) gives, in this case

$$y_n(x) = -\sum_{m=0}^{n-1} \binom{n-1}{m} T_{m+1}(x) \frac{\partial y_{n-m-1}(x)}{\partial x}$$

or, considering $p = m+1$,

$$y_n(x) = -\sum_{p=1}^{n} \binom{n-1}{p-1} T_p(x) \frac{y_{n-p}(x)}{\partial x}$$

or

$$y_n(x) = -T_n(x) - \sum_{p-1}^{n-1} \binom{n-1}{p-1} T_p(x) \frac{\partial y_{n-p}(x)}{\partial x} \quad . \tag{1.7.22}$$

The inverse transformation follows from (1.7.14) or, more directly, from (1.7.21).

In fact, in notation of (1.7.4),

$$F(x(y;\epsilon);\epsilon) = x$$

$$F^{(0)} = y, \quad F^{(n)}(y) = X^{(n)}(y)$$

and we have

$$x = y + \sum_{n=1}^{\infty} \frac{\epsilon^n}{n!} X^{(n)}(y).$$

The recurrence relation (1.7.21) gives, together with (1.7.8)

$$\begin{aligned} X_{j,k}(y) &= -\sum_{m=1}^{j} \binom{j-1}{m-1} T_m(y) \frac{\partial}{\partial y} X_{j-m,k}(y) \\ X^{(n)}(y) &= T_n(y) - \sum_{j=1}^{n-1} \binom{n-1}{j} X_{j,n-j}(y) \end{aligned} \tag{1.7.23}$$

with

$$X_{o,k} = X^{(k)}(y).$$

Applications of the above results will be given, explicitly in the next chapter, when the problem of integration of non-linear systems will be dealt with.

NOTES

Lindstedt has been given credit for developing a perturbation method which avoids secular and mixed secular terms in the perturbed harmonic oscillators. He described such a method on several occasions but always thought that the perturbing forces ought to be either odd or even functions of the angle variable involved. Such restriction was shortly after shown not to be necessary by Poincaré. In his celebrated "Méthodes Nouvelles", vol. 2, he developed a canonical analog of Lindstedt method which, even after a superficial look, proves to be a very elaborate generalization. However, it is obvious that the main idea of Poincaré's development comes from Delaunay and some remarks of Tisserand on Delaunay's Lunar Theory. One might in fact go back to Euler's second lunar theory. He obviously had learned a great deal, in between his Lunar Theories, about the development of frequencies of a perturbed system in power series of the small parameter of the problem. Such theories clearly had a great influence in Poincaré's work. The merit of von Zeipel was mainly the application of Poincaré's method to the theory of motion of a well defined system, although the systematic separation of terms of different period, in the development of perturbations, is an important point. Especially when one considers the fact that on several occasions short period phenomena are of no interest, but only long period or secular ones. The Averaging Methods in general, say as discussed by Cesari in his book, had been quite popular in Celestial Mechanics but with no mention to the convergence problem. Perhaps, a big hangover from Poincaré's definite statements on the divergence of Lindstedt's series. Such series were, and still are, used to produce quite accurate prediction of the position of Celestial bodies. Krylov and Bogoliubov did give some bounds in the truncation errors which, as a consequence of new efforts in celestial mechanics, in the sixties, were reviewed by Kyner. Strangely enough there is a great gap in the western literature in problems related to linear and nonlinear oscillations, a field very rich of references in the Soviet literature essentially from Liapunov until about 1950. Celestial Mechanics had been worked down to the bones by means of the available tools of classical analysis by the end of the last century and nonlinear circuit theory and mechanical systems did not seem to be palatable to western mathematicians. The masterpiece work of Cesari in 1940 was not immediately recognized, but there stood the first proof of convergence of an averaging method for a large variety of problems. Important works followed more than a decade after, by Gambill and Hale. The works of Birkhoff, Siegel and Wintner were more mathematically oriented toward qualitative properties of Dynamical Systems. The simultaneous analysis of Birkhoff's extensive analysis on the restricted problem of three bodies and of Strömgren's numerical experiments was undertaken only recently and summarized in the master work of Szebehely. Moulton and MacMillan should be considered among the scientists who had such capability of analysis of association between theories and numbers. And also Adams and Darwin. The method of Poisson for the variation of integrals of motion is something else that was overlooked for a long time. In the modern literature it is revived again by Kurth in 1959 and mentioned, under different names and aspects, by Danby and Brouwer-Clemence. Lately, stemming from nowhere, it produced a great deal of papers under the name of Universal Variables in the Newtonian problem of two bodies. The use of vector and matrix algebra and calculus is also still very rare, in books written basically more than a century after such tools were given a final form. Siegel's and Abraham's book show the process of evolution from classical to modern mathematical representation of exactly the same things. The definition of Lagrange's and Poisson's matrices is seldom found anywhere, and one has to refer to works on Quantum Mechanics and Field Theory. The proof of the symplectic condition for a canonical transformation is greatly simplified by the use of matrix notation. The connection between nonlinear circuit analysis and nonlinear mechanics methods and the classical averaging methods of Celestial Mechanics was clearly shown by Cesari in 1959. Equivalence statements between the KBM and von Zeipel's methods were first given in 1961 by Burstein and Sovolev. The efforts for a better theory of artificial satellites were certainly responsible for new researches in analytic and geometric theories. After 1960 it is obvious that a two way flux was established between researches in nonlinear oscillations and in Celestial Mechanics. Milestones were set by Moser, Hale and

Diliberto and, in the Soviet Union, by Kolmogorov, Arnol'd and Merman. A true departure from elaborations on works by Poincaré and Birkhoff was introduced by Hori with the application of Lie's Canonical Mappings. Lie's series appeared only one year before in Leimanis' book on Rigid Bodies Motion, but with no reference to perturbation techniques. The extension to non-canonical systems was presented by Hori at a Summer Institute in 1970 and, independently, worked out by Kamel. The extension is, nevertheless, not essential since any system can be written in Hamiltonian form as first shown by Dirac. The fact had been long known to researchers in Optimization and Control, although the great majority of Applied Mathematicians in other fields had and have not been aware of this important fact. Earliest references, to our knowledge, are the works of Miner, Tapley and Powers in 1967 and 1969. Finally, the operations with formal series as it is done with convergent series is well justified, e.g. by the work of Cartan. This is not the earliest reference, but is surely one of the best.

REFERENCES

1. Abraham, R., 1967, "Foundations of Mechanics", W. A. Benjamin Inc., Philadelphia.

2. Arnol'd, V. I., 1963, "Small Denominators and Problems of Stability of Motion in Classical and Celestial Mechanics", Uspeki Mat. Nauk USSR 18, 91-192.

3. Birkhoff, G. D., 1915, "Proof of Poincaré's Geometric Theorem", Trans. Amer. Math. Soc., 14, 14-22.

4. ________, 1915, "The Restricted Problem of Three Bodies", Rend. Circ. Mat. Palermo, 39, 1-115.

5. ________, 1927, "Dynamical Systems", Am. Math. Soc. Colloq. Pub., IX, Providence, R. I.

6. Breves, J. A., 1968, "A new proof of the conditions for a canonical transformation", Seminars, Univ. of São Paulo (in Cel. Mech., 6, No. 1, 1972).

7. Brouwer, D. and Clemence, G. M., 1961, "Methods of Celestial Mechanics", Academic Press, New York.

8. Brown, E. W., 1931, "Elements of Theory of Resonance", Rice Inst. Publ. 19, Houston.

9. Burstein, E. L. and Solovev, L. S., 1961, Dokl. Akad. Nauk USSR, 139, 855-858.

10. Campbell, J. A. and Jefferys, W. H., 1970, "Equivalence of the Perturbation Theories of Hori and Deprit", Cel. Mech., 2, 467.

11. Cartan, E., 1963, "Elementary Theory of Analytic Functions of One or Several Complex Variables", Addison-Wesley, Reading, Massachusetts.

12. Cesari, L., 1940, "Sulla Stabilità delle Soluzioni dei Sistemi di Equazioni Differenziali Lineari a Coefficienti Periodici", Atti Accad. Ital. Mem. Clas. Fis. Mat. e Nat., 11, 633-692.

13. ________, 1959, "Asymptotic Behavior and Stability Problems in Ordinary Differential Equations", Springer-Verlag, Berlin (Second Ed., 1963, Acad. Press, New York).

14. Danby, J. M. A., 1962, "Fundamentals of Celestial Mechanics", MacMillan, New York.

15. Darwin, G. H., 1911, "Scientific Papers", Cambridge Univ. Press, London.

16. Delaunay, C. E., 1860-1867, Mèm. Acad. Sci. Paris, 28, 29 (entire volumes).

17. Deprit, A., 1969, "Canonical Transformations Depending on a Small Parameter", Cel. Mech., 1, 12-30.

18. Diliberto, S. P. and Hufford, G., 1956, "Perturbation Theorems for Nonlinear Ordinary Differential Equations", Ann. Math. Studies 36, 207-236.

19. Diliberto, S. P., 1960 and 1961, "Perturbation Theorems for Periodic Surfaces. I and II", Rend. Circ. Mat. Palermo, 9, 265-299 and 10, 111.

20. Diliberto, S. P., Kyner, W. T. and Freund, R. F., 1961, "The Application of Periodic Surface Theory to the Study of Satellite Orbits", Astron. J., 66, 118-127.

21. Diliberto, S. P., 1967, "New Results on Periodic Surfaces and the Averaging Principle", U.S.-Japanese Semin. on Diff. Func. Equas., pp. 49-87, W. A. Benjamin, Inc., Philadelphia.

22. Dirac, P. A. M., 1958, "Generalized Hamiltonian Dynamics", Proceed. Roy. Soc. London, A246, 326-332.

23. Euler, L., 1772, "Theoria Motus Lunae, Novo Methodo", 775 pp., Petrop. (in Brown, E. W., "Lunar Theory", Dover Public., New York, 1960).

24. Gambill, R. A., 1954, "Criteria for Parametric Instability for Linear Differential Systems with Periodic Coefficients", Riv. Mat. Univ. Parma, 5, 169-181.

25. ________, 1955, Ibidem, 6, 37-43.

26. ________, 1956, Ibidem, 6, 311-319.

27. ________ and Hale, J. K., 1956, "Subharmonics and Ultraharmonics Solutions of Weakly Nonlinear Systems", J. Rat. Mech. Anal., 5, 353-398.

28. Giacaglia, G. E. O., 1964, "Notes on von Zeipel's Method", GSFC-NASA Publ. X-547-64-161, Greenbelt, Md.

29. ________, 1965, "Evaluation of Methods of Integration by Series in Celestial Mechanics", Ph.D. Dissertation, Yale University, New Haven.

30. Hale, J. K., 1954, "On the boundedness of the solution of linear differential systems with periodic coefficients," Riv. Mat. Univ. Parma, 5, 137-167.

31. ________, 1954, "Periodic Solutions of Nonlinear Systems of Differential Equations", Riv. Mat. Univ. Parma, 5, 281-311.

32. ________, 1958, "Sufficient Conditions for the Existence of Periodic Solutions of First and Second Order Differential Equations", J. Math. Mech., 7, 163-172.

33. ________, 1961, "Integral Manifolds of Perturbed Differential Equations", Ann. Math., 73, 496-531.

34. Hori, G., 1966, "Theory of General Perturbations with Unspecified Canonical Variables", Publ. Astron. Soc. Japan, 18, 287-296.

35. ________, 1970, "Lie Transformations in Nonhamiltonian Systems", Lecture Notes, Summer Institute in Orbital Mechanics, The Univ. of Texas at Austin, May 1970.

36. ________, 1971, "Theory of General Perturbations for Noncanonical Systems", Publ. Astron. Soc. Japan, 23, 567-587.

37. Kamel, A. A., 1970, "Perturbations Methods in the Theory of Nonlinear Oscillations", Cel. Mech., 3, 90-106.

38. Kolmogorov, N. A., 1954, "General Theory of Dynamical Systems and Classical Mechanics", Proc. Int. Cong. Math., Amsterdam, 1, 315-333, Noordhoff (1957).

39. Krylov, N. and Bogoliubov, N. N., 1934, "The Applications of Methods of Nonlinear Mechanics to the Theory of Stationary Oscillations", Publ. Ukranian Acad. Sci., No. 8, Kiev.

40. ________, 1947, "Introduction to Nonlinear Mechanics", Annals. Math. Studies, 11, Princeton Univ. Press, Princeton, New Jersey.

41. Kurth, R., 1959, "Introduction to the Mechanics of the Solar System", Pergamon Press, New York (Chapt. III, Section 3).

42. Kyner, W. T., 1964, "Qualitative Properties of Orbits about an Oblate Planet" Comm. Pure Appl. Math., 17, 227-231.

43. ________, 1967, "Averaging Methods in Celestial Mechanics", Proc. Intern. Astron. Union Symp., No. 25, Thessaloniki, Greece, 1964. Acad. Press, New York.

44. Leimanis, E., 1965, "The General Problem of Motion of Coupled Rigid Bodies about a Fixed Point", Springer-Verlag, New York (pp. 121-128).

45. Lie, M. S., 1888, "Theorie der Transformationgruppen", Teubner, Leipzig (vol. 1, I).

46. Lindstedt, A., 1882, "Beitrag zur Integration der Differentialgleichungen der Storungtheorie", Abh. K. Akad. Wiss. St. Petersburg, 31, No. 4 (See also: 1883, Comptes Rendus Acad. Sci. Paris, 3, and 1884, Bull. Astron., 1, 302).

47. MacMillan, W. D., 1920, "Dynamics of Rigid Bodies", Dover Publications, New York (pp. 403-413).

48. Merman, G. A., 1961, "Almost Periodic Solutions and the Divergence of Lindstedt's Series in the Planar Restricted Problem of Three Bodies", Bull. Inst. Theor. Astron. Leningrad, 8, 5.

49. Mersman, W. A., 1970, "A new algorithm for the Lie Transformation", Cel. Mech., 3, 81-89.

50. _________, 1970, "A Unified Treatment of Lunar Theory and Artificial Satellite Theory" in "Periodic Orbits, Stability and Resonances", Ed. G. E. O. Giacaglia, D. Reidel Pub. Co., Dordrecht, Holland.

51. _________, 1971, "Explicit Recursive Algorithm for the Construction of Equivalent Canonical Transformations", Cel. Mech., 3, 384-389.

52. Miners, W. E., Tapley, B. D. and Powers, W. F., 1967, "The Hamilton-Jacobi Method Applied to the Low-Thrust Trajectory Problem", Proc. 18th Cong. Intern. Astronaut. Fed., Belgrade, Yugoslavia.

53. Moulton, F. R. et al., 1920, "Periodic Orbits", Carnegie Inst. Wash. Publ., No. 161, Washington, D. C.

54. Moser, J., 1955, "Nonexistence of Integrals for Canonical Systems of Differential Equations", Comm. Pure Appl. Math., 8, 409-436.

55. _________, 1962, "On Invariant Curves of Area-Preserving Mappings of an Annulus", Nachr. Akad. Wiss. Gottingen Math. Phys. Kl. II, 1-20.

56. _________, 1968, "Lectures on Hamiltonian Systems", Mem. Amer. Math. Soc., 81, pp. 1-60.

57. Poincaré, H., 1886, "Sur la Méthode de M. Lindstedt", Bull. Astron., 3, 57.

58. _________, 1892-93-99, "Les Méthodes Nouvelles de la Mécanique Céleste", (3 vols.) Reprint by Dover Publ., New York, 1957.

59. _________, 1909-12 , "Leçons de Mécanique Céleste", Gauthier-Villars, Paris (vol. 2, part 2).

60. _________, 1912, "Sur un Théorème de Géometrie", Rend. Circ. Mat. Palermo 33, 375-407.

61. Poisson, S. D., 1834, "Mémoire sur la Variation des Constantes Arbitraires", J. École Polyt., 3, 266-344 (See also Ref. 41).

62. Powers, W. F. and Tapley, B. D., 1969, "Canonical Transformation Applications to Optimal Trajectory Analysis", AIAA Journal, 7, 394-399.

63. Shniad, H., 1969, "The Equivalence of von Zeipel Mappings and Lie Transforms", Cel. Mech., 2, 114-120. (For the background and earlier proof of the main theorem in Shniad's paper see: Lanczos, C., "The Variational Principles of Mechanics", Univ. of Toronto Press, Toronto, chapt. VII).

64. Siegel, C. L., 1941, "On the Integrals of Canonical Systems", Ann. Math., 42, 806-822.

65. ________, 1956, "Vorlesungen über Himmelsmechanik", Springer-Verlag, Berlin.

66. Strömgren, E., (see Ref. 67, pp. 550-555).

67. Szebehely, V., 1967, "Theory of Orbits - The Restricted Problem of Three Bodies", Acad. Press, New York.

68. Tisserand, F., 1896, "Traité de Mécanique Céleste", (4 vols.) Gauthier-Villars, Paris, (see also 1868, Thèse de Doctorat, J. de Liouville).

69. Wintner, A., 1941, "The Analytical Foundations of Celestial Mechanics", Princeton Univ. Press, Princeton, New Jersey.

CHAPTER II

PERTURBATION METHODS FOR HAMILTONIAN SYSTEMS. GENERALIZATIONS

1. Introduction.

This chapter is devoted to two main goals. First introduce the reader to known methods of canonical perturbations, describe them in a heuristic way and give examples so as to motivate the theorems presented in Chapters III and IV. Second, present some basic results about iterative procedures of fundamental importance on methods of averaging. Major contributors to this area are Lindstedt (1884), Poincaré (1893), Whittaker (1916), Siegel (1941), Krylov (1947), Bogoliubov (1945), Kolmogorov (1953), Arnol'd (1963), Diliberto (1961), Pliss (1966), Kyner (1961), Moser (1962), Hale (1961) with several overlappings in results. Many of these results have been unified and consolidated in celebrated books by Siegel (1956), Wintner (1947), Newytskii-Stepanov (1960), Cesari (1963), Hale (1969), Abraham (1967), Birkhoff (1927), Bogoliubov-Mitropolskii (1961), Lefschetz (1959), Minorsky (1962), Sansone-Conti (1964), Sternberg (1970).

It is a recognized fact, although several times not mentioned, that the averaging methods were introduced by Lindstedt (1882), though it is not clear whether his ideas stemmed from the efforts of Euler (1750) in the solution of the problem of motion of the moon. In linear periodic systems, an averaging method leads directly and essentially to the determination of Floquet's characteristic exponents. In non-linear systems, when they possess a Hamiltonian character, to the separation of the associate Hamilton-Jacobi equation and therefore the specification of the action and angle variables. In general non-linear and non-Hamiltonian systems, an averaging method leads to separability in an extended space, which can be called the cotangent space of the original system space. In regard to Hamiltonian systems, it has been an accepted and recognized result the fact that in general, they are not integrable. Nevertheless, such notion should be considered with care, depending on the definition of integrability. In fact, if the Hamiltonian is at least C^2 in a certain open region D of the phase space, there exists and is unique a solution corresponding to any initial point in D. In this respect, the system is certainly integrable. On the other hand, the word integrability is, in Hamiltonian systems, often associated with the idea of separability, so that an integrable system is a Stäckel's (or in particular, Liouville's) system. The two concepts can be associated by recalling the fact that if a solution exists and is unique for a time $0 \leq t < T$, then the motion in phase space is area preserving (or, the divergent of a Hamiltonian flow is zero). It is also true that such flow is canonical so that any point P(t) of the solution $(0 \leq t < T)$ is related to the initial point P(0) by a canonical transformation, which for t sufficiently small is C^2 and invertible. It follows that, in terms of the initial conditions, taken as a particular set of canonical variables, the system must necessarily be separable, for the Hamiltonian is reduced to a constant. Of course, such type of separability can only be achieved after the solution is known explicitly as a function of time and of the initial conditions, so that no help can come from such results. However, it serves to indicate the connection between the two concepts of integrability mentioned above.

As far as periodic linear systems are concerned we know, under quite general conditions, that the solution exists and has a well defined form as given by Floquet's theory.

For non-linear systems in general, integrability can only be understood as existence and uniqueness of solution. However, a connection with the idea of separability can be established by the "Hamiltonianization" of the system in the cotangent space, as will be shown later.

Most of the results concerning non-integrability are based on the existence of integrals in the vicinity of singular points (Siegel, 1941) or on the reducibility to Birkhoff's normal form by power series or on the convergence of iterative procedures. The negation of the above results does not evidently imply non-integrability. It was proved by Birkhoff that a normal form for Hamiltonian systems obtained by means of a series cannot in general be achieved. If the averaging methods are a translation, into some different language, of Birkhoff's normalization, then we cannot, in general, conclude on the divergence of these since we know that manipulation of a series does change its convergence character. Indeed, we shall formulate, as an example, an averaging method equivalent to a normalization and we shall expect divergence in general. On the other hand, averaging methods can be generalized, redefined, restated, and the perturbations subjected to such conditions that, such methods may converge at least for a certain set of initial conditions. In specific examples, adelphic integrals defined by formal series (Contopoulos, 1966, 1967) have shown remarkable character of true integrals of motion when submitted to a numerical verification for very long periods of time. The method of surface of section (Poincaré, 1893) has served an invaluable service in the search of possible integrals and has shown that integrals (not necessarily uniform or globally valid) may exist for systems notably defined as non-integrable (Bozis, 1970).

2. Convergence of a Classical Method of Iteration.

If one limits the time interval properly, it can be shown that under quite general conditions, the simplest method of successive approximation of solution by series, converges. In fact, we have the following results (MacMillan, 1912).

Let us initially consider a system of n equations in $x_1, x_2, \ldots, x_n$, depending on a parameter ϵ,

$$F_i(x;\epsilon) = 0; \quad i = 1,2,\ldots,n, \tag{2.2.1}$$

where x is the set $(x_1, x_2, \ldots, x_n)$. Further, suppose

a) $F_i(0;0) = 0; \quad i = 1,2,\ldots,n.$

b) $J = \det \dfrac{\partial(F_1, F_2, \ldots, F_n)}{\partial(x_1, x_2, \ldots, x_n)} \neq 0, \quad \text{for} \quad x = 0,\ \epsilon = 0.$

c) $\dfrac{\partial F_i}{\partial \epsilon} \neq 0, \quad \text{for some} \quad i, \quad \text{at} \quad x = 0.$

It follows that the functions F_i can be developed about the point $x = 0$, $\epsilon = 0$, in powers of x and ϵ. Then, one can easily prove that, if the F_i are analytic in their arguments in a certain region $(x;\epsilon)$, the series

$$x_j = \sum_{s=1}^{\infty} \epsilon^s a_{js} \tag{2.2.2}$$

obtained by successive approximations converge uniformly (in ϵ). The a_{js} are obtained by substituting the x_j into the expansions of F_i and equating coefficients of the same powers in ϵ. The proof of this can be found in any standard book of Analysis (e.g. Goursat, 1959). For the purpose of later use some details are needed. The expansion of $F_i(x;\epsilon)$ gives

$$F_i(x;\epsilon) = \left(\frac{\partial F_i}{\partial \epsilon}\right)_o \epsilon + \sum_{j=1}^{n} \left(\frac{\partial F_i}{\partial x_j}\right)_o x_j$$

$$+ \frac{1}{2} \sum_{j=0}^{n} \sum_{k=0}^{n} \left(\frac{\partial^2 F_i}{\partial x_j \partial x_k}\right)_o x_j x_k + \dots = 0$$

where, for uniformity of notation $x_o \equiv \epsilon$. The above expansion can be written

$$\sum_{j=1}^{n} \left(\frac{\partial F_i}{\partial x_j}\right)_o x_j \equiv b_o^{(i)} x_o + \sum_{k,j=0}^{n} b_{jk}^{(i)} x_j x_k$$
$$+ \sum_{\ell,k,j=0}^{n} b_{\ell kj}^{(0)} x_j x_k x_\ell + \dots \; . \tag{2.2.3}$$

If the series (2.2.2) are substituted into (2.2.3), the comparison of coefficients of same powers in ϵ (or x_o) gives

$$\sum_{j=1}^{n} \left(\frac{\partial F_i}{\partial x_j}\right)_o a_{j1} = b_o^{(i)}$$
$$\sum_{j=1}^{n} \left(\frac{\partial F_i}{\partial x_j}\right)_o a_{j2} = \sum_{j=0}^{n} \sum_{k=0}^{n} b_{jk}^{(i)} a_{j1} a_{k1} \tag{2.2.4}$$

$$\sum_{j=1}^{n} \left(\frac{\partial F_i}{\partial x_j}\right)_o a_{jp} = \Phi_{ip}(b_{j_o j_1 \dots j_k}, a_{rs}) \tag{2.2.5}$$

Therefore, at every step, the a_{jp} are computed from a given system of n equations whose right-hand sides are known if all previous approximations a_{j1}, $a_{j2},\ldots,a_{j,p-1}$ are known. The determinant of the system is not zero by hypothesis (b). From a formal point of view, equations (2.2.4) are totally similar to the sequence of linear inhomogeneous partial differential equations one encounters in the averaging method of Lindstedt-Poincaré or else in Lie's series asymptotic solutions.

Now consider the case in which $J = 0$ and assume at least one of its first minors is not zero. For instance, suppose $\partial F_i/\partial x_1 = 0$. Then, (n-1) of the equations (2.2.4) can be solved in terms of $x_2, x_3,\ldots,x_n$, as power series in x_0 and x_1. If the results are substituted in the n-th of the equations (2.2.4), an equation in x_0 and x_1 will then result. Since the coefficient of the first power of x_1 will be zero, then the solution of x_1 in terms of power series of x_0 will necessarily contain fractional powers of this parameter. This is a direct consequence of Weierstrass theorem on the factorization of a power series. The use of the "eliminating determinants" defined by Caley (1848) allows the solution in the case where all the first minors are zero. MacMillan (1912) further developed the method. The appearance of fractional powers in these cases has a direct consequence on the appearance of fractional powers in asymptotic series solutions to be developed later in problems of resonance.

Next, consider a system of differential equations

$$\dot{x}_i = \epsilon f_i(x;\epsilon,t) \tag{2.2.6}$$

for $i = 1,2,\ldots,n$, where f_i are analytic in $(x;\epsilon,t)$ for $x \in D$ (a given open region of R^n), $0 \leq \epsilon \leq 1$, $t \in R$, and regular at $x_i = \alpha_i$ $(i = 1,2,\ldots,n)$, $\epsilon = 0$, for all values of $t \in [0,T]$. The functions f_i are developable in power series of $\xi_i = x_i - \alpha_i$ and ϵ.

These series are convergent, provided in the interval $[0,T]$

$$|x_i - \alpha_i| \leq m_i \quad (i = 1,2,\ldots,n),$$

and $0 < \epsilon \leq \epsilon_o \leq 1$.

The expansion of (2.2.6) gives

$$\dot{\xi}_i = \epsilon[f_i(\alpha;0,t) + \sum_{j=0}^{n} (\frac{\partial f_i}{\partial x_j})_o \xi_j + \frac{1}{2} \sum_{j,k=0}^{n} (\frac{\partial f_i}{\partial x_j \partial x_k})_o \xi_j \xi_k + \ldots]$$

where $\xi_o = x_o = \epsilon$, $\alpha_o = 0$ and the subscript zero means that x_i are replaced by the α_i. We can actually write

$$\dot{\xi}_i = \epsilon[\phi_o^{(i)} + \sum_{j=0}^{n} \phi_j^{(i)} \xi_j + \sum_{j,k=0}^{n} \phi_{jk}^{(i)} \xi_j \xi_k + \ldots] \qquad (2.2.7)$$

where the ϕ's are functions of the α's and t. The goal is to obtain the ξ_i as power series in ϵ, with coefficients functions of time and of the constants α_i, that is

$$\xi_i = \sum_{k=0}^{\infty} \xi_k^{(i)} \epsilon^k \qquad (2.2.8)$$

where

$$\xi_k^{(i)} = \xi_k^{(i)}(\alpha;t), \quad i = 1,2,\ldots,n.$$

If (2.2.8) are substituted into (2.2.7) and coefficients of the same power of ϵ compared in both sides, there results a system of differential equations

$$\dot{\xi}_1^{(i)} = \phi_o^{(i)}$$

$$\dot{\xi}_2^{(i)} = \sum_{j=1}^{n} \phi_j^{(i)} \xi_1^{(j)}$$

$$\dot{\xi}_3^{(i)} = \sum_{j=1}^{n} \phi_j^{(i)} \xi_2^{(j)} + \sum_{k,j=1}^{n} \phi_{jk}^{(i)} \xi_1^{(j)} \xi_1^{(k)}$$

$$\dot{\xi}_p^{(i)} = \sum_{j=1}^{n} \phi_j^{(i)} \xi_p^{(j)} + F_p^{(i)} \qquad (2.2.9)$$

for $p = 1,2,3,\ldots$. The functions $F_p^{(i)}$ depend on the solution of all approximations up to stage p-1, so that at every stage of such sequence of approximations

$$\dot{\xi}_p^{(i)} = \phi_p^{(i)}(\xi_k^{(j)}, t, \beta_\ell) \quad \text{for} \quad i = 1,2,\ldots,n; \quad j = 1,2,\ldots,n;$$

$$k = 1,2,\ldots,p-1; \quad \ell = 1,2,\ldots,n(p-1).$$

The $\xi_p^{(i)}$ are obtained by quadratures. The constants of integration β are not arbitrary. In fact, if one stops at the p-th stage included, the solutions will depend on the initial constants $\alpha_1, \alpha_2, \ldots, \alpha_n$ and on np constants β. One might prefer the choice of setting all the β's equal to zero or the choice of defining them in a convenient way. In the second case they will be functions of the α's. If the constants α_k are initial conditions, that is, $x_i(0) = \alpha_i$, then the constants β's should be chosen so as to make all the $\xi_p^{(i)}$ vanish at $t = 0$. We now show that the series ξ_i obtained in this way are convergent in $[0,T]$ provided ϵ is sufficiently small.

Without loss of generality one can assume that the right members of (2.2.7) are convergent for $|\xi_i| \leq 1$, $\epsilon \leq 1$ in $[0,T]$. If this is not true, a change of scale for ξ_i and ϵ will always make this assumption possible. It follows that all coefficients ϕ in the right members of (2.2.7) are bounded and less than a positive number M, i.e..

$$|\phi_{j_1 j_2 \ldots j_k}^{(i)}| \leq M$$

for $i = 1,2,\ldots,n$ and $k = 1,2,3,\ldots$. The concept of majorant series can now be used. In fact, consider the equations

$$\dot{\eta}_i = \frac{M\epsilon}{(1-\epsilon)(1-\sum_{j=1}^{n} \eta_j)}, \quad (i = 1,2,\ldots,n). \tag{2.2.10}$$

The right-hand members can be expanded in power series of ϵ and η_j, with $|\epsilon| < 1$, $|\Sigma\, \eta_j| < 1$. Every coefficient is positive and greater than the corresponding coefficient in (2.2.7), in view of the foregoing hypotheses. Equations (2.2.10) can be solved by the method of successive approximations just described. It follows that the right-hand members of the equations (2.2.9) will be less than the corresponding ones for (2.2.10). Thus, if the solution of (2.2.10) converges, the solution of (2.2.9) also converges. But (2.2.10) can be integrated in closed form. If the initial values are all zero (as for the ξ_i if the α_i are initial conditions), it must result that

$$\eta_1 = \eta_2 = \dots = \eta_n = \eta$$

or

$$\dot{\eta} = \frac{M\epsilon}{(1-\epsilon)(1-n\eta)}$$

and, therefore,

$$\eta = \frac{1}{n}[1 - (1 - \frac{2Mn\epsilon t}{1-\epsilon})^{1/2}] \tag{2.2.11}$$

which satisfies the condition $\eta = 0$ for both $t = 0$ and $\epsilon = 0$. The expansion of (2.2.11) in power series of ϵ is convergent provided

$$|\frac{2Mn\epsilon t}{1-\epsilon}| < 1$$

in $[0,T]$, that is, provided

$$|\epsilon| < \frac{1}{1+2nMT} = \epsilon_0. \tag{2.2.12}$$

Since the method of successive approximations given is unique, it must coincide with the expansion of (2.2.11). Thus, the series for ξ_i are convergent in $[0,T]$ if $|\epsilon| < \epsilon_0$, where M is the upper bound for the coefficients of (2.2.7). It is seen that, for T large enough, the series only converges, in general, for $\epsilon \to 0$. The above estimate cannot be considered the best possible, so that the term "in general" is kept in for there are actual situations where the method described converges for ϵ small enough, but not zero, as $T \to \infty$.

Consider now the system of differential equations

$$\dot{x}_i = g_i(x;t) + \epsilon f_i(x;\epsilon,t), \quad (i = 1,2,\ldots,n). \tag{2.2.13}$$

Substituting the α_i of the previous method by the solutions $x_{io}(t)$ of (2.2.13) for $\epsilon = 0$, and defining

$$\xi_i = x_i - x_{io},$$

in the same way it is found that the coefficients $\xi_p^{(i)}$ satisfy differential equations of the type

$$\dot{\xi}_p^{(i)} = \sum_{j=1}^{n} \phi_j^{(i)} \xi_p^{(j)} + \Phi_p^{(i)}(\xi_k^{(\ell)};t) \tag{2.2.14}$$

where $k = 1,2,\ldots,p-1$; $\ell = 1,2,\ldots,n$; $i = 1,2,\ldots,n$; $p = 1,2,3,\ldots$. It is a remarkable property that the $\phi_j^{(i)}$ are independent of the particular p, as before, so that the homogeneous solutions of (2.2.14) are the same for any p. They are functions of the time explicitly and of $x_{io}(t)$, these last being functions of a set of n integration constants. As far as the integration constants for (2.2.14) they can be chosen so as to make the $\xi_p^{(i)} = 0$ at $t = 0$, and, using the terminology of Celestial Mechanics, in this case the solutions ξ_i and x_{io} are osculating at $t = 0$. There are other ways in which the constants of integration can be chosen, but this requires a modification due to the fact that the expansions are not done in the neighborhood of $\xi_i = 0$ (at $t = 0$).

Picard's classical method of approximations allows to show that the solution of a system

$$\dot{\xi}_i = \sum_{j=1}^{n} \phi_{ij}(t)\xi_j + k_i(t), \quad (i = 1,2,\ldots,n),$$

is dominated in the interval $[0,T]$ by the solution of the system

$$\dot{\eta}_i = M \sum_{j=1}^{n} \eta_j + M, \quad (i = 1,2,\ldots,n),$$

where M is the upper bound of the $\phi_{ij}(\epsilon)$ and $k_i(t)$ in the interval $[0,T]$. Then, in a similar way as was done before, one proves that, by using the majorant functions defined by

$$\dot{\eta}_i = M \frac{(\epsilon+\eta_1 + \dots + \eta_n)}{1-(\epsilon+\eta_1 + \dots + \eta_n)} ,$$

the series ξ_i are convergent in $[0,T]$ if

$$|\epsilon| < \exp[-MnT] \tag{2.2.15}$$

where M is the upper bound for the coefficients of (2.2.13) as power series of ξ_j and ϵ. The limitation one obtains in this case is much stronger, as T becomes large, than in the previous case. These cases have been discussed in details by Moulton and others (1920). However, as we shall see later, (2.2.15) may not be the best estimate for this case.

3. Secular Terms. Lindstedt's Device.

The above described methods have the classical characteristic of leading to secular terms, that is, series solutions where the $\xi_p^{(i)}$ contain terms which are linear (at least) in t. If such phenomenon could be avoided, and, more specifically, one could get $\xi_p^{(i)}(t)$ bounded for all t (say, almost periodic or periodic) the rate of convergence would certainly be improved and in special situations, as will be seen in the next chapter, actual convergence for all t can be obtained for sufficiently small ϵ.

At this moment we apply the method described in the previous section to the simple pendulum, show the appearance of secular terms and introduce Lindstedt's device in this particular application. For simplicity we shall assume that the initial conditions correspond to the libration case of the pendular motion, that is, oscillations of finite amplitude around the stable equilibrium solution. The equation of motion can be written as

$$\ddot{\theta} = -\omega_0^2 \sin\theta \tag{2.3.1}$$

where $\omega_0^2 = g/\ell$. Consider the convergent expansion of $\sin\theta$ in powers of θ and the change of variable $\theta = \sqrt{\epsilon}\, x$, so that (2.3.1) becomes

$$\ddot{x} + \omega_o^2 x = -\omega_o^2 \sum_{n=1}^{\infty} (-1)^n \frac{\epsilon^n x^{2n+1}}{(2n+1)!} \quad . \tag{2.3.2}$$

For $\epsilon = 0$ (infinitesimal oscillations), the solution is

$$x_o(t) = A \sin (\omega_o t + \alpha) \tag{2.3.3}$$

and let us consider the series

$$\xi = x - x_o = \sum_{m=1}^{\infty} \epsilon^m \xi_m(t)$$

or, the solution in the vicinity of $x_o(t)$, as given by

$$x = x_o(t) + \sum_{m=1}^{\infty} \epsilon^m \xi_m(t). \tag{2.3.4}$$

The method just described, substitutes (2.3.4) for x into (2.3.2) and equate coefficients of same powers of ϵ. As the first few approximations we find

$$\ddot{\xi}_1 + \omega_o^2 \xi_1 = \frac{1}{3!} \omega_o^2 x_o^3$$

$$\ddot{\xi}_2 + \omega_o^2 \xi_2 = \frac{1}{2!} \omega^2 x_o^2 \xi_1 - \frac{1}{5!} \omega_o^2 x_o^5$$

$$\ddot{\xi}_3 + \omega_o^2 \xi_3 = \frac{1}{2!} \omega_o^2 (x_o^2 \xi_2 + x_o \xi_1^2) - \frac{1}{4!} \omega_o^2 x_o^4 \xi_1 + \frac{1}{7!} \omega_o^2 x_o^7$$

$$\ddot{\xi}_4 + \omega_o^2 \xi_4 = \frac{1}{3!} \omega_o^2 (3x_o^2 \xi_3 + 6x_o \xi_1 \xi_2 + \xi_1^3) - \frac{1}{4!} \omega_o^2 (x_o^4 \xi_2 + 2x_o^3 \xi_1^2)$$

$$+ \frac{1}{6!} \omega_o^2 x_o^6 \xi_1 - \frac{1}{9!} \omega_o^2 x_o^9 \tag{2.3.5}$$

Let us analyze the solution for ξ_1 which makes $\xi_1 = \dot{\xi}_1 = 0$ at $t = 0$. With a proper choice for the unit of time we shall consider $\omega_o = 1$ without loss in generality. We also have that the particular solution of

$$\ddot{z} + z = a \sin p(t + \alpha)$$

is

$$z = \frac{a}{1-p^2} \sin p(t + \alpha), \quad p \neq 1,$$

$$z = -\frac{1}{2} a\, t \sin(t + \alpha), \; p \neq 1,$$

and of

$$\ddot{z} + z = a \cos p(t + \alpha)$$

is

$$z = \frac{a}{1-p^2} \cos p(t + \alpha), \quad p \neq 1,$$

$$z = \frac{1}{2} a\, t \sin(t + \alpha), \quad p = 1.$$

It easily follows that

$$\xi_1 = B \sin(t + \beta) - \frac{1}{16} A^3 t \sin(t + \alpha) + \frac{1}{192} A^3 \sin 3(t + \alpha) \qquad (2.3.6)$$

where B, β are given by

$$B \sin \beta = -\frac{A^3}{192} \sin 3\alpha,$$

$$B \cos \beta = -\frac{A^3}{64} \cos 3\alpha + \frac{A^3}{16} \sin \alpha.$$

In this particular example it is seen that a secular term appears in (2.3.6), that is, in the first approximation. (Actually, that is more often called a mixed secular term.) Evidently, the appearance of t outside trigonometric functions makes it quite difficult to have convergence of the above process for $t \to \infty$. The constants of integration B, β cannot in any way be used to cancel the troublesome term. The solution proposed by Lindstedt (1882) is to assume a reference solution, that is, a function $x_o(t)$ which is a modification of the zero order solution as far as the frequency is concerned. In fact, we consider

$$x_o(t) = A \sin(\omega t + \alpha) \qquad (2.3.7)$$

where we assume

$$\omega^2 = \omega_o^2 + \epsilon\omega_1 + \epsilon^2\omega_2 + \dots$$

or, taking $\omega_o = 1$ as before,

$$\omega^2 = 1 + \epsilon\omega_1 + \epsilon^2\omega_2 + \dots \tag{2.3.8}$$

where $\omega_1, \omega_2, \dots$ are constants (depending on A, α) to be conveniently chosen. By writing the equation (2.3.2) as

$$\ddot{x} + \omega^2 x - \epsilon\omega_1 x - \epsilon^2\omega_2 x - \dots = -\sum_{n=1}^{\infty} (-1)^n \epsilon^n \frac{x^{2n+1}}{(2n+1)!}$$

with the "zero" order solution

$$x_o(t) = A \sin(\omega t + \alpha)$$

where ω is given by (2.3.8), and unknown a priori, we obtain, as before

$$\ddot{\xi}_1 + \omega^2 \xi_1 - \omega_1 x_o = \frac{1}{3!} x_o^3$$

$$\ddot{\xi}_2 + \omega^2 \xi_2 - \omega_1 \xi_1 - \omega_2 x_o = \frac{1}{2!} x_o^2 \xi_1 - \frac{1}{5!} x_o^5$$

or

$$\ddot{\xi}_1 + \omega^2 \xi_1 = \omega_1 x_o + \frac{1}{3!} x_o^3$$

$$\ddot{\xi}_2 + \omega^2 \xi_2 = \omega_1 \xi_1 + \omega_2 x_o + \frac{1}{2!} x_o^2 \xi_1 - \frac{1}{5!} x_o^5$$

and the right-hand members are evidently odd functions of $(\omega t + \alpha)$, that is, sine series in $(\omega t + \alpha)$. In the equation for ξ_p the corresponding unknown approximation ω_p has to be determined so that secular (or mixed secular, in this case) terms should be avoided. The first order equation is

$$\ddot{\xi}_1 + \omega^2 \xi_1 = \omega_1 A \sin(\omega t + \alpha) + \frac{1}{8} A^3 \sin(\omega t + \alpha)$$
$$- \frac{1}{24} A^3 \sin(3\omega t + 3\alpha)$$

so that, defining

$$\omega_1 = -\frac{1}{8}A^2$$

the resonant forcing term is eliminated and the solution is

$$\xi_1 = B\sin(\omega t + \beta) + \frac{1}{192}A^3 \sin(3\omega t + 3\alpha)$$

where B, β can be defined by

$$B\sin\beta + \frac{1}{192}A^3 \sin 3\alpha = 0$$

$$B\cos\beta + \frac{1}{64}A^3 \cos 3\alpha = 0$$

that is,

$$\xi_1 = \dot{\xi}_1 = 0 \quad \text{at} \quad t = 0.$$

It is easily seen that to any order of approximation the equation to be integrated is

$$\ddot{\xi}_p + \omega^2\xi_p = \omega_p x_0 + A_1^p(A,\omega_1,\omega_2,\ldots,\omega_{p-1}) \sin(\omega t + \alpha)$$
$$+ \sum_{j=1}^{n_p} A_j^p(A,\omega_1,\omega_2,\ldots,\omega_{p-1}) \sin[(2j+1)(\omega t + \alpha)]$$

and the solution is found by setting

$$\omega_p = -A_1^p(A,\omega_1,\omega_2,\ldots,\omega_{p-1})$$
$$\xi_p = \sum_{j=1}^{n_p} \frac{A_j^p}{\omega^2[1-(2j+1)^2]} \sin[(2j+1)(\omega t + \alpha)].$$

It follows that the frequency ω is determined step by step and the solution is expressed as a purely periodic function of t, that is,

$$x = x_0(t) + \sum_{p=1}^{\infty} \epsilon^p \sum_{j=1}^{n_p} \frac{A_j^p}{\omega^2[1-(2j+1)^2]} \sin[(2j+1)(\omega t+\alpha)].$$

In this specific example, since the original equation can be integrated exactly, the convergence of the above procedure can be proved directly as long as the initial conditions are such that an oscillatory motion is verified. The series above diverges in case where the actual motion is a circulation. The case of asymptotic motion cannot, as far as we know, be dealt with an approximation of series. The circulation case can be made convergent by assuming a different change of variable. In fact, in this case, the angle θ increases steadily with time beside undergoing fluctuations. The steady increase with time must be taken care of by assuming

$$\theta = \alpha t + \sqrt{\epsilon}\, x$$

where

$$\alpha = \alpha_0 + \epsilon\alpha_1 + \epsilon^2\alpha_2 + \ldots \quad ,$$

$$x = x_0(t) + \epsilon\xi_1 + \epsilon^2\xi_2 + \ldots \quad ,$$

$$x_0 = A\, \sin(\omega t + \beta)$$

and

$$\omega^2 = \omega_0^2 + \epsilon\omega_1 + \epsilon^2\omega_2 + \ldots \quad .$$

When dealing with the canonical equivalent of Lindstedt's method we shall indicate as both cases (libration and circulation) can be treated in a unique fashion. This is possible by introducing elliptic functions with modulus of any value. The asymptotic case, then, will be given by a limiting case of the global solution. The possibility of such global solutions has been studied in details by Garfinkel and others (1971).

4. Poincaré's Method (Lindstedt's Method).

The method of successive adjustment of the frequencies of the system, for which we gave an example in the last section, applies to any system of ordinary differential equations which can be written in normal form and satisfies certain conditions of regularity at least locally. It is however desirable if such regularity extends over a certain domain. In this case we may assume the system to have the form

$$\dot{x}_i = f_i(x;t,\epsilon), \quad (i = 1,2,3,\ldots,n)$$

or, in vector form,

$$\dot{x} = f(x;t,\epsilon). \tag{2.4.1}$$

By the well known transformation to Dirac's cotangent space, system (2.4.1) can be brought into a canonical form, by defining the associate generalized momentum vector y $(y_i;\ i = 1,2,\ldots,n)$ and the Hamiltonian

$$H = f^T(x;t,\epsilon)y. \tag{2.4.2}$$

The equations of motion are

$$\dot{x} = H_y^T = f(x;t,\epsilon)$$

$$\dot{y} = H_x^T = -f_x(x;t,\epsilon)y$$

and we assume that the system

$$\dot{\xi} = f(\xi;t,0)$$

$$\dot{\eta} = -f_\xi(\xi;t,0)\eta$$

is integrable in some domain D of the 2n-dimensional phase space $(\xi;\eta)$ and for $0 \le t \le T$. We assume $f(x;t,\epsilon)$ to be at least C^2 in D, continuous with respect to t in $[0,T]$ and analytic in ϵ for $0 \le \epsilon \le 1$. The same properties are, therefore, also verified by the function H.

By these means, very little restriction is set by assuming a given system of equations to be Hamiltonian and, henceforth, the importance of perturbation methods in Hamiltonian systems.

With the above considerations it seems logical to ask whether a better estimate than (2.2.15) can be obtained. The Hamiltonian of the system under consideration (2.2.13) is

$$H = \sum_{i=1}^{n} y_i g_i(x;t) + \epsilon \sum_{i=1}^{n} y_i f_i(x;\epsilon,t) = H_o + \epsilon H_1 \tag{2.4.3}$$

and the canonical conjugate equations

$$\dot{x}_i = \frac{\partial H}{\partial y_i} = g_i + \epsilon f_i \tag{2.4.4}$$

$$\dot{y}_i = - \frac{\partial H}{\partial x_i} = - \sum_j y_j \frac{\partial g_j}{\partial x_i} - \epsilon \sum_j y_j \frac{\partial f_j}{\partial x_i} .$$

For $\epsilon = 0$, we assume that the system

$$\dot{x}_i = g_i(x;t)$$

$$\dot{y}_i = - \sum_j \frac{\partial g_j}{\partial x_i} y_j \tag{2.4.5}$$

is integrable. In fact, the first set is integrable by hypothesis and the solution is $x_i = x_{io}(t)$. Substitution of this solution into the second set gives a linear system

$$\dot{y}_i = \sum_j a_{ij}(t) y_j$$

which is, evidently, integrable, for t in the interval of definition of $x_{io}(t)$. Let the solution of (2.4.5) be written as

$$x_i = x_{io}(\alpha;\ \beta;\ t)$$

$$y_i = y_{io}(\alpha;\ \beta;\ t)$$

with

$$x_{io}(\alpha;\ \beta;\ 0) = \alpha_i$$

$$y_{io}(\alpha;\ \beta;\ 0) = \beta_i$$

for $i = 1,2,\ldots,n$. It follows from Jacobi's theorem that the solution of system (2.4.4) can be written as

$$x_i = x_{io}(\alpha;\ \beta;\ t)$$

$$y_i = y_{io}(\alpha;\ \beta;\ t)$$

if α, β are functions of t satisfying the equations

$$\dot{\alpha}_i = \epsilon \frac{\partial H_1}{\partial \beta_i}$$
$$\dot{\beta}_i = -\epsilon \frac{\partial H_1}{\partial \alpha_i} \qquad (2.4.6)$$

for $i = 1,2,\ldots,n$. But system (2.4.6) is of the type studied earlier [Equation (2.2.6)] and the application of the method of successive approximations will give the convergence criterion

$$|\epsilon| < \frac{1}{1 + 4nM'T}$$

which, if $M' \simeq M$, is a better estimate than (2.2.15) for the system (2.2.13).

We now return to the main purpose of this section and outline the general principle and rationale of Lindstedt's device as explained by Poincaré in canonical language. Let us consider a conservative dynamical system defined by the Hamiltonian

$$H = H(y;x;\epsilon) \qquad (2.4.7)$$

where y, x are n-dimensional vectors defined in phase space of dimension $2n$, ϵ is a dimensionless constant parameter and H is real analytic in some domain D of the phase space and for ϵ in $[0,1]$. We stress the fact that any analytic system $\dot{z} = f(z;\epsilon)$ can be reduced to the Hamiltonian form above, by introducing the cotangent phase space. Hamilton's principal function $W(y;X;\epsilon)$ is defined by the partial differential equation

$$H(y; \frac{\partial W}{\partial y}; \epsilon) = K(X;\epsilon) \qquad (2.4.8)$$

where $K(X;\epsilon)$ is obviously the Hamiltonian of the system written in terms of the new variables $(Y;X)$ defined by

$$Y_k = \frac{\partial W}{\partial X_k} = Y_k(y;X;\epsilon),$$

$$x_k = \frac{\partial W}{\partial y_k} = x_k(y;X;\epsilon), \qquad (2.4.9)$$

for $k = 1,2,\ldots,n$. Under the conditions specified for H, a function W satisfying

Equation (2.4.8) certainly exists (in the Jacobi sense) since the system of differential equations generated by (2.4.7) has a unique solution in D. The solution is evidently an analytic function of ϵ and the n constants of integration $X_1, X_2, \ldots, X_n$, in D. We assume that the system of differential equations generated by $H(y;x;0) = H_o(y;x)$ is integrable in the Liouville sense, that is, there exist n first integrals of motion in D, uniform and independent. If $x_1', x_2', \ldots, x_n'$ are such integrals, that is,

$$x_k'(y;x) = \alpha_k$$

along the solutions of (2.4.7) for $\epsilon = 0$ and in D, in general, the angular variables y_k' canonically associated to the action variables x_k' have frequencies (in time) which are linearly independent over the set of integers and, therefore, the motion is quasiperiodic (almost periodicity would, in this case, correspond to a system with an infinite number of basic frequencies). In terms of these action-angle variables the Hamiltonian (2.4.7) can be written as $H'(y';x';\epsilon)$ with the obvious condition

$$H'(y';x';0) = H_o'(x').$$

It is therefore with no loss of generality that, under the assumption that $H_o(y;x)$ leads to integrability (in the above specified sense), it can be thought as being a function of the momenta (x) only. It is also logical to expect that almost everywhere in D the frequencies $\omega_k^o = \partial H_o/\partial x_k$ are linearly independent over the integers. This implies, in particular, that none of these frequencies are zero in D, or, more precisely, none of the x_k momenta are ignorable. The problem is now reduced to one for which $H(y;x;0)$ is independent of y and therefore Hamilton's principal function $W(y;X;0)$ is a generator for the identity transformation, that is,

$$W(y;X;0) = y \cdot X.$$

We assume that W is analytic with respect to ϵ at $\epsilon = 0$, and therefore, for ϵ sufficiently small,

$$W(y;X;\epsilon) = y \cdot X + \epsilon S(y;X;\epsilon), \qquad (2.4.10)$$

with

$$S(y,X;\epsilon) = S_1(y;X) + \epsilon S_2(y;X) + \dots \tag{2.4.11}$$

a convergent power series in ϵ.

It follows that (2.4.9) can be written as

$$Y_k = y_k + \epsilon \frac{\partial S}{\partial X_k} = y_k + \epsilon F_k(y;X;\epsilon)$$

and

$$x_k = X_k + \epsilon \frac{\partial S}{\partial y_k} = X_k + \epsilon G_k(y;X;\epsilon) \tag{2.4.12}$$

for $k = 1,2,\dots,n$ and ϵ sufficiently small. Mappings of the sort (2.4.12) have been extensively studied principally by Moser (1955, 1961, 1962, 1967).

Under the above conditions, it is possible to show that there exists a formal series (2.4.11) which solves (2.4.8) up to any order (power) of ϵ. We introduce the "average" value $< f >$ of a quasi-periodic function $f(y_1,y_2,\dots,y_n)$, with $y_k = \omega_k t + y_k^o$, ω_k constant and linearly independent over the integers, by

$$< f > = \lim_{T \to \infty} \frac{1}{T} \int_0^T f \, dt. \tag{2.4.13}$$

In a generalized sense, a quasi-periodic function f with the property $< f > = 0$, will be said to be purely quasi-periodic. Obviously, if f is a Fourier Series in the n angular variables $y_1,y_2,\dots,y_n$, $< f >$ is the constant term of the Fourier's series. On the other hand, in general, if $< f > = 0$ then

$$\lim_{T \to \infty} \int_0^T f \, dt = \text{finite} \tag{2.4.14}$$

which is an obvious consequence of (2.4.13) for f quasi-periodic and L_2 for $t \in R$. A function $F(t)$ satisfying the condition

$$\lim_{T \to \infty} F(t) = \text{finite} \tag{2.4.15}$$

will be said to be free from secular terms. Any primitive of an L_2 purely quasi-periodic function satisfies this properly. Under the integrability assumption of H_o, it follows that, in terms of the action-angle variables (y,x) the Hamiltonian $H(y;x;\epsilon)$ is quasi-periodic if, for example, it has a convergent multi-dimensional Fourier series in $y_1, y_2, \ldots, y_n$, for ϵ in $[0,1]$ and $(x;y)$ in D.

The formal series S and K are now obtained by direct substitution of (2.4.10) and (2.4.11) into (2.4.8), that is,

$$H(y; \frac{\partial W}{\partial y}; \epsilon) = H(y; X + \epsilon \frac{\partial S_1}{\partial y} + \epsilon^2 \frac{\partial S_2}{\partial y} + \ldots; \epsilon)$$

$$K(X;\epsilon) = K_o(X) + \epsilon K_1(X) + \epsilon^2 K_2(X) + \ldots$$

Expansion of the first of these by Taylor series (which by hypothesis converges) gives, symbolically,

$$\begin{aligned} H(y, \frac{\partial W}{\partial y}; \epsilon) &= \sum_{k=0}^{\infty} \frac{1}{k!} \frac{\partial^k H}{\partial x^k}\bigg|_{x=X} (\epsilon \frac{\partial S_1}{\partial y} + \epsilon^2 \frac{\partial S_2}{\partial y} + \ldots)^k \\ &= \sum_{k=0}^{\infty} \frac{1}{k!} \sum_{p=0}^{\infty} \epsilon^p \frac{\partial^k H_p}{\partial x^k}\bigg|_{x=X} (\epsilon \frac{\partial S_1}{\partial y} + \epsilon^2 \frac{\partial S_2}{\partial y} + \ldots)^k \\ &= H_o(X) + \epsilon \frac{\partial H_o}{\partial X} (\frac{\partial S_1}{\partial y})^T + \frac{\epsilon^2}{2!} (\frac{\partial S_1}{\partial y}) \frac{\partial^2 H_o}{\partial X \partial X} (\frac{\partial S_1}{\partial y})^T + \ldots \\ &\quad + \epsilon^2 \frac{\partial H_o}{\partial X} (\frac{\partial S_2}{\partial y})^T + \ldots + \epsilon H_1(y;X) + \epsilon^2 \frac{\partial H_1}{\partial X}(y;X) (\frac{\partial S_1}{\partial y})^T \\ &\quad + \ldots + \epsilon^2 H_2(y;X) + \ldots \quad . \end{aligned} \qquad (2.4.16)$$

Expressions up to any order of approximation were first obtained by Giacaglia (1963). Equating coefficients of same powers in ϵ, one gets, to any order of approximation, an equation of the type

$$\sum_{k=1}^{n} \frac{\partial H_o}{\partial X_k} \frac{\partial S_p}{\partial y_k} + \Phi_p(y;X) + H_p(y;X) = K_p(X) \qquad (2.4.17)$$

where $\partial H_k / \partial X_\ell$ stands for $\partial H_k / \partial x_\ell \big|_{x=X}$. For example,

$$\Phi_1(y;X) = 0$$

$$\Phi_2(y;X) = \frac{1}{2!} \sum_{k=1}^{n} \sum_{\ell=1}^{n} \frac{\partial^2 H_o}{\partial X_k \partial X_\ell} \frac{\partial S_1}{\partial y_k} \frac{\partial S_1}{\partial y_\ell} + \sum_{k=1}^{n} \frac{\partial H_1}{\partial X_k} \frac{\partial S_1}{\partial y_k}$$

and so on. In general, $\Phi_p(y;X)$ is a function of $S_1, S_2, \ldots, S_{p-1}, K_1, K_2, \ldots, K_{p-1}$, so that the solutions of equations (2.4.17) can only be obtained in succession. One way of defining $K_p(X)$ is by the use of an averaging procedure

$$K_p(X) = < \Phi_p(y;X) + H_p(y;X) > \tag{2.4.18}$$

where y_k is supposed to be given by a linear function of time $y_k = \omega_k^o t + y_k^o$, and all the ω_k^o linearly independent over the set of integers (that is, rationally independent). The resulting $K_p(X)$ is certainly independent of t. It follows that the function

$$F_p = \Phi_p + H_p - K_p = F_p(y;X) \tag{2.4.19}$$

is purely quasi-periodic in view of the hypotheses on $H(y;x;\epsilon)$. The p-th approximate to the generating function, S_p, is obtained from the linear equation

$$\sum_{k=1}^{n} \omega_k^o \frac{\partial S_p}{\partial y_k} + F_p(y;X) = 0$$

where $\omega_k^o = \partial H_o / \partial X_k$. It is now obvious that if every $\omega_k^o \neq 0$, S_p results to be a quasi-periodic function in $y_1, y_2, \ldots, y_n$ $(y_k = \omega_k^o t + y_k^o)$ free from secular terms, that is, for linearly independent ω_k^o over the integers,

$$S_p(y;X) = \sum_{k=1}^{n} \frac{1}{\omega_k^o} \int F_p(y;X)\, dy_k + G_p(X) \tag{2.4.20}$$

where $G_p(X)$ is arbitrary. Obviously if one of the ω_k^o is zero the formula does not apply, unless $F_p(y;X)$ is such that

$$\frac{\partial F_p}{\partial y_k} = 0 \tag{2.4.21}$$

for that particular y_k. It is easily seen that

$$\lim_{t \to \infty} S_p(y;X) = \text{finite} \tag{2.4.22}$$

for $y_k = \omega_k^o t + y_k^o$. All these relations are easily shown and, by recurrence, it follows that one can determine the formal series

$$y \cdot X + \epsilon S_1 + \epsilon^2 S_2 + \epsilon^3 S_3 + \ldots$$

and

$$K_o + \epsilon K_1 + \epsilon^2 K_2 + \ldots ,$$

where $\epsilon S_1 + \epsilon^2 S_2 + \epsilon^3 S_3 + \ldots$ satisfies the property of being quasi-periodic and free from secular terms. The case where some $\omega_k^o = 0$ or it is small, in some sense, will be studied in chapter 5, under the general problem of resonance. The system is formally solved up to any desired degree of approximation and the "solution", in the new variables $(Y;X)$ is

$$\begin{aligned} Y_k &= \omega_k t + Y_k^o \\ X_k &= X_k^o \end{aligned} \tag{2.4.23}$$

where

$$Y_k^o = \text{const.}$$

$$X_k^o = \text{const.} + O(\epsilon^{p+1})$$

$$\omega_k = \frac{\partial K_o}{\partial X_k} + \epsilon \frac{\partial K_1}{\partial X_k} + \ldots + \epsilon^p \frac{\partial K_p}{\partial X_k} = \text{const.} + O(\epsilon^{p+1}) \tag{2.4.24}$$

where $O(\epsilon^{p+1})$ is the factor of the first term neglected after the last approximates S_p and K_p have been obtained. By no means, it should be interpreted as the error or an approximation to the error bound of the solution. This might be so, eventually, only in the case of convergence of the method. The problem will be dealt with in the next two chapters. A rough estimate by Kyner (1963) shows that the error bound is equivalent to that obtained by Bogoliubov and Mitropolsky (1951) for the canonical averaging method of Krylov-Bogoliubov-Mitropolsky (KBM), and in fact, Poincaré's Method was shown to be equivalent to that of KBM by Burstein and Solovev (1961). Such error bound is proportional to ϵ for $t \sim 1/\epsilon$, at worst. Convergence of the method, under particular circumstances, will be given in Chapter 3.

From a purely formal point of view we obtain, from (2.4.21),

$$\begin{aligned} y_k &= \omega_k t + Y_k^o + \epsilon N_k(Y_1, Y_2, \ldots, Y_n;\ X_1^o, X_2^o, \ldots, X_n^o; \epsilon) \\ x_k &= X_k^o + \epsilon W_k(Y_1, Y_2, \ldots, Y_n;\ X_1^o, X_2^o, \ldots, X_n^o; \epsilon) \end{aligned} \tag{2.4.25}$$

where N_k, W_k are quasi-periodic in $Y_1, Y_2, \ldots, Y_n$ and free from secular terms. It is obvious that one of the major causes of error is in the frequency ω_k, since any error is linearly multiplied by time. In practical applications, the best way out, in lack of exact solution, is to use a numerical observed value of ω_k from the average $< y_k >$ with respect to t. Such average, if relations (2.4.25) hold, is obviously ω_k. This use of observational evidence eliminates the in-track error due to a miscalculation of the frequency ω_k.

5. Fast and Slow Variables.

The case of proper degeneracy (Arnold, 1963) is quite common in perturbation theory. Generally speaking, the problem is defined by non-independent frequencies of the unperturbed system. That is, given the Hamiltonian

$$H_o = H_o(x)$$

and the frequencies

$$\omega_j = \frac{\partial H_o}{\partial x_j}; \quad j = 1, 2, \ldots, n$$

one has degeneracy if the matrix

$$\left\{\frac{\partial \omega_j}{\partial x_k}\right\}; \quad j, k = 1, 2, \ldots, n \tag{2.5.1}$$

is singular. This definition includes cases of rational dependence and when some of the action variables are not present in $H_o(x)$, that is, at least one of the ω_j is identically zero. It also includes linear systems, that is, cases in which

$$H_o = \omega_1 x_1 + \omega_2 x_2 + \ldots + \omega_n x_n. \tag{2.5.2}$$

Let us consider, here, the case where the matrix (2.5.1) has at least a minor of order m $(0 < m \leq n)$ which is not zero. The unperturbed system is non-linear, integrable and defined by m independent frequencies, corresponding to a set of m independent angular variables $y_k = \omega_k(x)t + y_k^o$, $k = 1,2,\ldots,m$. There exists, in this case, a canonical transformation $(x,y) \rightarrow (x',y')$ such that, at least locally, the Hamiltonian H_o is a function of only m momenta x' and the corresponding matrix (2.5.1) is non-singular. It may be worth noting, however, that <u>if none of the x are absent in H_o, one may perform a transformation to a new Hamiltonian whose Hessian matrix (2.5.1) is non-singular.</u> In fact, consider in general the Hamiltonian

$$H = H(y;x;\epsilon) = H_o(x) + \epsilon H_1(y;x) + \ldots$$

and suppose $\omega_j = \partial H_o/\partial x_j \neq 0$, $j = 1,2,\ldots,n$. If a function $F = \Phi(H)$ can be found such that

$$F = F_o(x) + \epsilon F_1(y;x) + \ldots$$

and such that, being $\Omega_j = \partial F_o/\partial x_j$, the matrix

$$\left\{ \frac{\partial \Omega_j}{\partial x_k} \right\}$$

is non-singular, the <u>apparent</u> degeneracy is eliminated. The equations of motion are now

$$\dot{y}_j = \frac{1}{\alpha} \frac{\partial F}{\partial x_j}$$

$$\dot{x}_j = -\frac{1}{\alpha} \frac{\partial F}{\partial y_j}$$

where α is the constant defined in terms of the initial conditions by

$$\dot{\Phi}(H) = \dot{\Phi}(H(y_o;x_o;\epsilon)) = \dot{\Phi}(h) = \alpha$$

and h is the energy integral corresponding to the initial conditions $(y_o;x_o)$. Evidently α can be developed in a power series of ϵ [we suppose H real analytic in all arguments] and if Φ is analytic, the power series

$$\Phi(H) = F_0(x) + \epsilon F_1(y;x) + \epsilon^2 F_2(y;x) + \dots .$$

converges. This process does not apply in the linear case (2.5.2) since, as it is easily verified, whatever $\Phi(H)$ is, the Hessian of $F_0(x)$ is zero. It does apply, however, in other cases. An important example is, for instance when

$$H_o = \frac{1}{x_1^2} + x_2$$

a case of many applications in celestial mechanics (two-body problem in rotating coordinates, restricted three-body problem in rotating coordinates, etc.). Although the Hessian of H_o is zero (H_o is linear in x_2) one sees that there are several functions of H_o leading to an F_o for which the Hessian is not zero (e.g.; Poincaré, 1893). Excluded the linear case we are therefore left with the case in which some of the momenta are not present in H_o. Let $(x_{p+1}, x_{p+2}, \dots, x_n)$ be the ignorable momenta and consider the equations generated by

$$H = H_o(x_1, x_2, \dots, x_p) + \epsilon H_1(x_1, x_2, \dots, x_n;\ y_1, y_2, \dots, y_n) + \dots$$

that is,

$$\dot{y}_k = \partial H/\partial x_k$$

$$\dot{x}_k = -\partial H/\partial y_k = -\epsilon\ \partial H_1/\partial y_k .$$

It follows that, as a "zero approximation", the x_k are constant and the y_k are linear functions of time $(k = 1, 2, \dots, p)$ or are constant $(k = p+1, p+2, \dots, n)$. If these results are put back into the equations of motion and the average with respect to $y_1, y_2, \dots, y_p$ is considered, to "first order" one obtains

$$y_k = \omega_k(x)t + y_k^o$$

$$x_k = x_k^o$$

with

$$\omega_k = \omega_k^o + \epsilon\omega_k^1 , \quad (k = 1, 2, \dots, p),$$

$$\omega_j = \epsilon\omega_j^1 \quad , \quad (j = p+1, \dots, n).$$

This crude description motivates the fact the the angular variables $y_1, y_2, \ldots, y_p$ (whose associate momenta $x_1, x_2, \ldots, x_p$ are present in H_0) are called fast and the angular variables $y_{p+1}, y_{p+2}, \ldots, y_n$ (whose associate momenta are absent in H) are called slow. As a consequence, any function containing at least a fast variable is said to be short periodic and any function containing none of the fast variable, long periodic. Obviously, we are not seeking here precise definitions, but only a traditional explanation of a terminology.

The problem now is to see whether there are formal series, in this case, which solve the generating function of Poincaré's method. In general the answer is negative, unless a unique situation occurs. This is the subject of the present section.

The elimination of fast variables is accomplished by a generalization of Hamilton's problem, where we require the new Hamiltonian to contain only slow variables. More precisely, we construct a generating function, as a formal series

$$W(y;X;\epsilon) = y\cdot X + \epsilon S_1(y;X) + \ldots$$

as in (2.4.10), (2.4.11) and (2.4.12), and require the energy conservation law in the form

$$H(y;x;\epsilon) = K(Y_{p+1}, Y_{p+2}, \ldots, Y_n; X; \epsilon) \tag{2.5.3}$$

so that the system reduces to one with a number of degrees of freedom equal to n-p. This is always possible since at any stage m of approximation the equation to be integrated is

$$\sum_{k=1}^{p} \omega_k(X) \frac{\partial S_m}{\partial y_k} + F_m(y;X) + H_m(y;X)$$

$$= K_m(y_{p+1}, y_{p+2}, \ldots, y_n; X; \epsilon)$$

and K_m is defined by the average of $F_m + H_m$ over the fast variables. The new Hamiltonian is obtained as a formal series. Admitting such series to be convergent (at least over a finite interval of time) the problem is now reduced to the equations generated by the Hamiltonian

$$K = K_o(X) + \epsilon K_1(Y_{p+1}, Y_{p+2}, \ldots, Y_n; X)$$

$$+ \epsilon^2 K_2(Y_{p+1}, Y_{p+2}, \ldots, Y_n; X) + \ldots$$

$$= K(Y_{p+1}, Y_{p+2}, \ldots, Y_n; X; \epsilon) \qquad (2.5.4)$$

while the constant momenta $X_1, X_2, \ldots, X_p$ play the role of parameters. In case of convergence, the relations

$$x_k = X_k + \epsilon \frac{\partial S(y;X;\epsilon)}{\partial y_k} \qquad (2.5.5)$$

for $k = 1,2,3,\ldots,p$ represent first integrals of the original system, depending on p parameters $X_1, X_2, \ldots, X_p$ which can be given arbitrary values.

The elimination of the slow variables reduces now to a simple condition. In fact, in (2.5.4), $K_o(X)$ depends only on $X_1, X_2, \ldots, X_p$ and is therefore a constant of motion. The Hamiltonian can now be written as

$$\epsilon F = \epsilon F_1(q;p) + \epsilon^2 F_2(q;p) + \ldots$$

$$= \epsilon F(q;p;\epsilon) \qquad (2.5.6)$$

where $q = (Y_{p+1}, Y_{p+2}, \ldots, Y_n)$, $p = (X_{p+1}, X_{p+2}, \ldots, X_n)$ and the parameters $X_1, X_2, \ldots, X_p$ have been omitted. The equations of motion are simply

$$\dot{q}_k = \epsilon \frac{\partial F}{\partial p_k}$$

$$\dot{p}_k = -\epsilon \frac{\partial F}{\partial q_k} \qquad (2.5.7)$$

for $k = 1,2,\ldots,n-p$. If $n-p = 1$, the system has a single degree of freedom and the problem is theoretically solved. If $n-p \geq 2$ the integration by a method of successive approximations of the type under discussion can only be performed, obviously, if the dominant part of ϵF, that is, $\epsilon F_1(q;p)$ corresponds to an integrable system. From this point on, we have a repetition of the process of Poincaré described in the previous section. Useless to say, the problem can formally be completely reduced if $F_1(q;p)$ does not depend on any q and contains all of the p variables, that is, $X_{p+1}, X_{p+2}, \ldots, X_n$. The actual contribution of von Zeipel (1916) was to recognize the fact that, although the complete reduction of the system may not be possible, partial reduction is a certain step toward the solution of the problem.

Error estimates of the method have been obtained by Kyner (1966) and, in case of convergence, accelerated process of convergence have been introduced by Moser (1966) based on a Newton-type iterative process. This process, actually first suggested by Kolmogorov (1954), has been widely used by Arnol'd (1963) in several papers. In this respect, much will be said in the next chapter. Evidently, there are several situations where the error estimate $O(\epsilon^2)$ obtained by Kyner can be improved a lot. For instance, in the proof of convergence in the Twist Mapping of Moser (1962) better than quadratic convergence may be obtained so that the error decreases with a power of ϵ which is increasing as the iterations are accumulated. For this to be true, the mapping involved does not even have to be analytic but only finitely many times differentiable.

6. Generalization of the Averaging Procedure, Birkhoff's Normalization and Adelphic Integrals.

In most cases, when the averaging method is applied, it is a basic hypothesis to assume that the Hamiltonian be multi-periodic in the angle variables, say $y_1, y_2, \ldots, y_n$. As seen in section 4 of this chapter, quasi-periodicity can be assumed as a slight generalization of the assumption of multi-periodicity, when a proper definition of average is introduced. Such hypotheses are a reminiscence of the special fields where the methods have been developed: celestial mechanics and oscillations in mechanical and electrical systems.

In order to introduce a more general approach to the problem, where the above mentioned hypotheses are not verified we initially consider a simple example. Let the Hamiltonian be given and such that

$$H(y_1,y_2,x_1,x_2) = H_o + H_1 + H_2 + \dots$$

where

$$H_o = \frac{1}{2} A_{11}(x_1^2 + y_1^2) + \frac{1}{2} A_{22}(x_2^2 + y_2^2) + A_{12}(x_1x_2 + y_1y_2)$$

$$H_p = H_p(y_1,y_2,x_1,x_2) \ , \quad p = 1,2,\dots$$

where H_p are homogeneous polynomials of degree $p+2$. The solution of the "dominant" part of the problem is immediate if one can eliminate the part $(x_1x_2 + y_1y_2)$. This can, in general, be accomplished quite easily by a linear canonical transformation $(y;x) \to (\eta;\xi)$

$$x_j = \sum_{k=1}^{2} a_{jk}\xi_k$$

$$\eta_j = \sum_{k=1}^{2} a_{kj}y_k$$

where, for example, one can take

$$a_{12} = A_{12}$$

$$a_{22} = A_{22} - A_{11}$$

$$a_{11} = (1 + a_{12}a_{21})/a_{22}$$

$$a_{21} = (A_{12}a_{22} - A_{22}a_{12})/(A_{22}a_{12}^2 + A_{11}a_{22}^2 - 2A_{12}a_{12}a_{22})$$

excluded the case $A_{11} = A_{22}$, where the above transformation is singular. This particular case is, of course, much more easily solved. The Hamiltonian is brought to the form

$$H = H_o + H_1 + H_2 + \dots$$

where $H_o = A_1(\xi_1^2 + \eta_1^2) + A_2(\xi_2^2 + \eta_2^2)$ and $H_1, H_2, \dots$ are again homogeneous polynomials of degree $3,4,\dots$ in $\xi_1,\xi_2,\eta_1,\eta_2$. Also

$$A_1 = \frac{1}{2}(A_{11}a_{11}^2 + A_{22}a_{21}^2 + 2a_{11}a_{21}A_{12}),$$

$$A_2 = \frac{1}{2}(A_{11}a_{12}^2 + A_{22}a_{22}^2 + 2a_{12}a_{22}A_{12}).$$

The solution of Hamilton's equation

$$A_1[(\frac{\partial W}{\partial \eta_1})^2 + \eta_1^2] + A_2[(\frac{\partial W}{\partial \eta_2})^2 + \eta_2^2] = F_o(\alpha_1,\alpha_2)$$

is immediate. With the "natural" choice

$$F_o = A_1\alpha_1^2 + A_2\alpha_2^2 ,$$

we find $S = S_1 + S_2$, where

$$(\frac{\partial S_k}{\partial \eta_k})^2 + \eta_k^2 = \alpha_k^2, \quad k = 1,2$$

and therefore

$$\alpha_k^2 = \xi_k^2 + \eta_k^2,$$

$$\beta_k = (\xi_k^2 + \eta_k^2)^{1/2} \arcsin (\eta_k/\alpha_k)$$

for $k = 1,2$. The inverse transformation is

$$\eta_k = \alpha_k \sin(\beta_k/\alpha_k) \quad ,$$

$$\xi_k = \alpha_k \cos(\beta_k/\alpha_k) \quad , \quad k = 1,2.$$

The dominant part of the Hamiltonian is reduced to

$$H_o = F_o = A_1\alpha_1^2 + A_2\alpha_2^2$$

while the complete Hamiltonian will in general be made up by terms

$$\alpha_1^p\alpha_2^q \begin{matrix}\cos\\ \sin\end{matrix} (m \frac{\beta_1}{\alpha_1} + n \frac{\beta_2}{\alpha_2}). \tag{2.6.1}$$

The zero-th order solution

$$\alpha_k = \text{const.} \quad ,$$

$$\eta_k = (-A_k\alpha_k)t + \eta_k^o, \quad (k = 1,2)$$

shows that Poincaré's method will produce mixed secular terms due to differentiations with respect to α_1 or α_2 in the generating function of the method (containing necessary terms of the form (2.6.1)). The solution to the question is actually simpler, at least in the formal sense. In fact, suppose the Hamiltonian contains the variables $(x;y)$ in the combinations $x_1^2 + y_1^2$ and $x_2^2 + y_2^2$ only, i.e.,

$$H = H(x_1^2 + y_1^2, x_2^2 + y_2^2).$$

In this case, since

$$\dot{x}_j = \frac{\partial H}{\partial(x_j^2 + y_j^2)} 2y_j,$$

$$\dot{y}_j = - \frac{\partial H}{\partial(x_j^2 + y_j^2)} 2x_j,$$

it follows that

$$x_j^2 + y_j^2 = \text{const.} = c_j^2$$

and therefore

$$x_j = c_j \cos(\omega_j t + \sigma_j)$$

$$y_j = c_j \sin(\omega_j t + \sigma_j)$$

where

$$\omega_j = -2 \frac{\partial H}{\partial(x_j^2 + y_j^2)} = \text{const.}$$

and c_j, σ_j are arbitrary. This is analogous to Whittaker's (1937) remark that if the Hamiltonian is function of the variables $\omega_j = x_j y_j$ only, then the ω_j are constant. The same remark applies, of course, to any combinations of the associate coordinate and momenta. These considerations lead naturally to the question whether, assuming H_0 say to have the form

$$A_1(x_1^2 + y_1^2) + A_2(x_2^2 + y_2^2),$$

it is possible to reduce all the Hamiltonian to a function of the combinations $x_1^2 + y_1^2$ and $x_2^2 + y_2^2$. The answer to this question is affirmative in the sense that, at least formally, the reduction can in general be obtained by a series of homogeneous polynomials in the variables involved, although the convergence of these series, as such, has never been investigated. The equivalence to the problem of Birkhoff's normalization is, nevertheless, evident.

Consider, then, the dominant part H_o of the Hamiltonian to be a function only of $x_1^2 + y_2^2$ and $x_2^2 + y_2^2$. The higher order parts of the Hamiltonian are functions of the variables $(x;y)$ say in the combinations

$$w_1^p w_2^q \, u_1^m u_2^n$$

where

$$w_1 = x_1 x_2 + y_1 y_2$$

$$w_2 = x_1 y_2 - x_2 y_1$$

$$u_1 = x_1^2 + y_1^2$$

$$u_2 = x_2^2 + y_2^2$$

This is, for instance, the case of Celestial Mechanics when Poincaré's variables are used (e.g. Brouwer and Clemence, 1961). Generally, one can assume the higher order parts of H to be homogeneous polynomials of increasing degree in x_1, y_1, x_2, y_2. The elimination of all terms except the combinations u_1, u_2 from H_1 can be accomplished by means of a generating function

$$S = x_1' y_1 + x_2' y_2 + S_1 + S_2 + \dots$$

so that one finds

$$H_1(x';y) + \sum_{k=1}^{2} \frac{\partial S_1}{\partial y_k} \frac{\partial H_o}{\partial x_k'} = H_1'(x';y) + \sum_{k=1}^{2} \frac{\partial S_1}{\partial x_k'} \frac{\partial H_o'}{\partial y_k}$$

where primes indicate new variables and new Hamiltonian. Now, since $H_o = A_1 u_1 + A_2 u_2$, the function H_1' is defined by that part of H_1, if any, containing purely the combinations u_1 and u_2, which we call H_{1s}. The remaining terms, called H_{1p}, will allow for the determination of S_1. It follows that, since

$H_o'(x',y) = H_o(x',y),$

$$\sum_{k=1}^{2} \left(\frac{\partial H_o}{\partial x_k'} \frac{\partial S_1}{\partial y_k} - \frac{\partial H_o}{\partial y_k} \frac{\partial S_1}{\partial x_k'}\right) = -H_{1p}(x';y)$$

where, in H_{1p}, any term in u_1 and/or u_2 is necessarily factored by a term in w_1 or w_2. Now, considering the form of H_o, one has

$$\frac{\partial H_o}{\partial x_i'} = 2\alpha_i x_i', \quad \frac{\partial H_o}{\partial y_i} = 2\alpha_i y_i$$

where

$$\alpha_i = \left(\frac{\partial H_o}{\partial u_i}\right)_{x=x'} .$$

Therefore, the equation for S_1 becomes

$$2 \sum_i \alpha_i \left(x_i' \frac{\partial S_1}{\partial y_i} - y_i \frac{\partial S_1}{\partial x_i'}\right) = -H_{1p}(x';y).$$

On the other hand, considering the definition of w_k and u_k $(k = 1,2)$, it follows that

$$x_1' \frac{\partial S_1}{\partial y_1} - y_1 \frac{\partial S_1}{\partial x_1'} = w_2' \frac{\partial S_1}{\partial w_1'} - w_1' \frac{\partial S_1}{\partial w_2'} ,$$

$$x_2' \frac{\partial S_1}{\partial y_2} - y_2 \frac{\partial S_1}{\partial x_2} = w_1' \frac{\partial S_1}{\partial w_2'} - w_2' \frac{\partial S_1}{\partial w_2'}$$

where

$$w_1' = x_1'x_2' + y_1y_2,$$

$$w_2' = x_1'y_2 - x_2'y_1 .$$

The equation for S_1 becomes

$$2(\alpha_1-\alpha_2)\left(w_2' \frac{\partial S_1}{\partial w_1'} - w_1' \frac{\partial S_1}{\partial w_2'}\right) = -H_{1p}(w_1',w_2',u_1',u_2')$$

$$= \Phi_{1p}(w_1',w_2'),$$

where the dependence on u_1', u_2' is omitted and is not relevant to the subsequent discussion (as long as isolated dependence of u_1', u_2' does not occur). The solution of this last equation is obtained by introducing the auxiliary variables of integration

$$z_1 = w_1'^2 - w_2'^2, \quad z_2 = w_1'^2 + w_2'^2.$$

With this substitution, one finds

$$S_1 = \frac{1}{4(\alpha_1-\alpha_2)} \int^{z_1} \frac{\Phi^*_{1p}(z_1,z_2)}{(z_2^2-z_1^2)^{1/2}}\, dz_1 + \Phi_1(z_2)$$

where Φ_1 is an arbitrary function of z_2, u_1', u_2'. The method does not apply when $\alpha_1 = \alpha_2$, that is

$$\frac{\partial H_o}{\partial u_1} = \frac{\partial H_o}{\partial u_2}.$$

This is, obviously, a case of internal resonance of the linear approximation, which is exceptional. A similar treatment and overall discussion holds to any order of approximation. The Hamiltonian is, at least formally, reduced to

$$H' = H_o' + H_1' + \dots = H'(u_1', u_2')$$

so that

$$u_1' = x_1'^2 + y_1^2 = \text{const.},$$

$$u_2' = x_2'^2 + y_2^2 = \text{const.}$$

The relations between primed and unprimed variables are obtained by

$$y_k = y_k' - \frac{\partial S_1}{\partial x_k'} - \frac{\partial S_2}{\partial x_k'} - \dots ,$$

$$x_k = x_k' + \frac{\partial S_1}{\partial y_k} + \frac{\partial S_2}{\partial y_k} + \dots .$$

<u>The foregoing considerations establish a clear connection between Poincaré's problem</u>

and Birkhoff's normalization. (Birkhoff, 1927; Siegel, 1956). The problems are actually identical in scope and such identity has been shown in specific applications, quite recently, by Deprit (1969, 1971). It is a well known fact that the series introduced by Birkhoff are generally divergent, although exceptional cases exist. New results connected with such problems are rare and the theorems of Kolmogorov and Moser could apply due to the non-linearity of the equations generated by H_0. The next connection of importance is with the concept of Adelphic Integrals introduced by Whittaker (1937). Recently, the definition and series approximation given by Whittaker, have been explored in specific examples by Contopoulos (1963) who, by the way, has shown that such integrals, supposed only formal results, do hold, in practice, for a very long interval of time, specifically as long as a computer could handle the integration with reasonable confidence in the accuracy of the results. The motivation for the question is: can we find, for a conservative system, some other integral which is independent from the energy integral? Evidently there are systems where such is the case and, in fact, by definition, an integrable system with n degrees of freedom has n such integrals. Although a well known result of Poincaré indicates that dynamical systems are non-integrable, such result relies on the existence of uniform (with respect to a certain parameter) integral. In the vicinity of singular points, Siegel (1941) has also shown the non-existence of analytic integrals and Moser (1955) the non-existence of differentiable integrals. Noneless, integrals may exist for specific values of parameters appearing in the equations, for specific values of initial conditions, or other exceptional cases as, for instance, just continuous integrals. We shall give, at the end of this section, an example of such exceptions.

Let $F(y;x;t)$ be a (differentiable in D) integral of a conservative system defined by the Hamiltonian $H(y;x)$, supposed to be C^2 in a certain domain D of the $2n$-dimensional phase space $y = (y_1, y_2, \ldots, y_n)$, $x = (x_1, x_2, \ldots, x_n)$. It is well known that F being independent of H (here, not a function of H alone), the following condition

$$(F,H) + \frac{\partial F}{\partial t} = 0 \qquad (2.6.2)$$

is necessary and sufficient. In explicit form, the Poisson parenthesis is, here

$$(F,H) = \sum_{k=1}^{n} \left(\frac{\partial F}{\partial y_k}\frac{\partial H}{\partial x_k} - \frac{\partial F}{\partial x_k}\frac{\partial H}{\partial y_k}\right).$$

If F is explicitly time independent, the condition is simply $(F,H) = 0$.

Now consider the case in which H depends on a dimensionless parameter ϵ, $|\epsilon| \in [0,1]$, and such that it is developable in Taylor series in the vincinity of $\epsilon = 0$, for $|\epsilon| < \epsilon_o$, and also, such that

$$H(y;x;\epsilon) = H_o(x) + \epsilon H_1(y;x) + \epsilon^2 H_2(y;x) + \dots . \tag{2.6.3}$$

Finally, suppose F to be time independent and an analytic function (in the real sense) of ϵ, for $|\epsilon| < \epsilon_c$. Then

$$F(y;x;\epsilon) = F_o(y;x) + \epsilon F_1(y;x) + \epsilon^2 F_2(y;x) + \dots \tag{2.6.4}$$

and we require $F_k(y;x)$, $k = 0,1,2,\dots$, to be differentiable in D. If F is an integral for all ϵ, $|\epsilon| < \epsilon_o$ say, then one must have $(F_o, H_o) = 0$, or, more explicitly

$$\sum_{k=1}^{n} \frac{\partial F_o}{\partial y_k}\frac{\partial H_o}{\partial x_k} = 0. \tag{2.6.5}$$

It is evident that any $F_o(x)$ satisfies (2.6.5) and also, being $F_o^*(x,y)$ a solution of (2.6.5) then $F_o^*(x,y) + F_o^{**}(x)$ also is, whatever $F_o^{**}(x)$ may be. We shall exclude cases of resonance, in this case, situations where the functions $\omega_k^o = \partial H_o/\partial x_k$ are dependent, or, in particular, linearly dependent over the set of integers, for $x \in D$. Actually we shall assume the infinitely many conditions

$$\left|\sum_{k=1}^{n} j_k \omega_k^o\right| > K(\omega^o)\left[\sum_{k=1}^{n} |j_k|\right]^{-n-1} \tag{2.6.6}$$

for all integers not all zero j_k and a convenient constant K. Cases of resonance or near resonance have been discussed, in details, in the problem of Adelphic Integrals, by Contopoulos (1968, 1970). For systems with $n > 2$ degrees of freedom not even a heuristic solution of the problem is available in the literature, although it can be produced with no major difficulties. The above conditions exclude particular solutions (or "near" solutions) of the type

$$F_o = \sum_{k=1}^{n} p_k y_k ,$$

where p_k are integers such that

$$\sum_{k=1}^{n} j_k p_k = 0.$$

Then one has the following lemma. Lemma 1. "The function F_o is an arbitrary function of $x_1, x_2, \ldots, x_n$ and of any linear form $\alpha_1 y_1 + \alpha_2 y_2 + \ldots + \alpha_n y_n$ where α_k are real non-rational numbers such that

$$\alpha_1 \omega_1^o + \alpha_2 \omega_2^o + \ldots + \alpha_n \omega_n^o = 0."$$

We note that since the solution of the system generated by H_o is

$$x_k = \text{const.}$$

$$y_k = \omega_k^o(x)t + y_k^o,$$

any function of $\alpha_1 y_1 + \alpha_2 y_2 + \ldots + \alpha_n y_n$ reduces to an absolute constant. For this reason, we shall consider the solution $F_o = F_o(x)$ for (2.6.5). This is, in fact, obvious, since F_o has to be an integral of the system generated by H_o and therefore, a function of the n integrals $x_1, x_2, \ldots, x_n$ of that system.

Lemma 2. "If $F_o = F_o(x)$ and $F_1(y,x)$ is 2π-periodic with respect to $y_1, y_2, \ldots, y_n$, with zero average, then (H_1, F_1) is 2π-periodic in $y_1, y_2, \ldots, y_n$ and has zero average, provided $H_1(y,x)$ is 2π-periodic in $y_1, y_2, \ldots, y_n$."

In fact, the condition $(F,H) = 0$ leads to the sequence of conditions

$$(F_p, H_o) + (F_{p-1}, H_1) + (F_{p-2}, H_2) + \ldots + (F_o, H_p) = 0 \qquad (2.6.7)$$

for $p = 1,2,3,\ldots$. For $p = 1$, we have

$$(F_1 H_o) + (F_o, H_1) = 0$$

or

$$\sum_{j=1}^{n} \omega_j^o \frac{\partial F_1}{\partial y_j} = \sum_{j=1}^{n} P_j(x) \frac{\partial H_1}{\partial y_j} \tag{2.6.8}$$

where

$$P_j(x) = \frac{\partial F_o}{\partial x_j} .$$

The right-hand member of (2.6.8) certainly is 2π-periodic in each y_k and has zero average. The same will be true for $F_1(y;x)$ provided one disregards any arbitrary function of x in the solution, and the ω_j^o satisfy (2.6.6). Now let $\theta = p_1y_1 + p_2y_2 + \dots + p_ny_n$ be any argument in the Fourier series of $H_1(y;x)$, with $p_1, p_2, \dots, p_n$ integers, not all zero. In view of the linearity of (2.6.8) one can reason with that single argument. Thus, eliminating arbitrary functions of x, we have, for that argument

$$F_1 = \frac{\sum_{j=1}^{n} p_j P_j}{\sum_{j=1}^{n} p_j \omega_j^o} (-A \sin\theta + B \cos\theta) \tag{2.6.9}$$

where we have defined $H_1 = A \cos\theta + B \sin\theta + \dots$. The factor of the right-hand member of (2.6.9) is a function of x, let say, $C(x)$, and in view of (2.6.6), is not large (obviously we need the constant $K(\omega^o)$ to be $O(1)$ with respect to ϵ).

It follows that

$$(F_1, H_1)_\theta = \{\sum_{j=1}^{n} p_j \frac{\partial C}{\partial x_j}\} \{\frac{B^2 - A^2}{2} \sin 2\theta + AB \cos 2\theta\}$$

which proves the lemma since terms independent of θ can only be produced by trigonometric functions of the same argument.

Consider (2.6.7) for $p = 2$. The function F_2 is defined by

$$(F_2, H_o) + (F_1, H_1) + (F_o, H_2) = 0$$

or

$$\sum_{k=1}^{n} \omega_k^0 \frac{\partial F_2}{\partial y_k} = \sum_{k=1}^{n} P_k(x) \frac{\partial H_2}{\partial y_k} - (F_1,H_1)$$

and it follows that, disregarding arbitrary functions of x, F_2 is also a 2π-periodic function of $y_1,y_2,\ldots,y_n$.

In general, however, it is not true that F_p will be 2π-periodic in the angular variables $y_1,y_2,\ldots,y_n$. This is verified only under very special conditions. The most important example is when H is a cosine series in the angles $y_1,y_2,\ldots,y_n$. In this case, it is easily seen that F is also a cosine series. Therefore, any function obtained from a Poisson's Parenthesis is a sine series, and cannot contain any constant term. This is easily seen by writing

$$(F,H) = \sum_{k=1}^{n} \left(\frac{\partial F}{\partial y_k}\frac{\partial H}{\partial x_k} - \frac{\partial F}{\partial x_k}\frac{\partial H}{\partial y_k}\right)$$

and observing that, in each binomial, one has the product of a sine series by a cosine series.

The same is true also when H is a sine series. In problems of Celestial Mechanics, when Newtonian forces are considered, these conditions are satisfied.

The convergence of such method of approximation has been proved by Whittaker (1916) for some special classes of problems with two degrees of freedom, namely, in the vicinity of an equilibrium point of the general elliptic type and as long as the normal frequencies ω_1, ω_2 are irrational one to the other and for deviations sufficiently small from equilibrium. Although Whittaker felt very strongly in favor of the convergence for more general systems, he pointed out the fact that such adelphic integrals could not generally be uniformly convergent for any value of the independent variable and with respect to all values of the constants of integration or the parameters of the problem in any interval. This last consideration follows clearly from the fact that, as the ratio ω_1/ω_2 changes from an irrational to a rational value, the series defining the adelphic integrals take a completely different form. The same situation occurs in the application of averaging methods with respect to the type of motion defined by H_0 (the reference solution). In non-linear oscillations, the normal modes depend on the initial conditions and therefore, it seems natural to conclude that, as far as the initial conditions are concerned, convergence in any domain of the phase space is not possible. This is, in fact, the essential reasoning behind Poincaré's Theorem on the divergence of series in Celestial Mechanics (Poincaré, 1898, Vol. II). There is, of course, one case where convergence is a fact not even in question: the obvious situation where the series terminates. Even if an integral exists, in the form of a polynomial in ϵ, there remains the problem of what should be the zero-th approximation F_o. The difference between obtaining a series (eventually divergent) and a polynomial, may depend on the choice of $F_o(x)$. If a general principle for such choice could be found, we would have a criterion for the existence of integrals which are polynomials of certain physical parameters. For instance, consider the case

$$H = H_o(x) + H_1(y;x)$$

quite common in problems of perturbations. In this case, the equation defining F_p is

$$(F_p, H_o) + (F_{p-1}, H_1) = 0$$

for $p = 1,2,3,\ldots$. Evidently, if $F_k (k \geq p)$ are identically zero, it follows that $F = F_o + F_1 + \ldots + F_{p-1}$ where

$$(F_o, H_o) = 0$$

$$(F_1, H_o) + (F_o, H_1) = 0$$

$$(F_2, H_o) + (F_1, H_1) = 0$$

$$(F_{p-1}, H_o) + (F_{p-2}, H_1) = 0$$

$$(F_{p-1}, H_1) = 0.$$

The last condition implies that F_{p-1} is an integral of the system generated by H_1. This is a necessary condition for the integral F to be a polynomial of degree $p-1$ in ϵ. Evidently, for this to happen it is sufficient that F_{p-1} be equal to, or a function of, H_1. This is the case for instance of Kovalevskaya's integral for the motion of a symmetric top under the influence of gravity. For this motion we introduce Andoyer's variables (1926)

$$L = p_\psi = G \cos b$$

$$p_\theta = G \sin b \sin(\ell - \psi)$$

$$p_\phi = H = G \cos I$$

where ϕ, ψ, θ are the usual Euler angles as defined in Goldstein (1951), G is the magnitude of the angular momentum, I is the inclination of the invariable plane (normal to the angular momentum vector) with respect to the inertial equatorial plane, b is the inclination of the body principal inertial equatorial plane with respect to the invariable plane and ℓ the angle between the body x-axis and the interception of the body (x, y) plane with the invariable plane. Let h be the angle between the inertial X axis and the interception of the invariable and (X, Y) planes and let g the angle between the interceptions of the invariable plane with the planes (X, Y) and (x, y). Then the quantities $(L, G, H; \ell, g, h)$ are canonically conjugate (e.g., Deprit, 1966) and the kinetic energy is

$$\mathscr{H}_0 = \frac{1}{2}\left(\frac{1}{A} \sin^2 \ell + \frac{1}{B} \cos^2 \ell\right)(G^2 - L^2) + \frac{1}{2C} L^2$$

where A, B, C are the principal moments of inertia (w.r.t. x, y, z respectively). If one assumes $A = B$,

$$\mathscr{H}_0 = \frac{1}{2}\left(\frac{1}{C} - \frac{1}{A}\right)L^2 + \frac{1}{2A} G^2$$

while the potential, by proper choice of the axes, can be written

$$\begin{aligned} w\mathscr{H}_1 = w\{x_G[&\sin I \sin g \cos \ell + (\sin b \cos I \\ &+ \cos b \sin I \cos g) \sin \ell] + z_G[\cos b \cos I \\ &- \sin b \sin I \cos g]\} \end{aligned}$$

where w is the weight of the top and x_G, $y_G = 0$, z_G are the coordinates of the center of mass in the body system. We have, of course, the integrals $\mathcal{H}_0 + w\mathcal{H}_1 = E$ (energy) and $H = G \cos I = H_0$ (since h is ignorable).

Consider an integral $F(L, G, H, \ell, g, h)$ of the system, such that, $F = F_0 + wF_1 + \ldots$

$$F_0 = \psi(L,G)$$

$$F_k = F_k(L, G, \ell, g) \quad (k = 1,2,\ldots)$$

where we have assumed h to be cyclic, for obvious reasons, and $H = H_0$ is simply a parameter not shown explicitly. We shall write

$$\mathcal{H}_0 = \frac{a}{2} L^2 + \frac{b}{2} G^2$$

$$\mathcal{H}_1 = A^\circ \sin(\ell+g) + B^\circ \sin(\ell-g) + C^\circ \sin \ell + D^\circ \cos g + E^\circ$$

where

$$A^\circ = x_G(L+G)(G^2-H^2)^{1/2}/2G^2 \quad ,$$

$$B^\circ = x_G(L-G)(G^2-H^2)/^{1/2}2G^2 \quad ,$$

$$C^\circ = x_G H(G^2-L^2)^{1/2}/G^2 \quad ,$$

$$D^\circ = -z_G(G^2-L^2)^{1/2}(G^2-H^2)^{1/2}/G^2 \quad ,$$

$$E^\circ = z_G L H G^{-2} \quad .$$

The conditions for F to be an integral are

$$(\mathcal{H}_0, F_k) + (\mathcal{H}_1, F_{k-1}) = 0 \tag{2.6.10}$$

for $k = 1,2,\ldots$. We shall leave $\psi(L,G) = F_0$ undefined and try to determine under what conditions in ψ and the physical parameters, the series for F terminates. We obtain from (2.6.10)

$$aL \frac{\partial F_k}{\partial \ell} + bG \frac{\partial F_k}{\partial g} = \{A^\circ \cos(\ell+g) + B^\circ \cos(\ell-g) + C^\circ \cos \ell\} \frac{\partial F_{k-1}}{\partial L}$$

$$+ \{A^\circ \cos(\ell+g) - B^\circ \cos(\ell-g) - D^\circ \sin g\} \frac{\partial F_{k-1}}{\partial G}$$

$$- \{A_L^o \sin(\ell+g) + B_L^o \sin(\ell-g) + C_L^o \sin \ell + D_L^o \cos g + E_L\} \frac{\partial F_{k-1}}{\partial \ell}$$

$$- \{A_G^o \sin(\ell+g) + B_G^o \sin(\ell-g) + C_G^o \sin \ell + D_G^o \cos g + E_G\} \frac{\partial F_{k-1}}{\partial g} \tag{2.6.11}$$

For $k = 1$, we find

$$F_1 = \frac{A^o(\psi_L + \psi_G)}{aL + bG} \sin(\ell+g) + \frac{B^o(\psi_L - \psi_G)}{aL - bG} \sin(\ell-g)$$

$$+ \frac{C^o\psi_L}{aL} \sin \ell + \frac{D^o\psi_G}{bG} \cos g + E' = A' \sin(\ell+g)$$

$$+ B' \sin(\ell-g) + C' \sin \ell + D' \cos g + E' \tag{2.6.12}$$

where E' is an arbitrary function of L, G. The function F_1 has the same form as H_1. In fact, this is necessary since, taking $\psi = H_o$, it must result $F_1 = H_1$ + arbitrary function of L, G. It is also clear that there exists no $\psi \not\equiv 0$ such that $F_1 = 0$. For $k = 2$, Eq. (2.6.11) gives (Giacaglia, 1967):

$$F_2 = A_{0,1} \cos g + A_{0,2} \cos 2g + A_{1,-1} \cos(\ell-g)$$

$$+ A_{1,0} \cos \ell + A_{1,1} \cos(\ell+g) + A_{1,2} \cos(\ell+2g)$$

$$+ A_{2,-2} \cos(2\ell-2g) + A_{2,-1} \cos(2\ell-g) + A_{2,0} \cos 2\ell$$

$$+ A_{2,1} \cos(2\ell+g) + A_{2,2} \cos(2\ell+2g) + B_{1,-2} \sin(\ell-2g)$$

$$+ B_{1,-1} \sin(\ell-g) + B_{1,0} \sin \ell + B_{1,1} \sin(\ell+g)$$

$$+ B_{1,2} \sin(\ell+2g) + E'' \tag{2.6.13}$$

where E'' is an arbitrary function of L, G and $A_{k,j}$, $B_{k,j}$ are given functions of ψ_L, ψ_G, A^o, B^o, C^o, D^o, E', L, G, a, b and their derivatives. If one imposes the condition $F_2 = 0$, all coefficients must be identically zero and we find

$$E'' = E'_L = E'_G = 0$$

and, being k a nonzero constant,

$$A' = kA^o$$

$$B' = kB^o$$

$$C' = kC^o$$

$$D' = kD^o$$

so that $F_o = k\mathscr{A}_o$ and $F_1 = k\mathscr{A}_1$. This shows that every differentiable integral (valid for all values of w) and of the form $F_o + wF_1$ is necessarily proportional to $H = H_o + wH_1$. From (2.6.11), for $k = 3$, we find

$$F_3 = \sum_{k=-3}^{3} \sum_{j=-3}^{3} [A'_{k,j} \cos(k\ell+jg) + B'_{k,j} \sin(k\ell+jg)] \tag{2.6.14}$$

with k,j not simultaneously zero and $A'_{k,j}$, $B'_{k,j}$ functions of ψ_L, ψ_G, A^o, B^o, C^o, D^o, E', E'', L, G, a, b and their derivatives. Setting equal to zero all coefficients of this trigonometric polynomial, we find

$$\begin{aligned} &\text{I)} \quad a = b \ (A = 2C) \\ &\text{II)} \quad D^o = E^o = E' = 0 \ (z_G = 0) \\ &\text{III)} \quad \psi = (G^2 - L^2)^2/A^4 \\ &\text{IV)} \quad E'' = 2x_G^2 [G^2(L^2 + H^2) - 2L^2H^2]/A^2G^4. \end{aligned} \tag{2.6.15}$$

With these conditions, it follows from (2.6.12) that

$$F_1 = \frac{2x_G}{A^3} (1 - \frac{1}{G^2}) (G^2 - H^2)^{1/2} [(G-L) \sin(\ell+g) - (G+L) \sin(\ell-g)] + \frac{4x_G H^2}{A^3 G^2} (G^2 - L^2)^{3/2} \sin \ell$$

or

$$F_1 = \frac{4x_G}{A^3} G^2 \sin^2 b [\sin I (\sin g \cos \ell - \cos b \cos g \sin \ell) - \cos I \sin b \sin \ell]. \tag{2.6.16}$$

From (2.6.13) we find

$$F_2 = \frac{4x_G^2}{A^2} \{\frac{1}{2} (\frac{L^2+H^2}{G^2} - 2 \frac{L^2H^2}{G^4}) - \frac{1}{2} (1 - \frac{H^2+L^2}{G^2} + \frac{L^2H^2}{G^4}) \cos 2g$$

$$+ 2 \frac{LH}{G^2} (1 - \frac{L^2}{G^2})^{1/2} (1 - \frac{H^2}{G^2})^{1/2} \cos g\}$$

or

$$F_2 = \frac{4x_G^2}{A}(1 - \cos^2 b \cos^2 I - \sin^2 b \sin^2 I \cos^2 g$$

$$+ 2 \sin b \cos b \sin I \cos I \cos g).$$

It is easily seen that $F_3, F_4, \ldots$ are all zero, so that we have established the integral

$$F = F_o + wF_1 + w^2F_2$$

which is Kowalevskaya's integral (e.g., Leimanis, 1958). In fact, writing F in terms of p, q, r (components x,y,z of the rotation vector) and of Euler's angles, we find

$$\text{I)} \quad F_o = A^{-4}G^4 (1 - \frac{L^2}{G^2})^2 = A^{-4}G^4 \sin^4 b = (p^2 + q^2)^2$$

$$\text{II)} \quad F_1 = -4x_G A^{-1}[(p^2 - q^2) \sin \psi \sin \theta + 2pq \cos \psi \sin \theta]$$

$$\text{III)} \quad F_2 = 4x_G^2 A^{-2}(1 - \cos^2\theta) = 4x_G^2 A^{-2} \sin^2\theta .$$

Using Leimanis's notation,

$$\mu = wx_G A^{-1}$$

$$\xi = \sin \psi \sin \theta$$

$$\eta = \cos \psi \sin \theta$$

$$\zeta = \cos \theta$$

it follows

$$F = (p^2 - q^2 - 2\mu\xi)^2 + (2pq - 2\mu\eta)^2.$$

We have thus found an integral, valid for any value of w (which here takes the place of ϵ) but under the restriction $A = B = 2C$. Of course, for more general situations Arnol'd (1963) has shown that the system is integrable for a sufficiently small value of w, i.e., has shown stability of the fast top.

7. The Solution of Poincaré's Problem in Poisson's Parentheses. Elimination of Secular Terms from Adelphic Integrals.

In this section we shall indicate how to solve Poincaré's problem using Poisson's Parentheses and, at the same time, how to eliminate secular terms in the construction of Adelphic Integrals. We shall deal specifically with a case of degeneracy in which the dominant part of the Hamiltonian depends on a single action variable and the perturbation is 2π-periodic in the angle variables. As we have seen, this situation introduces series difficulties in Poincaré's method, difficulties which led von Zeipel to the already described generalization. Also, as we have seen, Poincaré's method constructs n formal integral whose zero-th order approximations are the action variables, constants of the unperturbed case. The other n formal integrals are essentially the constants of integration for the angle variables when all of these have ultimately been eliminated from the Hamiltonian. The process we are going to discuss is essentially that introduced by Whittaker, although the elimination of secular terms in the procedure was introduced by Giacaglia (1965). With the usual notation, the recurrence relations are

$$(H_o,F_k) = -\sum_{j=0}^{k-1} (H_{k-j},F_j) = -\psi_k(y;x) \tag{2.7.1}$$

where ψ_k is known when all the $k-1$ preceeding approximates are known. For $k = 0$, $\psi_k = 0$. Also, assuming $H_o = H_o(x)$,

$$\sum_{j=1}^{n} \omega_j^o \frac{\partial F_k}{\partial y_j} = \psi_k(y;x) \tag{2.7.2}$$

with the condition that every $F_k(y;x)$ should have no secular term, in the sense that the substitution $y_j = \omega_j^o\tau + y_j$ should give

$$\lim_{\tau \to \infty} F_k(y(\tau);x) = \text{bounded}. \qquad (2.7.3)$$

Nevertheless, since ψ_k is obtained by multiplication of trigonometric series it will contain terms that are functions only the x-type variable, so that, upon integration of (2.7.2), condition (2.7.3) will not be verified in general. The unwanted "secular behavior" can be eliminated by introducing an averaging procedure to be briefly described hereafter. We shall consider the highly degenerate case where H_o depends on one of the momenta only, say $H_o = H_o(x_1)$. Also, following Poincaré's results, we try to obtain integrals which, for $H = H_o$, reduce to the momenta, that is,

$$F_j = x_j + \epsilon \, \Delta F_j(y;x) \qquad (2.7.4)$$

for $j = 1,2,3$. The question remains if this choice will lead to the integrals

$$x_j' = x_j + \epsilon W_j(y;x) \qquad (2.7.5)$$

given by Poincaré's method. Since, by hypothesis, F_j and x_j' are integrals, the function

$$F_j - x_j' = \epsilon(\Delta F_j - W_j)$$

is also an integral. Now ΔF_j and W_j are not integrals (because x_j is not), so that $\Delta F_j \equiv W_j$. It follows that, if the process converges for ϵ in some interval, the two methods lead to the same result, although the use of Poisson parentheses gives explicit forms and add extra features to the solution. We let therefore

$$F = x_1 + F_1(y;x) + F_2(y;x) + \ldots \qquad (2.7.6)$$

$$H = H_o(x_1) + H_1(y;x) + H_2(y;x) + \ldots \qquad (2.7.7)$$

where H satisfies the foregoing conditions. The first order equation $(k = 1)$ from (2.7.1) gives

$$\omega_1^o \frac{\partial F_1}{\partial y_1} = - \frac{\partial H_1}{\partial y_1}$$

so that

$$F_1 = - \frac{1}{\omega_1^o} H_{1p}(y;x) + F_{1s}(y_2,y_3,\ldots,y_n;x) \tag{2.7.8}$$

where H_{1p} is defined by the operation of subtracting from H_1 the average with respect to y_1. In general

$$f_p(y;x) = f(y;x) - \lim_{T \to \infty} \frac{1}{T} \int_0^T f\,(t,y_2,\ldots,y_n;x)\,dt$$

or, for the multiperiodic case under consideration,

$$f_p(y;x) = f(y;x) - \frac{1}{2\pi} \int_0^{2\pi} f(y_1,y_2,\ldots,y_n;x)\,dy_1.$$

The index s indicates absence of y_1 and F_{1s} is, evidently, arbitrary. The second order approximation gives

$$\omega_1^o \frac{\partial F_2}{\partial y_1} = \sum_i \left(\frac{\partial H_1}{\partial x_i}\frac{\partial F_1}{\partial y_i} - \frac{\partial H_1}{\partial y_i}\frac{\partial F_1}{\partial x_i}\right) - \frac{\partial H_2}{\partial y_1} = \psi_{2p} + \psi_{2s}$$

where ψ_{2p} and ψ_{2s} are known and given by

$$\psi_{2p} = \sum_i \left[\left(\frac{\partial H_1}{\partial x_i}\frac{\partial F_1}{\partial y_i}\right)_p - \left(\frac{\partial H_1}{\partial y_i}\frac{\partial F_1}{\partial x_i}\right)_p \right] - \frac{\partial H_2}{\partial y_1}$$

$$\psi_{2s} = \sum \left[\left(\frac{\partial H_1}{\partial x_i}\frac{\partial F_1}{\partial y_i}\right)_s - \left(\frac{\partial H_1}{\partial y_i}\frac{\partial F_1}{\partial x_i}\right)_s \right].$$

If F_2 has to be free from secular terms, ψ_{2s} must vanish, giving the condition

$$\psi_{2s} = (H_{1s},F_{1s}) - \frac{1}{2\omega_1^{o2}} \frac{\partial\omega_1^o}{\partial x_1} [\frac{\partial}{\partial y_1} H_{1p}^2]_s = 0$$

and since the last term on the right is zero, F_{1s} is defined by

$$(H_{1s}, F_{1s}) = 0 \tag{2.7.9}$$

for which we need only a particular solution, the simplest of which, in this case is $F_{1s} = 0$. Consider for simplicity the case

$$H = H_0 + H_1, \tag{2.7.10}$$

so that the second order approximation is given by

$$\omega_1^0 \frac{\partial F_2}{\partial y_1} = \psi_{2p} = (H_{1p}, F_{1s}) - \frac{1}{\omega_1^0}(H_{1s}, F_{1p})$$

$$- \frac{1}{2\omega_1^{02}} \frac{\partial \omega_1^0}{\partial x_1} [\frac{\partial}{\partial y_1} H_{1p}^2]_p$$

and since $[\frac{\partial}{\partial y_1} H_{1p}^2]_p = \frac{\partial}{\partial y_1} H_{1p}^2$ it follows that

$$F_{2p} = - \frac{1}{\omega_1^{02}} \int (H_{1s}, F_{1p}) dy_1 - \frac{1}{2\omega_1^{03}} \frac{\partial \omega_1^0}{\partial x_1} (H_{1p}^2)_p \tag{2.7.11}$$

and

$$F_2 = F_{2p} + F_{2s}$$

where $F_{2s}(y_2, y_3, \ldots, y_n; x)$ is arbitrary. Under hypothesis (2.7.10) the third approximation gives

$$\omega_1^0 \frac{\partial F_3}{\partial y_1} = (H_1, F_2) = \psi_{3p} + \psi_{3s}$$

where

$$\psi_{3s} = (\bar{H}_{1s}, F_{2s}) + (H_{1s}, F_{2p})_s$$

and imposing the condition $\psi_{3s} = 0$, defines the arbitrary function F_{2s} by

$$(H_{1s}, F_{2s}) = - \frac{1}{2\pi} \int_0^{2\pi} (H_{1s}, F_{2p}) dy_1 = \phi_{3s}(y_2, \ldots, y_n; x)$$

where ϕ_{3s} is known. The homogeneous characteristics of $F_{1s}, F_{2s}, \ldots, F_{ks}$ are the same for any k, and given by

$$\frac{dy_2}{\frac{\partial H_{1s}}{\partial x_2}} = \frac{dy_3}{\frac{\partial H_{1s}}{\partial x_3}} = \cdots = \frac{dy_n}{\frac{\partial H_{1s}}{\partial x_n}} = \frac{dx_2}{-\frac{\partial H_{1s}}{\partial y_2}} = \frac{dx_3}{-\frac{\partial H_{1s}}{\partial y_3}} = \cdots = \frac{dx_n}{-\frac{\partial H_{1s}}{\partial y_n}} = d\tau$$

where τ is an auxiliary parameter. The solution for F_{ks} $(k = 1,2,\ldots)$ will thus depend on the solution of the system

$$\frac{dy_j}{d\tau} = \frac{\partial H_{1s}}{\partial x_j}$$

$$\frac{dx_j}{d\tau} = -\frac{\partial H_{1s}}{\partial y_j} \qquad (j = 2,3,\ldots,n)$$

This corresponds to a dynamical system with $n-1$ degrees of freedom and whose Hamiltonian is H_{1s}. Nevertheless it should be noted that one needs only a particular solution (in the Jacobi sense) of such system. Of course, if for some values of $(x;y)$, one or more of the partials $\partial H_{1s}/\partial x_k$ is zero or small (say as small as ϵ), the solution will contain <u>singularities</u> or <u>small divisors</u>, and the method cannot proceed. One of the ways to handle this situation is suggested by the considerations pertinent to resonance and to be described in chapter V. Here we limit our discussion to the particular case where the derivative $\partial H_{1s}/\partial x_2 \simeq O(\epsilon^{1/2})$ and we plainly assume the expansion

$$F = F_0 + \epsilon^{1/2}F_1 + \epsilon F_2 + \epsilon^{3/2}F_3 + \cdots .$$

From the fundamental relation $(F,H) = 0$ it follows that, by equating terms of the same order in ϵ,

$$(H_0,F_0) = 0 \quad ,$$

$$(H_0,F_1) = 0 \quad ,$$

$$(H_0,F_2) + \Big(\frac{\partial H_1}{\partial x_1}\frac{\partial F_0}{\partial y_1} + \frac{\partial H_{1p}}{\partial x_2}\frac{\partial F_0}{\partial y_2} + \cdots + \frac{\partial H_1}{\partial x_n}\frac{\partial F_0}{\partial y_n}$$

$$- \frac{\partial H_1}{\partial y_1}\frac{\partial F_0}{\partial x_1} - \frac{\partial H_1}{\partial y_2}\frac{\partial F_0}{\partial x_2} - \cdots - \frac{\partial H_1}{\partial y_n}\frac{\partial F_0}{\partial x_n}\Big) = 0$$

$$(H_o,F_3) + \frac{\partial H_{1s}}{\partial x_2}\frac{\partial F_o}{\partial y_2} + (\frac{\partial H_1}{\partial x_1}\frac{\partial F_1}{\partial y_1} \; \frac{\partial H_{1p}}{\partial x_2}\frac{\partial F_1}{\partial y_2} + \dots + \frac{\partial H_1}{\partial x_n}\frac{\partial F_1}{\partial y_n}$$

$$- \frac{\partial H_1}{\partial y_1}\frac{\partial F_1}{\partial x_1} - \frac{\partial H_1}{\partial y_2}\frac{\partial F_1}{\partial x_2} - \dots - \frac{\partial H_1}{\partial y_n}\frac{\partial F_1}{\partial x_n}) = 0,$$

and so forth. If, again, $H_o = H_o(x_1)$, $F_o = x_1$, it follows that

$$\omega_1^o \frac{\partial F_1}{\partial y_1} = 0 \quad \therefore \quad F_1 = F_{1s}(y_2, y_3, \dots, y_n; x)$$

and

$$\omega_1^o \frac{\partial F_2}{\partial y_1} = - \frac{\partial H_1}{\partial y_1} = \psi_2$$

$$\omega_1^o \frac{\partial F_3}{\partial y_1} = (H_1, F_1) - \frac{\partial H_{1s}}{\partial x_2}\;\frac{\partial F_1}{\partial y_2} = \psi_3$$

$$\omega_1^o \frac{\partial F_k}{\partial y_1} = (H_1, F_{k-2}) + \frac{\partial H_{1s}}{\partial x_2}\frac{\partial}{\partial y_2}(F_{k-3}-F_{k-2}) = \psi_k$$

for $k = 4,5,6,\dots$. Now F_{1s} is arbitrary and can be taken equal to zero, so that, automatically, one gets $\psi_3 = 0$ and therefore

$$\omega_1^o \frac{\partial F_3}{\partial y_1} = 0$$

or

$$F_3 = F_{3s}(y_2, y_3, \dots, y_n;\; x_1, x_2, \dots, x_n).$$

On the other hand

$$F_2 = - \frac{1}{\omega_1^o} H_{1p} + F_{2s}$$

so that

$$\psi_{4s} = (H_{1s}, F_{2s}) + (H_{1p}, F_{2p})_s - \frac{\partial H_{1s}}{\partial x_2}\frac{\partial F_{2s}}{\partial y_2}$$

which should be zero. Since

$$(H_{1p}, -\frac{1}{\omega_1^o} H_{1p})_s = -\frac{1}{\omega_1^{o\,2}} \frac{\partial\omega_1^o}{\partial x_1} (H_{1p} \frac{\partial H_{1p}}{\partial y_1})_s = 0$$

it follows that

$$\psi_{4s} = (H_{1s}, F_{2s}) - \frac{\partial H_{1s}}{\partial x_2} \frac{\partial F_{2s}}{\partial y_2} = 0$$

and $F_{2s} = 0$ satisfies, in particular, the requirements. In any event, the characteristics (up to any order) are

$$\frac{dy_3}{\frac{\partial H_{1s}}{\partial x_3}} = \cdots = \frac{dy_n}{\frac{\partial H_{1s}}{\partial x_n}} = \frac{dx_3}{-\frac{\partial H_{1s}}{\partial y_3}} = \cdots = \frac{dx_n}{-\frac{\partial H_{1s}}{\partial y_n}} = d\tau$$

with the required disappearance of the small divisor $\partial H_{1s}/\partial x_2$. If $F_{2s} = 0$, F_2 is completely defined and F_4 is given by

$$F_4 = -\frac{1}{\omega_1^{o\,2}} \int [(H_{1s}, H_{1p}) - \frac{\partial H_{1s}}{\partial x_2} \frac{\partial H_{1p}}{\partial y_2}]\, dy_1$$
$$- \frac{1}{2\omega_1^{o\,3}} \frac{\partial \omega_1^o}{\partial x_1} (H_{1p}^2)_p$$

and so forth. At every stage of the approximation, the characteristics are the same and do not present any singularity. It is also clear that the method can be applied equivalently to cases in which more than one derivative $\partial H_{1s}/\partial x_k$ is small.

Suppose now $F_o = x_2$ so that F will correspond to x_2' of the Poincaré problem. In this case

$$\omega_1^o \frac{\partial F_1}{\partial y_1} = -\frac{\partial H_1}{\partial y_2},$$

so that,

$$F_1 = -\frac{1}{\omega_1^o} \int \frac{\partial H_1}{\partial y_2} dy_1 + \psi_1(y_2, y_3, \ldots, y_n; x).$$

However, the integrand $\partial H_1/\partial y_2$ may contain terms which are independent from y_1 and, therefore, F_1 will have a secular increase in y_1. Such secular parts will be

$$F_{1s} = -\frac{1}{\omega_1^o}\int \frac{\partial H_{1s}}{\partial y_2}\, dy_1 + \psi_1(y_2, y_3, \ldots, y_n; x)$$

which cannot be zero unless H_{1s} does not depend on y_2. Therefore, one is forced to deviate from the assumption $F_o = x_2$ and assume a more general form

$$F_o = F_o(y_2, y_3, \ldots, y_n; x).$$

If it is possible to choose F_o so that secular terms are not present in the higher approximations, one can at least obtain a formal integral, eventually convergent. The equation for F_1 is obtained from

$$(H_o, F_1) + (H_1, F_o) = 0$$

or

$$\omega_1^o \frac{\partial F_1}{\partial y_1} = (H_1, F_o).$$

The "secular" part of the right hand member should be zero, that is,

$$(H_{1s}, F_o) = 0 \qquad (2.7.12)$$

since F_o does not contain y_1 by hypothesis. This hypotheses is easily justified by the condition $(H_o, F_o) = 0$, with $H_o = H_o(x_1)$. An immediate solution of (2.7.12) is to assume

$$F_o = k\, H_{1s}$$

where k is a constant. From the point of view of Hamilton-Jacobi theory, it is clear that this choice is suggested by the equivalent situation in Poincaré's method (Giacaglia, 1965, p. 16).

<u>The interesting physical feature of this process is that the "secular" part of H_1 becomes the zero order approximation of an integral of motion.</u> The

interpretation of this fact lies in the conservation of the energy of the system under canonical transformations. Also, there is a close connection, at this point, with perturbation methods based on Lie Series Transforms to be discussed later in this chapter.

Now consider the original system

$$\dot{y}_k = \frac{\partial H}{\partial x_k}, \quad \dot{x}_k = -\frac{\partial H}{\partial y_k}$$

where $H = H(y;x)$, $k = 1,2,\ldots,n$. Let $t = y_{n+1}$, so that

$$\dot{y}_\alpha = \frac{\partial \mathscr{H}}{\partial x_\alpha}, \quad \dot{x}_\alpha = -\frac{\partial \mathscr{H}}{\partial y_\alpha} \tag{2.7.13}$$

where $\mathscr{H} = H + x_{n+1}$, $x_{n+1} = \text{const} = \beta$, $\alpha = 1,2,\ldots,n+1$. The angle variables of the system are, according to Poincaré's method

$$y'_\alpha = \omega_\alpha t + y'_{\alpha o}$$

where $y'_{\alpha o}$ are absolute constants and

$$\omega_\alpha = -\frac{\partial x_{n+1}}{\partial x'_\alpha} = \omega_\alpha^0 + \epsilon\omega_\alpha^1 + \epsilon^2\omega_\alpha^2 + \ldots$$

and the ω_α^k are functions of $x'_1, x'_2, \ldots, x'_n$. In particular

$$\omega_\alpha^0 = \frac{\partial H_o}{\partial x'_\alpha}.$$

so that

$$y'_\alpha = \omega_\alpha^0 t + \epsilon\nu_\alpha(x';\epsilon)t + \beta_\alpha.$$

On the other hand

$$y'_\alpha = \frac{\partial W}{\partial x'_\alpha} = y'_\alpha(x';y;\epsilon) = y_\alpha + \epsilon\mu_\alpha(x';y;\epsilon).$$

Comparison of the last two relations gives

$$\beta_\alpha = y_\alpha - \omega_\alpha^0 t + \epsilon(\mu_\alpha - \nu_\alpha t).$$

On the other hand, the β_α are constants of the system (2.7.13), and can be written as

$$\beta_k = y_k - \omega_k^o y_{n+1} + \epsilon\theta_k(x';y;t,\epsilon) \tag{2.7.14}$$

and the zero order part of such integrals can be taken as

$$F_{ko} = y_k - \omega_k^o y_{n+1} \quad (k = 1,2,\ldots,n). \tag{2.7.15}$$

Poisson's condition is now written in the form

$$\sum_{\alpha=1}^{n+1} \left(\frac{\partial \mathscr{H}}{\partial x_\alpha}\frac{\partial F}{\partial y_\alpha} - \frac{\partial \mathscr{H}}{\partial y_\alpha}\frac{\partial F}{\partial x_\alpha}\right) = 0.$$

If $H_o = H_o(x)$, the zero order approximation would be given by

$$\sum_{k=1}^{n} \frac{\partial H_o}{\partial x_k}\frac{\partial F_o}{\partial y_k} = \frac{\partial F_o}{\partial y_{n+1}}$$

and a particular solution is

$$F_o = y_1 - \frac{\partial H_o}{\partial x_1} y_{n+1} = y_1 - \omega_1^o y_{n+1}$$

which is of the form (2.7.15) for $k = 1$.

The question arises whether the formal series obtained in this form have some meaning, since, in the present case, linear terms in time cannot be eliminated. But the same question is present in Poincaré's method, where the frequencies $\omega_k = \omega_k^o + \epsilon\omega_k^o + \ldots$ are indeed obtained, in practical cases, only up to a certain degree of approximation p. This fact, as mentioned before, is reflected in the conclusion that, even if the series converge, in practical cases the solution cannot be valid for an interval of time which, at best if $O(\epsilon^{-p})$.

Writing the condition as

$$(F,H) + \frac{\partial F}{\partial t} = 0$$

the "integrals" F corresponding to (2.7.14) are formally obtained as follows. We suppose

$$F_o = y_1 - \omega_1^o t$$

and

$$H = H_o(x_1) + H_1(y;x).$$

The recurrence relations for F_k are

$$(F_k, H_o) + \frac{\partial F_k}{\partial t} = -(F_{k-1}, H_1)$$

or

$$\omega_1^o \frac{\partial F_k}{\partial y_1} + \frac{\partial F_k}{\partial t} = (H_1, F_{k-1}).$$

For $k = 1$,

$$\omega_1^o \frac{\partial F_1}{\partial y_1} + \frac{\partial F_1}{\partial t} = \frac{\partial H_1}{\partial x_1} + \frac{\partial \omega_1^o}{\partial x_1} t \frac{\partial H_1}{\partial y_1},$$

for $k = 2$,

$$\omega_1^o \frac{\partial F_2}{\partial y_1} + \frac{\partial F_2}{\partial t} = (H_1, F_1),$$

and so forth. The solution for F_1 is found to be

$$F_1 = \frac{1}{\omega_1^o} \int \frac{\partial H_1}{\partial x_1} dy_1 + \frac{1}{\omega_1^{o2}} \frac{\partial \omega_1^o}{\partial x_1} \int \frac{\partial H_1}{\partial y_1} y_1 \, dy_1$$

$$- \frac{1}{\omega_1^{o2}} \frac{\partial \omega_1^o}{\partial x_1} y_1 \int \frac{\partial H_1}{\partial y_1} dy_1 + t \frac{1}{\omega_1^o} \frac{\partial \omega_1^o}{\partial x_1} \int \frac{\partial H_1}{\partial y_1} dy_1$$

$$+ \psi_1(y_2, y_3, \ldots, y_n; x)$$

with ψ_1 arbitrary. On the other hand, one has

$$\frac{\partial H_1}{\partial y_1} y_1 = \frac{\partial}{\partial y_1} (H_1 y_1) - H_1,$$

so that, if H_1 is 2π-periodic in every variable $y_1, y_3, \ldots, y_n$, we obtain

$$\int \frac{\partial}{\partial y_1}(H_1 y_1)\,dy_1 - \int H_1 dy_1 = -\int H_{1p} dy_1 .$$

Hence

$$F_1 = \frac{1}{\omega_1^o}\int \frac{\partial H_1}{\partial x_1} dy_1 - \frac{1}{\omega_1^{o2}} \frac{\partial \omega_1^o}{\partial x_1}\int H_{1p} dy_1 + t\,\frac{1}{\omega_1^o}\frac{\partial \omega_1^o}{\partial x_1} H_{1p} + \psi_1 .$$

The only undesirable term is the first in the right hand member, from which secular terms in y_1 may arise. They are precisely

$$\frac{1}{\omega_1^o}\int \frac{\partial H_{1s}}{\partial x_1} dy_1 = \frac{1}{\omega_1^o}\frac{\partial H_{1s}}{\partial x_1} y_1 .$$

The function

$$F_1 - \frac{1}{\omega_1^o}\frac{\partial H_{1s}}{\partial x_1} y_1 = \frac{1}{\omega_1^o}\left(1 + \frac{\partial \omega_1^o}{\partial x_1} t\right) H_{1p}$$

$$- \frac{1}{\omega_1^{o2}}\frac{\partial \omega_1^o}{\partial x_1}\int H_{1p} dy_1 + \psi_1$$

is multiperiodic in $y_1, y_2, \ldots, y_n$ and secular in t, this second characteristic being unavoidable and indeed necessary. The situation suggests therefore a modification of the function F_o as follows. We consider

$$F_o = y_1 - \omega_1^o t + \psi_o(y_2, y_3, \ldots, y_n; x)$$

which, evidently, is a solution for

$$(F_o, H_o) + \frac{\partial F_o}{\partial t} = 0.$$

Then, the equation for F_1 becomes

$$\omega_1^o - \frac{\partial F_1}{\partial y_1} + \frac{\partial F_1}{\partial t} = \frac{\partial H_1}{\partial x_1} + t\,\frac{\partial \omega_1^o}{\partial x_1}\frac{\partial H_1}{\partial y_1} + (H_1, \psi_o)$$

whose solution is the same as before, with the addition of the term

$$\frac{1}{\omega_1^o} \int (H_1, \psi_o) dy_1.$$

The part of this integral which contains secular terms in y_1 will be zero if, and only if,

$$\frac{\partial H_{1s}}{\partial x_1} + (H_{1s}, \psi_o) = 0.$$

The last equation defines the way in which the arbitrary function ψ_o should be chosen. The solution of this partial differential equation is equivalent to the integration of the characteristics

$$\frac{dy_k}{d\tau} = \frac{\partial H_{1s}}{\partial x_k}, \quad \frac{dx_k}{d\tau} = - \frac{\partial H_{1s}}{\partial y_k} \tag{2.7.16}$$

where τ is any parameter and $k = 2,3,\ldots,n$, whereas x_1 has to be treated as a constant parameter. If y_k, x_k are obtained from these as functions of τ, then H_{1s} is expressed as a function of τ, and ψ_o is obtained from

$$\psi_o = -\int \frac{\partial H_{1s}}{\partial x_1} d\tau.$$

After the integration is performed, ψ_o is again set in terms of $y_2, y_3, \ldots, y_n$; $x_1, x_2, x_3, \ldots, x_n$. The addition of ψ_o to F_o shall have the effect of changing the reference frequency ω_o^1, which, in other terms, is simply Lindstedt's device.

With this, we have established a clear connection between the definition of an Adelphic Integral and the formal integration of a Hamiltonian system by Poincaré's method. Such connection as we shall see next, establishes a fundamental bridge toward the methods using Lie Series Transforms and on Auxiliary System.

8. Perturbation Techniques Based on Lie Transforms.

This section is devoted to a, as brief as possible, view of perturbation methods introduced first by Hori (1966). As we have seen, it is perfectly justified to assume Hori's generator S to depend on the parameter ϵ and, therefore, define a canonical transformation by

$$y_j = \eta_j + \sum_{n=1}^{\infty} \frac{\epsilon^n}{n!} D_S^{n-1} \frac{\partial S}{\partial \xi_j}$$
$$x_j = \xi_j - \sum_{n=1}^{\infty} \frac{\epsilon^n}{n!} D_S^{n-1} \frac{\partial S}{\partial \eta_j} \tag{2.8.1}$$

for $j = 1,2,\ldots,n$, where y_j are coordinates, x_j momenta, and η_j, ξ_j the corresponding new variables, and

$$S = S(\eta;\xi;\epsilon).$$

The image of any function $f(y;x;\epsilon)$ into the new phase space $(\eta;\xi)$, via the generator S, is given by

$$f(y;x;\epsilon) = \sum_{n=0}^{\infty} \frac{\epsilon^n}{n!} D_S^n f(\eta;\xi;\epsilon), \tag{2.8.2}$$

where, we recall the definitions

$$D_S^0 f = f$$
$$D_S^1 f = (f,S) = \sum_{k=1}^{n} \left(\frac{\partial f}{\partial \eta_k}\frac{\partial S}{\partial \xi_k} - \frac{\partial f}{\partial \xi_k}\frac{\partial S}{\partial \eta_k}\right)$$
$$D_S^n f = D_S^1(D_S^{n-1} f), \qquad n = 1,2,\ldots .$$

Obviously, all functions involved f, S must be, at least, infinitely many time differentiable and the series above should converge for ϵ sufficiently small.

Now consider the original system of differential equations to be defined by the Hamiltonian

$$H = H(y;x;\epsilon).$$

which, for simplicity, we assume to be analytic in the $2n+1$ arguments for $(y;x) \in D$ and $0 \leq \epsilon < \epsilon_0$. The equations are

$$\dot{y} = H_x, \quad \dot{x} = -H_y, \tag{2.8.3}$$

and we assume that the power series

$$H(y;x;\epsilon) = \sum_{k=0}^{\infty} \epsilon^k H_k(y;x) \tag{2.8.4}$$

is such that $H_0(y;x)$ is integrable in D, in the Liouville sense, that is, the system

$$\begin{aligned} \frac{d\eta_k}{d\tau} &= \frac{\partial H_0}{\partial \xi_k}(\eta;\xi) \\ \frac{d\xi_k}{d\tau} &= -\frac{\partial H_0}{\partial \eta_k}(\eta;\xi) \end{aligned} \tag{2.8.5}$$

for $k = 1,2,\ldots,n$ has the explicit solution

$$\begin{aligned} \eta_k &= \eta_k^*(\alpha_1,\alpha_2,\ldots,\alpha_n;\ \beta_1 + \omega_1\tau,\ \beta_2,\beta_3,\ldots,\beta_n) \\ \xi_k &= \xi_k^*(\alpha_1,\alpha_2,\ldots,\alpha_n;\ \beta_1 + \omega_1\tau,\ \beta_2,\beta_3,\ldots,\beta_n) \end{aligned} \tag{2.8.6}$$

where (α,β) are constants of integration and $\omega_1 = \omega_1(\alpha_1)$, by the usual specific choice of the energy integral dependence on only one of the α's, say α_1. The requirement that the jacobian matrix

$$\frac{\partial(\eta^*;\xi^*)}{\partial(\beta;\alpha)}$$

be non-singular, for sufficiently small τ, allows the inversion of the above relations as

$$\begin{aligned} \alpha_k &= \alpha_k(\eta;\xi), \quad k = 1,2,\ldots,n \\ \beta_1 + \omega_1\tau &= \beta_1^*(\eta;\xi) \\ \beta_k &= \beta_k(\eta;\xi), \quad k = 2,3,\ldots,n. \end{aligned} \tag{2.8.7}$$

Following Hori's definition we shall call (2.8.5) the <u>auxiliary system.</u> It should be kept in mind that, since H_0 is supposed to be integrable in the Liouville sense, there exists a canonical transformation, in particular (2.8.6) if $(\alpha;\beta)$ are action-angle variables, which reduces H_0 to a function only of the new momenta, in this case of α_1 only.

We now consider the problem of producing first integrals of motion of (2.8.3), independent of H. We consider a complete canonical transformation (2.8.1) with a generator

$$\epsilon S(\eta;\xi;\epsilon) = \sum_{k=1}^{\infty} \epsilon^k S_k(\eta;\xi) . \qquad (2.8.8)$$

The transformation being time independent, if $K(\eta;\xi;\epsilon)$ is the new Hamiltonian, it follows that

$$H(y;x;\epsilon) = K(\eta;\xi;\epsilon) \qquad (2.8.9)$$

where, in the left hand side the coordinates and momenta $(y;x)$ are supposed functions of $(\eta;\xi;\epsilon)$ through (2.8.1). According to (2.8.2), such transformation is obtained by direct application of S, if this is a known function, so that

$$K(\eta;\xi;\epsilon) = \sum_{n=0}^{\infty} \frac{\epsilon^n}{n!} D_S^n H(\eta;\xi;\epsilon) \qquad (2.8.10)$$

If the series on the right converges, as a power series in ϵ, we must assume that a similar convergent power series exists for K, that is,

$$K(\eta;\xi;\epsilon) = \sum_{n=0}^{\infty} \epsilon^n K_n(\eta;\xi). \qquad (2.8.11)$$

Making use of (2.8.4) and (2.8.8), the right hand side of (2.8.10) yields, equating coefficients of the same power in ϵ,

$$\begin{aligned} K_o(\eta;\xi) &= H_o(\eta;\xi) \\ K_p(\eta;\xi) &= (H_o, S_p) + F_p \qquad (2.8.12) \\ p &= 1,2,3,\ldots, \end{aligned}$$

where F_p is a function of H_o, $H_1,\ldots,H_{p-1}$, S_1, $S_2,\ldots,S_{p-1}$, and possible to be specified either directly or by recurrence. The specification of F_p is not important as far as the discussion of the method is concerned and the advantage of one or another form is pertinent to the specific problem under study. For $p \geq 1$, equation (2.8.12) represents a partial differential equation in S_p with the typical characteristic of averaging methods, that is, K_p is also unknown. The equation can be written as

$$\sum_{k=1}^{n} \left(\frac{\partial H_o}{\partial \eta_k} \frac{\partial S_p}{\partial \xi_k} - \frac{\partial H_o}{\partial \xi_k} \frac{\partial S_p}{\partial \eta_k}\right) + F_p(\eta_1,\ldots,\eta_n;\ \xi_1,\ldots,\xi_n) = K_p(\eta_1,\ldots,\eta_n;\ \xi_1,\ldots,\xi_n) \tag{2.8.13}$$

or, using the auxiliary system

$$-\frac{dS_p}{d\tau} + F_p'(\alpha;\beta_1 + \tau,\ \beta_2,\ldots,\beta_n) = K_p'(\alpha;\beta_1 + \tau,\ \beta_2,\ldots,\beta_n). \tag{2.8.14}$$

<u>The averaging principle in this method can be interpreted by imposing the condition that K_p should not depend on τ.</u> If, as usual, we assume $H(y;x;\epsilon)$ to be a 2π-periodic function of each y and because H_o is Liouville integrable, the y^* and x^* are quasiperiodic, or periodic, functions of τ, a classical result following from the general theory of action and angle variables. We generalize the average to a quasiperiodic function, as was discussed previously, by setting

$$K_p' = \lim_{T\to\infty} \frac{1}{T}\int_0^T F_p'(\alpha;\beta_1+\tau,\ \beta_2,\ldots,\beta_n)\,d\tau = K_p'(\alpha;\ -\ ,\ \beta_2,\beta_3,\ldots,\beta_n) = K_p(\eta;\xi) \tag{2.8.15}$$

the last transformation in (2.8.15) being obtained by means of (2.8.7). It follows that

$$\frac{dS_p}{d\tau} = F_p'(\alpha;\beta_1 + \tau,\ \beta_2,\ldots,\beta_n) - K_p'(\alpha;\ -\ ,\ \beta_2,\ldots,\beta_n)$$

or

$$S_p = \int(F_p'-K_p')\,d\tau = S_p'(\alpha;\beta_1 + \tau,\ \beta_2,\ldots,\beta_n) = S_p(\eta;\xi), \tag{2.8.16}$$

again making use of (2.8.7) to perform the last transformation. It is also obvious that, in view of the definition of K'_p,

$$\lim_{\tau \to \infty} S'_p(\alpha; \beta_1 + \tau, \beta_2, \ldots, \beta_n) = \text{finite}$$

and, under the foregoing hypotheses, S'_p is quasiperiodic (or periodic with no constant term) with respect to τ. By recurrence, or otherwise, one can show that the process can be repeated for any $p = 1,2,3,\ldots$ which proves that <u>there exists a formal series</u>

$$S = S_0(\eta; \xi) + \epsilon S_1(\eta; \xi) + \ldots \qquad (2.8.17)$$

<u>which reduces the Hamiltonian to</u>

$$K = K_0(\eta; \xi) + \epsilon K_1(\eta; \xi) + \ldots \qquad (2.8.18)$$

<u>with the property that, if $(\eta; \xi)$ are substituted by the solution of the auxiliary system, K does not depend explicitly on τ,</u> and therefore,

$$\frac{\partial K'}{\partial \tau} = \frac{dK'}{d\tau} = 0 \qquad (2.8.19)$$

where K' is defined by the formal series

$$K' = K'_0(\alpha; -, \beta_2, \ldots, \beta_n) + K'_1(\alpha; -, \beta_2, \beta_3, \ldots, \beta_n) + \ldots .$$

Obviously, one can write

$$\frac{dK'}{d\tau} = \sum_{k=1}^{n} \left(\frac{\partial K}{\partial \eta_k} \frac{\partial \eta_k}{d\tau} + \frac{\partial K}{\partial \xi_k} \frac{\partial \xi_k}{d\tau}\right)$$

$$= \sum_{k=1}^{n} \left(-\dot{\xi}_k \frac{d\eta_k}{d\tau} + \dot{\eta}_k \frac{d\xi_k}{d\tau}\right)$$

$$= \sum_{k=1}^{n} \left(- \frac{\partial K_0}{\partial \xi_k} \dot{\xi}_k - \frac{\partial K_0}{\partial \eta_k} \dot{\eta}_k\right) = - \frac{dK_0}{dt}$$

and in view of (2.8.14),

$$\frac{dK_0}{dt} = 0$$

so that

$$K_o(\eta;\xi) = \text{constant} = J_o . \tag{2.8.20}$$

We conclude that as a result of a Lie Transform, such that the new Hamiltonian does not depend on the auxiliary time τ, one obtains a new (formal) integral of motion, given by (2.8.20). The validity of this formal result can only be verified by analyzing the convergence of the method. Since it has been shown that Lie's Method and von Ziepel's Method are equivalent (Shniad) and that, if Kolmogorov's Method converges (under variable frequencies) so does von Ziepel's (Moser, 1966), the convergence, under sufficiently small and several time differentiable perturbations, of the Lie Transform Method, can be inferred indirectly. Again, such convergence cannot be uniform with respect to ϵ or the initial conditions. The advantage of the method outlined here is that only quadratures are involved, in opposition to Poincaré's Method where, in general, one has to deal with partial differential equations. Equivalently important advantages are, of course, the production of the transformation in explicit form (see 2.8.1), the ability of writing any function of $(y;x)$ in terms of $(\eta;\xi)$ by the direct use of the generator S (see 2.8.2) and the invariance of the method and resulting quantities with respect to canonical transformations, a fact which follows directly from the invariance of Poisson's parentheses with respect to such transformations.

We recall that a canonical transformation

$$\begin{aligned} Q &= Q(q;p;\tau) \\ P &= P(q;p;\tau) \end{aligned} \tag{2.8.21}$$

defined by a Lie generator $S(Q;P;\tau)$ can be defined by the solution of the system

$$\begin{aligned} \frac{dQ}{d\tau} &= \left(\frac{\partial S}{\partial P}\right)^T \\ \frac{dP}{d\tau} &= -\left(\frac{\partial S}{\partial Q}\right)^T \end{aligned} \tag{2.8.22}$$

for the initial conditions $(\tau = 0)$

$$Q(q;p;0) = q,$$
$$P(q;p;0) = p, \tag{2.8.23}$$

where τ is a parameter. The right hand members of (2.8.21) are supposed C^2 in all the $2n+1$ variables, in some domain of the phase space and τ restricted to some interval, say, $|\tau| \le \tau_0$. For the Poincaré generator $W(q;P;\tau)$ the same canonical transformation is given by

$$Q = q + \frac{\partial W}{\partial P^T}(q;P;\tau)$$
$$p = P + \frac{\partial W}{\partial q^T}(q;P;\tau) \tag{2.8.24}$$

under the condition

$$W(q;P;0) \equiv 0$$

which is equivalent to the initial conditions (2.8.23). It has been established that

$$S(Q;P;\tau) = \frac{\partial W}{\partial \tau}(q;P;\tau) \tag{2.8.25}$$

where Q is given by the first of (2.8.24). Assuming the expansions

$$W(q;P;\tau) = \sum_{n=1}^{\infty} W_n(Q;P)\tau^n \tag{2.8.26}$$

and

$$S(Q;P;\tau) = \sum_{n=0}^{\infty} S_{n+1}(Q;P)\tau^n,$$

equating coefficients of like powers in τ in (2.8.25), gives the relations among the W_k and the S_j, as obtained earlier.

Mersman (1971) produced Deprit's algorithm by setting $\tau = \epsilon$ in the above formalism. If S corresponds now to Lie's generator S of equation (1.5.7), to keep the notation used there one should substitute $S_{n+1}/n!$ for S_n in the expansion of (2.8.25) and obtain

$$S_1 = W_1$$

$$S_2 = 2W_2 - \sum_i \frac{\partial W_1}{\partial Q_i}\frac{\partial W_1}{\partial P_i}$$

$$S_3 = 6W_3 - \sum_i \; 2\frac{\partial W_1}{\partial Q_i}\frac{\partial W_2}{\partial P_i} + 2\frac{\partial W_2}{\partial Q_i}\frac{\partial W_1}{\partial P_i}$$

$$+ \sum_{i,j} \frac{\partial^2 W_1}{\partial Q_i \partial Q_j}\frac{\partial W_1}{\partial P_i}\frac{\partial W_1}{\partial P_j} + 2\frac{\partial^2 W_1}{\partial Q_i \partial P_j}\frac{\partial W_1}{\partial P_i}\frac{\partial W_1}{\partial Q_j}$$

and so on. Hori's formalism is also obtained from (2.8.25) by substituting $S(Q;P)$ for $S(Q;P;\tau)$ and setting $\tau = 1$ thereafter, that is, the expansions of (2.8.25) corresponding to (2.8.26) are

$$W(q;P) = \sum_{n=1}^{\infty} W_n(Q;P)$$

$$S(Q;P) = S_1(Q;P)$$

which, substituted into (2.8.25), or directly into the expansions following (1.6.10) give

$$W = S_1 + \frac{1}{2}\sum_i \frac{\partial S_1}{\partial Q_i}\frac{\partial S_1}{\partial P_j}$$

$$+ \frac{1}{6}\sum_{i,j} \left(\frac{\partial^2 S_1}{\partial Q_i \partial Q_j}\frac{\partial S_1}{\partial P_i}\frac{\partial S_1}{\partial P_j} + \frac{\partial^2 S_1}{\partial Q_i \partial P_j}\frac{\partial S_1}{\partial P_i}\frac{\partial S_1}{\partial Q_j}\right. \tag{2.8.27}$$

$$\left. + \frac{\partial^2 S_1}{\partial P_i \partial Q_j}\frac{\partial S_1}{\partial Q_i}\frac{\partial S_1}{\partial P_j}\right) + \cdots \quad .$$

The parameter ϵ is then introduced into W and S_1, as

$$W = W(Q;P;\epsilon)$$

$$S_1 = U(Q;P;\epsilon)$$

and one assumes the formal series

$$W = \sum_{n=1}^{\infty} W_n(Q;P)\epsilon^n,$$

$$U = \sum_{n=1}^{\infty} U_n(Q;P)\epsilon^n. \tag{2.8.28}$$

The inverse of (2.8.27) is found to be, by a way or another,

$$S_1 = W - \frac{1}{2}\sum_i \frac{\partial W}{\partial Q_i}\frac{\partial W}{\partial P_i}$$
$$+ \frac{1}{12}\sum_{i,j}\left(\frac{\partial^2 W}{\partial Q_i \partial Q_j}\frac{\partial W}{\partial P_i}\frac{\partial W}{\partial P_j} + 4\frac{\partial^2 W}{\partial Q_i \partial P_j}\frac{\partial W}{\partial P_i}\frac{\partial W}{\partial Q_j}\right.$$
$$\left. + \frac{\partial^2 W}{\partial P_i \partial P_j}\frac{\partial W}{\partial Q_i}\frac{\partial W}{\partial Q_j}\right) + \ldots \; . \tag{2.8.29}$$

Introducing (2.8.28) and equating like powers of ϵ, one finds from (2.8.27)

$$W_1 = U_1 ,$$
$$W_2 = U_2 + \frac{1}{2}\sum_i \frac{\partial U_1}{\partial Q_i}\frac{\partial U_1}{\partial P_i} , \tag{2.8.30}$$

----

or, from (2.8.29)

$$U_1 = W_1 ,$$
$$U_2 = W_2 - \frac{1}{2}\sum_i \frac{\partial W_1}{\partial Q_i}\frac{\partial W_1}{\partial P_i} , \text{ etc.} \tag{2.8.31}$$

The foregoing relations allow the translation of the perturbation method introduced by Hori (1966) and described at the beginning of this section, into Deprit's formalism.

As an example consider Duffing's Equation without damping, that is,

$$\ddot{u} + u + \epsilon\gamma u^3 = \epsilon B \cos \omega t \tag{2.8.32}$$

where $\epsilon \geq 0$, $\gamma \geq 0$, B, $\omega \neq 0$ are constant parameters. We consider the case when ω is not rational and moreover for $p \neq 0$, q integers, a relation

$$|p\omega - q| \geq K(p)\epsilon^{1/2} \tag{2.8.33}$$

is satisfied for a conveniently chosen $K(p)$, say $K(p) = p^{5/2-\sigma}$, $\sigma \geq 4$, integer. If (2.8.33) is not satisfied we do have a case of resonance and it will be discussed in the last chapter.

Introducing the homogeneous complete canonical transformation

$$u = (2p_1)^{1/2} \sin q_1$$

$$\dot{u} = (2p_1)^{1/2} \cos q_1$$

the equation (2.8.32) can be written

$$\dot{q}_1 = \frac{\partial H}{\partial p_1}, \qquad \dot{p}_1 = -\frac{\partial H}{\partial q_1} \tag{2.8.34}$$

where

$$H = p_1 + \epsilon\gamma p_1^2 \sin 4q_1 - \epsilon B\,(2p_1)^{1/2} \sin q_1 \cos \omega t.$$

Further, introducing the coordinate

$$q_2 = \omega t$$

with the conjugate momentum p_2, the system takes the form

$$\dot{q}_j = \frac{\partial K}{\partial p_j}, \quad \dot{p}_j = -\frac{\partial K}{\partial q_j} \tag{2.8.35}$$

for $j = 1,2$, and

$$K = p_1 + \omega p_2 + \epsilon(\gamma p_1^2 \sin 4q_1 - B(2p_1)^{1/2} \sin q_1 \cos q_2$$

$$= K_o(p_1,p_2) + \epsilon K_1(q_1,q_2,p_1, -)$$

The <u>auxiliary system</u> is defined by K_o and has the solution

$$q_1^o = \tau + \beta_1$$

$$q_2^o = \omega\tau + \beta_2$$

$$p_1^o = \alpha_1$$

$$p_2^o = \alpha_2$$

where α_1, α_2, β_1, β_2 are constants. Let the new Hamiltonian be

$$K^* = K_o^* + \epsilon K_1^* + \epsilon^2 K_2^* + \ldots$$

and the Lie generator

$$\epsilon S = \epsilon S_1 + \epsilon^2 S_2 + \ldots ,$$

with the condition that K^* should not depend on τ and therefore K_o^* is an integral of motion in the new coordinates and momenta q_1^*, q_2^*, p_1^*, p_2^*.

The equation

$$-\frac{dS_1}{d\tau} + K_1(q_1^o, q_2^o, p_1^o, -) = K_1^*$$

gives, under the condition that ω is not an integer

$$K_1^* = \frac{3}{8} \gamma p_1^2$$

$$S_1 = -\frac{1}{4} \gamma p_1^2 \sin 2q_1^* + \frac{1}{32} p_1^{*2} \sin 4q_1^*$$

$$+ \frac{B(2p_1^*)^{1/2}}{2(1+\omega)} \cos (q_1^* + q_2^*)$$

$$+ \frac{B(2p_1^*)^{1/2}}{2(1-\omega)} \cos (q_1^* - q_2^*) .$$

The second order approximation, using

$$-\frac{dS_2}{d\tau} + \frac{1}{2}(K_1 + K_1^*, S_1) + K_2 = K_2^*$$

where, in our case, $K_2 = 0$, gives

$$K_2^* = \frac{17}{64} \gamma^2 p_1^{*3} + \frac{B^2}{8(1-\omega^2)}$$

$$S_2 = -\frac{1}{2}\left(\frac{21}{32} \gamma^2 p_1^{*3} + \frac{B^2}{4(1-\omega^2)}\right) \sin 2q_1^*$$

$$- \frac{3}{128} \gamma^2 p_1^{*3} \sin 4q_1^* - \frac{7}{192} \gamma^2 p_1^{*3} \sin 6q_1^*$$

$$- \frac{B^2}{8\omega(1-\omega^2)} \sin 2q_2^* + \frac{B\gamma(2p_1^*)^{3/2}}{32(1-\omega^2)(1+\omega)} (13-\omega^2) \sin (q_1^* + q_2^*)$$

$$+ \frac{B\gamma(2p_1^*)^{3/2}}{32(1-\omega^2)(1-\omega)} (13-\omega^2) \sin (q_1^* - q_2^*)$$

$$- \frac{B\gamma(2p_1^*)^{3/2}}{128(1-\omega^2)(3+\omega)} (21-5\omega^2) \cos (3q_1^* + q_2^*)$$

$$- \frac{B\gamma(2p_1^*)^{3/2}}{128(1-\omega^2)(3-\omega)} (21-5\omega^2) \cos (3q_1^* - q_2^*)$$

$$+ \frac{B\gamma(2p_1^*)^{3/2}}{128(5+\omega)} \cos (5q_1^* + q_2^*) + \frac{B\gamma}{128(5-\omega)} \cos (5q_1^* - q_2^*)$$

$$- \frac{B^2}{16(1+\omega)^2} \sin (2q_1^* + 2q_2^*) - \frac{B^2}{16(1-\omega)^2} \sin (2q_1^* - 2q_2^*) .$$

With the current approximation the new Hamiltonian is given by

$$K^* = p_1^* + \omega p_2^* + \frac{3}{8} \epsilon\gamma p_1^{*2} + \frac{17}{64} \epsilon^2\gamma p_1^{*3} + O(\epsilon^3)$$

where we have neglected absolute constants. On the other hand K_0^* is an integral of motion, that is,

$$p_1^* + \omega p_2^* = \text{const.}$$

so that

$$K^* - p_1^* - \omega p_2^* = \frac{3}{8} \epsilon\gamma p_1^{*2} + \frac{17}{64} \epsilon^2\gamma p_1^{*3} + \ldots$$

is also an integral, so that the problem is, in principle, reduced to quadratures and, except for values of ω rational or "close" to rational, the general solution can be found. The relations between the two sets of variables $(q;p)$ and $(q^*;p^*)$ are given by (2.8.1), or in the present notation

$$q_j = q_j^* + \sum_{n \geq 1} \frac{\epsilon^n}{n!} D_S^{n-1} \frac{\partial S}{\partial p_j^*}$$
$$p_j = p_j^* - \sum_{n \geq 1} \frac{\epsilon^n}{n!} D_S^{n-1} \frac{\partial S}{\partial q_j^*} \qquad (2.8.36)$$

for $j = 1,2$. Obviously, since S does not depend on p_2^*, it follows that $q_2 = q_2^*$, that is, the transformation does not change the time $(q_2^* = \omega t)$. Since we have defined

$$\epsilon S = \epsilon S_1 + \epsilon^2 S_2 + \dots$$

if one sets

$$W = \epsilon S \tag{2.8.37}$$

the transformations can be written

$$\begin{aligned} q_j &= q_j^* + \sum_{n \geq 1} \frac{1}{n!} D_W^{n-1} \frac{\partial W}{\partial p_j^*} \\ p_j &= p_j^* - \sum_{n \geq 1} \frac{1}{n!} D_W^{n-1} \frac{\partial W}{\partial q_j^*} \end{aligned} \tag{2.8.38}$$

or to second order in ϵ,

$$q_j = q_j^* + \frac{\partial W_1}{\partial p_j^*} + \frac{\partial W_2}{\partial p_j^*} + \frac{1}{2} (\frac{\partial W_1}{\partial p_j^*}, W_1)$$

$$p_j = p_j^* - \frac{\partial W_1}{\partial q_j^*} - \frac{\partial W_2}{\partial q_j^*} - \frac{1}{2} (\frac{\partial W_1}{\partial q_j^*}, W_1)$$

where

$$W_1 = \epsilon S_1$$

$$W_2 = \epsilon S_2.$$

Clearly, assuming convergence of the method, the p_j^* are reduced to constants and the q_j^* to linear functions of time ($q_2^* = \omega t$). The frequency of the angle variable q_1^* is a power series in ϵ. To the second order,

$$q_1^* = (1 + \frac{3}{4} \epsilon \gamma p_1^* + \frac{51}{64} \epsilon^2 \gamma p_1^{*2} + \dots)t + \beta_1^*$$

where p_1^*, β_1^* are constants.

9. Perturbation Methods of Non-Hamiltonian Systems Based on Lie Transforms.

Hori (1970, 1971) and Kamel (1970) have developed, independently, methods of perturbations of non-linear systems in general, by generalizing the approach to Hamiltonian systems. Clearly, such generalization is not strictly necessary since, as mentioned before, any system can be reduced to Hamiltonian form by doubling its dimension and introducing Dirac's cotangent space. The price one has to pay by having twice the number of differential equations we started with, is more than compensate by the fact that only two functions are to be solved for, the new Hamiltonian and the generator of the transformation. The direct approach requires the dealing with as many unknowns as there are variables, in fact, by direct application of the results of section 1.7, twice as many, as will be clear in a moment. Here we follow closely the presentation given by Kamel (1970). Consider a system of n first order differential equations

$$\dot{x} = f(x;\epsilon) \tag{2.9.1}$$

and assume $f(x;\epsilon)$ real analytic in the $n+1$ variables $(x_1, x_2, \ldots, x_n, \epsilon)$ in some domain $\Omega\{x \in D \subset R^n, |\epsilon| < \epsilon_0\}$. The right-hand side of (2.9.1) can be expanded for ϵ sufficiently small in the convergent power series

$$\dot{x} = \sum_{k \geq 0} \frac{\epsilon^k}{k!} f^{(k)}(x) \tag{2.9.2}$$

where

$$f^{(k)}(x) = \left.\frac{\partial^k f}{\partial \epsilon^k}\right|_{\epsilon=0} .$$

The functions $f^{(k)}(x)$ are obviously real analytic in D. This condition can eventually be relaxed by attaching to the process to follow a smoothing operation at every stage of approximation but, for the general understanding of the method, this is not advisable. We shall not consider nonautonomous systems and the observation that such cases can be treated just as well by treating t as another x-type coordinate is not generally appropriate. Such is the case, for instance, when questionsof asymptotic behavior, stability and periodic solutions are dealt with.

If equation (2.9.1) or (2.9.2) cannot be integrated in general, one seeks a transformation to a new system of n variables ξ, say

$$x = x(\xi;\epsilon) \tag{2.9.3}$$

such that the differential equation in ξ

$$\dot{\xi} = g(\xi;\epsilon) \tag{2.9.4}$$

resulting from (2.9.3) and (2.9.1) be more easily treatable. Obviously, stated in this form, the problem is too general to define what should be the properties of that transformation. One way to look at it is, of course, to assume that for $\epsilon = 0$, the equation (2.9.1) has a known general solution, that is, the equation

$$\dot{y} = f(y;0) = f^{(0)}(y) \tag{2.9.5}$$

is integrable. We might then ask the question whether there exists a transformation (2.9.3) such that (2.9.1) is brought into the form (2.9.5), that is,

$$\dot{\xi} = f^{(0)}(\xi). \tag{2.9.6}$$

Since for $\epsilon = 0$ the transformation (2.9.3) is obviously the identity, again we are lead to the search of a near identity transformation

$$x = \xi + \epsilon h(\xi;\epsilon) \tag{2.9.7}$$

and assume $h(\xi;\epsilon)$ to be analytic in some domain of the $n+1$ variables $(\xi;\epsilon)$ containing $\epsilon = 0$. It is obviously invertible, therefore, near $\epsilon = 0$, for ϵ sufficiently small. So one writes

$$x = \xi + \sum_{k \geq 1} \frac{\epsilon^k}{k!} E_k(\xi) \tag{2.9.8}$$

and the transformed system of differential equations will be, in general,

$$\dot{\xi} = \phi(\xi;\epsilon) = \sum_{k \geq 0} \frac{\epsilon^k}{k!} \phi^{(k)}(\xi),$$

with

$$\phi^{(k)}(\xi) = \left.\frac{\partial^k \phi}{\partial \epsilon^k}\right|_{\epsilon=0}.$$

The problem is now, given the transformation (2.9.8), to obtain the functions $\phi^{(k)}(\xi)$ in (2.9.9) from the functions $f^{(k)}(x)$ in (2.9.2). Obviously this can be accomplished in several ways but a recursive algorithm like the one discussed in section 1.7 is recommended if high orders and systematic formalism are sought. Differentiation of (2.9.8) with respect to t gives

$$\dot{x} = \dot{\xi} + \sum_{k \geq 1} \frac{\epsilon^k}{k!} \frac{\partial E_k}{\partial \xi} \dot{\xi}$$

and introducing (2.9.2) and (2.9.9), one finds

$$\sum_{k \geq 0} \frac{\epsilon^k}{k!} f^{(k)}(x) = \sum_{k \geq 0} \frac{\epsilon^k}{k!} \phi^{(k)}(\xi)$$

$$+ \sum_{k \geq 1} \frac{\epsilon^k}{k!} \frac{\partial E_k}{\partial \xi} \sum_{j \geq 0} \frac{\epsilon^j}{j!} \phi^{(j)}(\xi) \qquad (2.9.10)$$

From relation (1.7.2) we now see that

$$f(x(\xi;\epsilon);\epsilon) = \sum_{n=0}^{\infty} \frac{\epsilon^n}{n!} f_n(\xi)$$

and recursive relations are available for the definition of $f_n(\xi)$, as for instance, equation (1.7.14) or (1.7.15) or the resulting relations in section (1.7). From (2.9.10) it now follows that

$$f_n(\xi) = \phi^{(n)}(\xi) + \sum_{m=1}^{\infty} \binom{n}{m} \frac{\partial E_m(\xi)}{\partial \xi} \phi^{(n-m)}(\xi). \qquad (2.9.11)$$

If one considers (1.7.22)

$$E_n(\xi) = -T_n(\xi) - \sum_{m=1}^{n-1} \binom{n-1}{m-1} T_m(\xi) \frac{\partial E_{n-m}(\xi)}{\partial \xi},$$

or, with notation (1.7.19),

$$E_n(\xi) = -T_n(\xi) - \sum_{m=1}^{n-1} \binom{n-1}{m-1} L_m \, E_{n-m}(\xi). \qquad (2.9.12)$$

We write the inverse of (2.9.8) as

$$\xi = x + \sum_{k \geq 1} \frac{\epsilon^k}{k!} X^{(k)}(x) \qquad (2.9.13)$$

so that

$$X^{(n)}(x) = T_n(x) - \sum_{m=1}^{n-1} \binom{n-1}{m} L_m X_{m,n-m}(x) \tag{2.9.14}$$

using the notation introduced in (1.7.20) and (1.7.21), that is,

$$\begin{aligned} X_{p,q}(x) &= - \sum_{m=1}^{p} \binom{p-1}{m-1} L_m X_{p-m,q}(x), \\ X_{0,q}(x) &= X^{(q)}(x). \end{aligned} \tag{2.9.15}$$

Finally, one finds

$$\phi^{(n)}(\xi) = f^{(n)}(\xi) + \sum_{j=1}^{n} \binom{n}{j} [f_{j,n-j}(\xi) - \frac{\partial E_j}{\partial \xi} \phi^{(n-j)}(\xi)] \tag{2.9.16}$$

which is the recurrence relation we have sought. Obviously, Equation (2.9.16) contains the coefficients T_n defining the mapping (2.9.8), that is, the coefficients $X^{(n+1)}(x)$ of the expansion

$$\frac{\partial \xi}{\partial \epsilon} = \sum_{n \geq 0} \frac{\epsilon^n}{n!} X^{(n+1)}(x)$$

from (2.9.13), and

$$\frac{\partial \xi}{\partial \epsilon} = T(\xi;\epsilon) = \sum_{n \geq 0} \frac{\epsilon^n}{n!} T_{n+1}(\xi)$$

as in (1.7.6), (1.7.7.) and (1.7.8). At each stage of the approximation, $T_n(\xi)$ has to be chosen properly so as to meet our special requirements, whenever necessary. Such unknown can be put in direct evidence in (2.9.16), by writing

$$\frac{\partial T_n}{\partial \xi} \phi^{(0)}(\xi) - \frac{\partial \phi^{(0)}(\xi)}{\partial \xi} T_n(\xi) = \phi^{(n)} - f^{(n)} + G_n(\xi) \tag{2.9.17}$$

where $G_n(\xi)$ depends on all previous approximations. In fact, Kamel finds

$$G_n(\xi) = \frac{\partial E_n^*(\xi)}{\partial \xi} \phi^{(0)}(\xi) - f^*_{n,0}(\xi) + \sum_{m=1}^{n-1} \binom{n}{m} [\frac{\partial E_m}{\partial \xi} \phi^{(n-m)}(\xi) - f_{m,n-m}(\xi)] \tag{2.9.18}$$

where

$$E_n^*(\xi) = E_n(\xi) \quad \text{for} \quad T_n = 0,$$

$$f_{n,o}^*(\xi) = f_{n,o}(\xi) \quad \text{for} \quad T_n = 0,$$

$$f_{p,q}(\xi) = -\sum_{m=1}^{p} \binom{p-1}{m-1} L_m f_{p-m,q},$$

$$f_{o,q}(\xi) = \phi^{(q)}(\xi).$$

A thorough development of the method has been given by Kamel (1970) and Henrard (1970) and more recently by Hori (1971). Kamel shows how the generalized Lie Transform approach contains in essence the important methods of two-variable expansions procedures and matching of asymptotic solutions due to Kevorkian (1966). This subject is not dealt with here since it is explored in detail in the work of Cole (1968). It is worth noting that Deprit's presentation of Lie Transforms generated by functions depending on a (small) parameter and applied to Hamiltonians also depending on that parameter can as well be analyzed by forgetting, a priori, the presence of a parameter as it has been shown earlier in these notes and following Mersman's work (1971). In like manner the foregoing formalism can be simplified by introducing operators and functions which are not functions of a parameter and, later, introduce power series in ϵ in all results. This was accomplished by Hori (1971) and here we limit ourselves to sketch his results. Consider n variables $(\xi_1, \xi_2, \ldots, \xi_n)$ and an operator $T_k(\xi)$, and let

$$D_\xi = \sum_{k=1}^{n} T_k(\xi) \frac{\partial}{\partial \xi_k} . \tag{2.9.19}$$

Consider the mapping

$$x_j = \xi_j + \sum_{p \geq 1} \frac{1}{p!} D_\xi^{p-1} T_j(\xi) \tag{2.9.20}$$

where

$$D_\xi^o = 1,$$

$$D_\xi^1 = D_\xi,$$

$$D_\xi^p = D_\xi D_\xi^{p-1},$$

which are the analog of (2.8.1) and subsequent definitions. In particular $T_k(\xi)$ plays the role of $\partial S/\partial\xi_k$. We also consider the mapping of a real analytic function $f(x)$ of n variables $(x_1,x_2,\ldots,x_n)$ into the ξ-space as given by

$$f(x) = f(\xi) + \sum_{p\geq 1} \frac{1}{p!} D_\xi^p f(\xi) \tag{2.9.21}$$

and, actually, (2.9.20) is a consequence of (2.9.21). We define the inverse transformation

$$T_k^{-1}(x) = T_k(\xi)\big|_{\xi=x} \tag{2.9.22}$$

and

$$D_x = \sum_{k=1}^{n} T_k^{-1}(x) \frac{\partial}{\partial x_k} \tag{2.9.23}$$

so that, the inverse mapping of (2.9.21) is

$$f(\xi) = f(x) + \sum_{p\geq 1} \frac{(-1)^p}{p!} D_x^p f(x), \tag{2.9.24}$$

a direct generalization of Lie's Transform. All of the above relations are actually contained in the previous formalism (ϵ dependent) and their proof is straightforward.

The equation

$$\dot{x}_k = f_k(x), \tag{2.9.25}$$

by means of the transformation (2.9.20) generated by T_k via the mapping (2.9.20), changes into

$$\dot{\xi}_k = \phi_k(\xi). \tag{2.9.26}$$

Making use of (2.9.24), the inverse of (2.9.20) is given by

$$\xi_j = x_j + \sum_{p\geq 1} \frac{(-1)^p}{p!} D_x^{p-1} T_j^{-1}(x). \tag{2.9.27}$$

Since from (2.9.25)

$$\frac{d}{dt} = \sum_{k=1}^{n} f_k \frac{\partial}{\partial x_k} \tag{2.9.28}$$

for any function $F(x)$, that is,

$$\frac{d}{dt} F(x) = \dot{F}(x) = \sum_{k=1}^{n} \frac{\partial F}{\partial x_k} \dot{x}_k = \left(\sum_{k=1}^{n} f_k \frac{\partial}{\partial x_k} \right) F,$$

the computation of $\phi_k(\xi)$ in (2.9.26) is obtained as follows. Differentiate (2.9.27) to get

$$\dot{\xi}_j = \dot{x}_j + \sum_{p \geq 1} \frac{(-1)^p}{p!} \frac{d}{dt} \{ D_x^{p-1} T_j^{-1}(x) \}$$

and introduce (2.9.25) and (2.9.23) to find

$$\dot{\xi}_j = f_j(x) + \sum_{p \geq 1} \frac{(-1)^p}{p!} \sum_{k=1}^{n} f_k(x) \frac{\partial}{\partial x_k} (D_x^{p-1} \; T_j^{-1}(x))$$

or, using (2.9.21) and (2.9.20),

$$\dot{\xi}_j = f_j(\xi) + \sum_{p \geq 1} \frac{1}{p!} D_\xi^p f_j(\xi) + \sum_{p \geq 1} \frac{(-1)^p}{p!} \sum_{k=1}^{n} f_k(\xi) \frac{\partial}{\partial \xi_k} (D_\xi^{p-1} T_j(\xi))$$

$$+ \sum_{p \leq 1} \frac{(-1)^p}{p!} \sum_{q \geq 1} \frac{1}{q!} D_\xi^q \left\{ \sum_{k=1}^{n} f_k(\xi) \frac{\partial}{\partial \xi_k} (D_\xi^{p-1} \; T_j(\xi)) \right\} = \phi_j(\xi) \; . \tag{2.9.29}$$

Now we consider the series

$$\begin{aligned} f_j &= f_j^{(0)} + f_j^{(1)} + \dots \\ \phi_j &= \phi_j^{(0)} + \phi_j^{(1)} + \dots \\ T_j &= T_j^{(0)} + T_j^{(1)} + \dots \end{aligned} \tag{2.9.30}$$

and we search for the operators T_j so that the ϕ_j take a desired form. Obviously, the equations

$$\dot{y}_k = f_k^{(0)}(y) \tag{2.9.31}$$

are supposed to have a well defined general solution. The decomposition (2.9.30) of f_j is intended not necessarily as a power series in some small parameter ϵ and also is not necessarily an infinite series. In fact, the normal case of a perturbed integrable system (2.9.31) will be $f_j^{(k)} = 0$ for $k \geq 2$, that is,

$$f_j = f_j^{(o)} + f_j^{(1)}.$$

By feeding the series (2.9.30) into (2.9.29) one obtains a recursive algorithm for the unknowns $\phi_j^{(k)}$ and $T_j^{(k)}$, by equating terms of same order. In this respect, the explicit use of a parameter ϵ to represent the orders is quite useful, though not necessary. That is, equating terms of same order can be translated into the easier language of equating coefficients of like powers of ϵ, assuming $f_j^{(k)} = O(\epsilon^k)$, $\phi_j^{(k)} = O(\epsilon^k)$, $T_j^{(k)} = O(\epsilon^k)$.

The first few approximations give, all functions intended to be in terms of the ξ variables,

$$f_j^{(o)} = \phi_j^{(o)}$$

$$\sum_{k=1}^{n} \{-f_k^{(o)} \frac{\partial T_j^{(1)}}{\partial \xi_k} + T_k^{(1)} \frac{\partial f_j^{(o)}}{\partial \xi_k}\} + f_j^{(1)} = \phi_j^{(1)}$$

$$\sum_{k=1}^{n} \{-f_k^{(o)} \frac{\partial T_j^{(2)}}{\partial \xi_k} + T_k^{(2)} \frac{\partial f_j^{(o)}}{\partial \xi_k}\} + \frac{1}{2!} \sum_{k=1}^{n} T_k^{(1)} \cdot$$

$$\cdot \frac{\partial}{\partial \xi_k} (f_j^{(1)} + \phi_j^{(1)}) - \frac{1}{2!} \sum_{k=1}^{n} \frac{\partial T_j^{(1)}}{\partial \xi_k} (f_k^{(1)} + \phi_k^{(1)}) + f_j^{(2)} = \phi_j^{(2)}$$

and, in general,

$$\sum_{k=1}^{n} \{-f_k^{(o)} \frac{\partial T_j^{(p)}}{\partial \xi_k} + T_k^{(p)} \frac{\partial f_j^{(o)}}{\partial \xi_k}\} + \ldots + f_j^{(p)} = \phi_j^{(p)} \tag{2.9.32}$$

which, in fact, is equivalent to the previous relation (2.9.17). Here we introduce the important notion of <u>auxiliary system</u> by defining

$$\frac{d\xi_j}{d\tau} = f_j^{(o)}(\xi) \tag{2.9.33}$$

with the general solution

$$\xi_j = \xi_j(\tau) \tag{2.9.34}$$

so that

$$\sum_{k=1}^{n} \{-f_k^{(o)} \frac{\partial T_j^{(p)}}{\partial \xi_k} + T_k^{(p)} \frac{\partial f_j^{(o)}}{\partial \xi_k}\} = - \frac{dT_j^{(p)}}{d\tau} + \sum_{k=1}^{n} T_k^{(p)} \frac{\partial f_j^{(o)}}{\partial \xi_k} .$$

The general equation (2.9.32) reduces to a linear system for the $T_j^{(p)}(\xi)$ at every stage of approximation, that is,

$$- \frac{dT_j^{(p)}}{d\tau} + \sum_{k=1}^{n} \frac{\partial f_j^{(o)}}{\partial \xi_k}(\xi(\tau))T_k^{(p)}(\tau) + F_j^{(p)}(\tau) = \phi_j^{(p)} \qquad (2.9.35)$$

where the ξ's are substituted by the solution (2.9.34) of the auxiliary system. It is clear that (2.9.35) is a straight generalization of (2.8.9). It is noted that, as in the usual averaging methods, $\phi_j^{(p)}$ should be chosen so as to avoid secular terms in $T_j^{(p)}(\tau)$, that is,

$$\lim_{\tau \to \infty} T_j^{(p)}(\tau) = \text{finite.}$$

The simplest case is when the $f_j^{(o)}$ are linear functions of the ξ's, so that (2.9.35) is a linear (non-homogeneous) system with constant coefficients, for any order of approximation. If this is not the case, say the $\partial f_j^{(o)} / \partial \xi_k \big|_{\xi=\xi(\tau)}$ are periodic or quasiperiodic functions of τ the integration of (2.9.36) is obviously not a trivial task. It is therefore advisable, in general, to produce a decomposition of the $f_j(\xi)$ such that $f_j^{(o)}(\xi)$ are linear.

Van der Pol Equation

As an example consider the equation

$$\ddot{x} + \epsilon(1-x^2)\dot{x} + x = 0$$

which can be written

$$\dot{x}_1 = x_2$$
$$\dot{x}_2 = -x_1 - \epsilon(1-x_1^2)x_2. \qquad (2.9.36)$$

Here we consider

$$f_1^{(0)} = x_2, \quad f_2^{(0)} = -x_1$$

$$f_1^{(1)} = f_1^{(2)} = \ldots = f_1^{(p)} = \ldots = 0$$

$$f_2^{(1)} = -\epsilon(1-x_1^2)x_2$$

$$f_2^{(2)} = f_2^{(3)} = \ldots = f_2^{(p)} = \ldots = 0.$$

The auxiliary system is

$$\frac{d\xi_j}{d\tau} = f_j^{(0)} = \phi_j^{(0)}$$

whose solution we write in the form

$$\xi_1 = \alpha \cos(\tau+\beta)$$
$$\xi_2 = -\alpha \sin(\tau+\beta) \tag{2.9.37}$$

where α, β are scalar constants. The first order equations become

$$-\frac{dT_1^{(1)}}{d\tau} + T_2^{(1)} = \phi_1^{(1)}$$

$$-\frac{dT_2^{(1)}}{d\tau} - T_1^{(1)} + \epsilon[1-\alpha^2\cos^2(\tau+\beta)]\alpha\sin(\tau+\beta) = \phi_2^{(1)}$$

or

$$\frac{d^2T_1^{(1)}}{d\tau^2} + T_1^{(1)} = \epsilon[(1-\frac{\alpha^2}{4})\alpha\sin(\tau+\beta) - \frac{\alpha^3}{4}\sin(3\tau+3\beta)]$$
$$-\frac{d\phi_1^{(1)}}{d\tau} - \phi_2^{(1)}.$$

In order to avoid singular terms, the term in $\sin(\tau+\beta)$ must be avoided in the equation for $T_1^{(1)}$ and a possible choice is

$$\frac{d^2T_1^{(1)}}{d\tau^2} + T_1^{(1)} = -\epsilon\frac{\alpha^3}{4}\sin(3\tau+3\beta),$$

$$\frac{d\phi_1^{(1)}}{d\tau} + \phi_2^{(1)} = \epsilon(1-\frac{\alpha^2}{4})\alpha\sin(\tau+\beta),$$

$$\phi_2^{(1)} = \frac{d\phi_1^{(1)}}{d\tau} = \frac{\epsilon}{2}(1-\frac{\alpha^2}{4})\alpha\sin(\tau+\beta)$$

so that

$$\phi_2^{(1)} = -\epsilon[1 - \frac{1}{4}(\xi_1^2 + \xi_2^2)]\xi_2$$

$$\phi_1^{(1)} = -\epsilon[1 - \frac{1}{4}(\xi_1^2 + \xi_2^2)]\xi_1$$

and therefore

$$T_1^{(1)} = \frac{\epsilon\alpha^3}{32} \sin 3(\tau+\beta) = \frac{\epsilon}{32} \xi_2(\xi_2^2 - 3\xi_1^2)$$

$$T_2^{(1)} = \frac{\epsilon}{2} \xi_1(-1 + \frac{7}{16} \xi_1^2 - \frac{5}{16} \xi_2^2).$$

The first order equations in the new variables are thus

$$\frac{d\xi_1}{dt} = \xi_2 - \epsilon[1 - \frac{1}{4}(\xi_1^2 + \xi_2^2)]\xi_1$$

$$\frac{d\xi_2}{dt} = -\xi_1 - \epsilon[1 - \frac{1}{4}(\xi_1^2 + \xi_2^2)]\xi_2$$

and, in fact, one easily verifies that the equation for ξ_2 is obtained from that of ξ_1 by the substitutions $\xi_2 \to -\xi_1$, $\xi_1 \to \xi_2$, provided the choice $\phi_2^{(1)} = d\phi_1^{(1)}/d\tau$ is made. If we let

$$u^2 = \xi_1^2 + \xi_2^2$$

it is found that

$$\frac{du^2}{dt} = -2\epsilon(1 - \frac{u^2}{4})u^2$$

and, therefore,

$$u^2 = \xi_1^2 + \xi_2^2 = \frac{4ke^{-\epsilon t}}{ke^{-\epsilon t} \pm 1}$$

and $\pm$ sign choice depending on the sign of the constant k, that is, on the initial conditions, since u^2 has to be positive.

For $\epsilon > 0$ we obtain the asymptotic behavior

$$u^2 \to 0 \quad \text{as} \quad t \to \infty$$

which is the well known damped motion toward a focus. If $\epsilon < 0$, $u^2 \to 4$ as $t \to \infty$, which is the limit cycle of the van der Pol equation. The fact that a first order theory (in ϵ) is able to give full information in the asymptotic behavior of the system is that, to any order, the equations for ξ_1, ξ_2 have the same character of the first order equations, that is,

$$\dot{\xi}_1 = [1 + \epsilon^2 f_2(u^2) + \epsilon^4 f_4(u^4) + \ldots]\xi_2$$

$$- \epsilon[1 - \frac{u^2}{4} + \epsilon^2 g_3(u^2) + \epsilon^4 g_5(u^2) + \ldots]\xi_1$$

$$\dot{\xi}_2 = \dot{\xi}_1(\xi_2 \to -\xi_1, \quad \xi_1 \to \xi_2),$$

so that the above described asymptotic properties are conserved.

NOTES

Integrability of a Dynamical System has been quite a controversial issue. One feels that, as far as Hamiltonian systems are concerned, separability of the Hamilton-Jacobi equation might be a good definition, although this is not the general opinion. Stäckel Theorem, unfortunately, does not give any indication on how to actually construct a coordinates system which separates the equation. The only thing we clearly know is that if there are n independent integrals for an n-dimensional system, then, according to Arnol'd, the invariant manifolds are tori and, on these, the motion is generally quasi-periodic. The existence of such manifolds for a certain class of systems is also conjectured by Diliberto under the name of periodic surfaces. The issue for non Hamiltonian systems is more complex, although, as for the example given at the end of Chapter 5, one may think of a generalized Birkhoff normalization, in case of disturbed harmonic oscillators. Many problems can actually be reduced to harmonic oscillators by a proper choice of variables and time. For instance, the Newtonian problem of two bodies, by making use of the Levi-Civita transformation

$$x = u^2 - v^2$$

$$y = 2uv$$

combined with the time transformation

$$d\tau = dt/r,$$

reduces to a simple harmonic motion. Other force laws have been recently considered by Giacaglia and associates, following methods introduced by Kustaanheimo.

We are also lead to the study of integrability of a system in the vicinity of a stable equilibrium solution, a subject where many efforts have been made by Siegel and Moser, as well as many others. Although convergence of normalization methods cannot be established, it is obvious from the results of Contopoulos, Barbanis and Bozis that, under quite general circumstances, other integrals (or quasi-integrals) may exist both in normal and resonant systems. Evidence of existence of integrals has also been established by means of the method of Surface of Section by Hènon and associates.

As far as methods of successive approximations are concerned, to produce series solutions of a system, any simple method will do, and convergence in a properly bounded interval of time can be achieved. The question could also be answered by simply applying Picard's method of iterations, which, in fact, has been done by several researchers, especially where numerical techniques are involved.

Given a system depending on a small parameter, the way the solution goes in terms of powers of such parameter, is set by how close one is to a singular (equilibrium) point of the system and on the stability character of such singular point. Properties of this sort were studied originally by Birkhoff for the behavior of area preserving mappings in the vicinity of fixed points. More recent and important results are due to Moser and Gelfand - Lidskii. A typical example of the change in behavior of expansions with respect to a parameter in the vicinity of an equilibrium point can be seen in the Restricted Problem of three bodies at the five Euler-Lagrange solutions. Such expansions can go in powers of $\epsilon^{1/3}$, $\epsilon^{1/2}$ or ϵ, as recently shown by Szebehely and associates (1970). The method of successive approximations by MacMillan given in section 2 can be changed easily into an averaging method, for the Eq. (2.2.6) or its expanded form (2.2.7). Such a method was given by Cesari and lately by Hale. The appearance of secular terms in a solution, as given in the example at the beginning of section 3, led Lindstedt to the introduction of the averaging methods. In several problems, a poor choice of a reference solution decides on the success of the subsequent approximations, in the same fashion as the wrong choice of coordinates decides on the integrability (separability) of a system. The Hamiltonianization of a system, originally due to Dirac, is only practical in cases where system (2.4.5) has constant coefficients (excluding exceptional cases), that is, the g_i are linear, with constant coefficients, in the components x_j of x. If this is not the case, the definition of the reference solution from (2.4.5) might be a very difficult task. As far as Poincaré's method (which he calls Lindstedt's Method), it has been called von Zeipel's methods mainly because it was through his work on Asteriods that Brouwer obtained a spectacular solution for the problem of artificial satellites of the earth in 1959. The averaging methods entered with full power in Celestial Mechanics beginning about that same time. It is interesting to observe that Poincaré's Method was only exceptionally used in Nonlinear Mechanics, including the Russian Literature, before that time. Also, equation (2.4.8) indicates that, except for the averaging operation, all these methods, in conservative systems, are just a solution of Hamilton-Jacobi's equation by successive approximations. The main disadvantage is that the relations between original and new variables, generated by W (Eq. 2.4.10), are implicit and their inversion has only been recently fully solved by the introduction of Lie's Series. The fact that, if the average of a quasi-periodic function is zero, the integral of such function is bounded, can also be verified if one assumes a certain irrationality condition among the basic frequencies of the corresponding Fourier series ω_1, $\omega_2, \ldots, \omega_n$; precisely,

$$\left| \sum_{j=1}^{n} p_j \omega_j \right| \geq K \left| \sum_{j=1}^{n} p_j \right|^{-\sigma}$$

for some positive constants K and $\sigma > n-1$. If such conditions are not verified (they are not for a set of ω's of zero measure), then the integral of a zero average quasi-periodic function may not be bounded due to presence of small divisors, as discussed by Moser, in the theory of quasi-periodic motions. From the purely geometric point of view, Moser has made important steps on the study of area preserving mappings which are "close" to the identity (see 2.4.12). His work has the obvious influence of Birkhoff and Siegel. The expansions involved in actual calculations and decurring from (2.4.16) are actually tedious and incredibly long. The recent introduction of automatic symbol processors in fast electronic calculators has nevertheless eliminated most of the practical difficulties. Important results have been announced by Kovalevsky, Chapront and Deprit, in typical problems of Celestial Mechanics. We are not aware of analogous developments in Nonlinear Mechanics and Circuit Theory.

Degenerate systems, as defined by Arnol'd, are unfortunately very common in actual problems, thus the importance of the understanding of their behavior under perturbations. The essential geometric difficulty lies in the fact that the Invariant Manifolds of the Unperturbed Problem have a lower dimension than those of the perturbed one. Also, linear perturbed systems are much more sensible to resonance conditions and very difficult to describe. The stress given to the definition of fast and slow variables is justified by the fact that the former correspond usually to small amplitude oscillations and do not affect the latter which are associated with large scale deviations, with respect to the unperturbed system, over a long time. In many instances averaging is understood simply as a process of elimination of time when it appears explicitly in the equations. It is achieved simply by taking the average of the right-hand members of the differential equations. This is, in fact, the first step in the KBM Method. Such a procedure is explored in many ways by Hale (section V.3, pp. 171-208, 1969) who studies the deviation, as time goes to infinity, from a given non-autonomous system

$$\dot{x} = \epsilon f(t,x,\epsilon) \tag{A}$$

and the averaged system

$$\dot{x} = \epsilon f_0(x) \tag{B}$$

where

$$f_0(x) = \lim_{T \to \infty} \frac{1}{T} \int_0^T f(t,x,o)\,dt.$$

Hale obtains conditions for the existence of periodic solutions, as well as their stability character. The starting proposition is that, under quite general conditions, there exists a transformation

$$x = y + \epsilon u(t,y,\epsilon) \tag{C}$$

such that the equation (A) above is reduced to

$$\dot{y} = \epsilon f_0(y) + \epsilon F(t,y,\epsilon) \tag{D}$$

with $F(t,y,0) = 0$. One sees clearly that the near identity transformation (C) produces a system (D) which differs from the averaged system (B) by a quantity $O(\epsilon^2)$ at least. The error estimate by Kyner is actually derived from this basic result.

From a sophisticated point of view, Moser in 1970 studied the topology of Kepler's Motion, the singularities of the manifold of the state of motion and introduces the concept of averaging on manifolds, avoiding the explicit use of coordinates. His intrinsic representation applies special techniques to the vector field defined by the Keplerian motion. The regularization process used by Moser to study orbits in the vicinity of the origin ($r = 0$) is due to Levi-Civita and generally known as the inversion transformation. It is not usually applied in global studies since it introduces new singularities at points where the velocity of the particle is zero.

Assuming the right-hand sides of the differential equations to be periodic in time, Laricheva obtains much better error bounds for the averaged equations of Celestial Mechanics than those given by Bogoliubov and Mitropolski. In his work "Theory of Orbits about an Oblate Plant", Kyner in 1963 gives an excellent description of the averaging methods as well as the connection, in that particular example, with Diliberto's theory of periodic surfaces. In the case, they happen to be, as expected, two-dimensional tori, since the field of the Planet is supposed to have rotational symmetry. He also applies a technique developed by Hale in the book "Nonlinear Oscillations", in order to obtain conditions of periodicity and also develop approximate solutions.

As far as the application of Poincaré's Method to Hamiltonian systems, when the Hamiltonian is a power series in both coordinates and momenta, as in the example at the beginning of section 6, it was described by Giacaglia in 1965. The problem arises naturally in small oscillations and, in Celestial Mechanics, in the use of Poincaré's variables and problems of resonance. In this way one provides a certain generalization in the concept of Birkhoff's Normalization, by assuming, in principle, any combination of coordinates and momenta and, second, by giving a more systematic way of producing the Normalization. The application of Lie's series by Deprit is an example, however, on how complex the actual development of the Method might become. The characteristic exponents are better obtained, in this case, by using Cesari's method developed in 1940, as was shown by Giacaglia in the libration cases of the Elliptic Restricted Problem in 1971. Obviously, after the characteristic exponents are obtained to some order as power series in the small parameter of the problem, Lyapunov's transformation easily reduces the problem to the integration of a linear system whose coefficients are constant within that same order. The problem of small divisors in Poincaré's Method is here translated into a problem of parametric resonance.

The construction of integrals of motion via a successive approximation to the Poisson condition, undertaken by Contopoulos in several works, shows very well the change in the form of such integrals (or quasi-integrals) when a region of resonance is crossed. Since, in the limit, the resonance points are at least as dense as the rational numbers on the line, one expects a very wild behavior of the integrals, changing from one form to another, infinitely many times, in every finite interval of frequencies, defined by the small parameter of the problem and/or by the initial conditions. This fact will not prevent the convergence for a specific value of the frequencies, in fact, over a set of values with non zero measure. Such integrals however cannot be analytic, nor can their series by uniformly convergent or continuous. However, the number of discontinuities is countable and with zero measure. All these considerations and conjectures are intimately connected with Moser's and Kolmogorov's theories.

The construction of Kovalevskaya's Integral we have given in section 6 is a rare example of a series which terminates and, obviously, must be an exceptional situation. It is nevertheless an indication of the danger in defining a system integrable or nonintegrable for all possible situations.

Lie Transform techniques are quite popular at present and they actually represent a real breakthrough from Classical Methods. At least one can say they were not known to Poincaré, a thing hard to discover in perturbation theories. The credit for this new method goes to Hori. Later works and modified algorithms should only be considered as refinements or different forms of the same basic idea. One of the best examples of applications of the method has been given by Deprit et al. for the main problem of earth's artificial satellites. Also, a recent application to the motion of a rigid body under the influence of central gravitation has been given by Giacaglia et al. Several examples are also treated by Choi and Tapley. The example we have given for the solution of van der Pol equation is extended to third order by Hori in his recent paper on the subject of non Hamiltonian systems, and is the best reference on the actual use of the method for non Hamiltonian systems. The Hamiltonianization of van der Pol equation

$$\ddot{x} = -\epsilon(1-x^2)\dot{x}-x$$

is readily obtained by defining $x = y_1$, $\dot{x} = y_2$, so that

$$\dot{y}_1 = y_2$$

$$\dot{y}_2 = -\epsilon(1-y_1^2)y_2-y_1$$

and the Hamiltonian is

$$H = x_1\dot{y}_1 + x_2\dot{y}_2 = H_o + H_1,$$

where

$$H_o = x_1y_2 - x_2y_1,$$

$$H_1 = -\epsilon(1-y_1^2)x_2y_2.$$

The equations of motion are

$$\dot{y}_k = H_{x_k}, \quad \dot{x}_k = -H_{y_k}$$

and the auxiliary system is defined by

$$K_o = \xi_1\eta_2 - \xi_2\eta_1,$$

that is

$$\frac{d\xi_1}{d\tau} = \xi_2, \quad \frac{d\xi_2}{d\tau} = -\xi_1,$$

$$\frac{d\eta_1}{d\tau} = \eta_2, \quad \frac{d\eta_2}{d\tau} = -\eta_1$$

with the obvious solution

$$\xi_1^0 = \alpha_1 \sin(\tau + \beta_1)$$

$$\xi_2^0 = \alpha_1 \cos(\tau + \beta_1)$$

$$\eta_1^0 = \alpha_2 \sin(\tau + \beta_2)$$

$$\eta_2^0 = \alpha_2 \cos(\tau + \beta_2).$$

From the first order equation

$$- \frac{dS_1}{d\tau} + H_1 = K_1$$

we obtain

$$K_1 = \lim_{T \to \infty} \frac{1}{T} \int_0^T H_1(\tau) d\tau$$

$$= \lim_{T \to \infty} \frac{1}{T} \int_0^T [-\epsilon(1-\eta_1^{0\,2})\eta_2^0 - \eta_1^0] d\tau$$

$$= - \frac{\epsilon\alpha_1\alpha_2}{2} (1 - \frac{\alpha_2^2}{4}) \cos(\beta_1 - \beta_2),$$

$$S_1 = \int [H_1(\tau) - K_1] d\tau.$$

For a complete solution up to third order the paper by Choi is suggested.

Finally, for a detailed and excellent description of the averaging methods both from the point of view of Krylov-Bogoliubov and of Poincaré, as well as the meaning of neglecting high order terms, we refer to the classical work of Musen, and to the extensive work of Volosov.

REFERENCES

1. Abraham, R., 1967, "Foundations of Mechanics", W. A. Benjamin, Inc., Philadelphia.

2. Andoyer, H., 1926, "Cours de Mécanique Céleste", (Vol. I), Gauthier-Villars, Paris.

3. Arnol'd, V. I., 1963, "Proof of A. N. Kolmogorov's Theorem on the Conservation of Quasiperiodic Motions under Small Perturbations of the Hamiltonian", Uspekhi Mat. Nauk USSR, 18, 13-40.

4. ____________, 1963, "Small Denominators and Problems of Stability of Motion in Classical and Celestial Mechanics", Uspekhi Mat. Nauk USSR, 18, 91-192.

5. Barbanis, B., 1966, "The Topology of the Third Integral", Intern. Astrom. Union Symp. No. 25, pp. 19-25, Academic Press, New York.

6. Birkhoff, G. D., 1927, "Dynamical Systems", Am. Math. Soc. Colloq. Public. IX, Providence, Rhode Island.

7. Bogoliubov, N. N., 1945, "On some statistical methods in mathematical physics", (Paper) Izv. Akad. Nauk USSR, Moscow.

8. ______________ and Mitropolskii, Y. A., 1951, "Asymptotic Methods in the Theory of Nonlinear Oscillations", Gordon and Breach, New York.

9. Bozis, G., 1966, "A New Integral in the Restricted Problem of Three Bodies", Doctoral Thesis, Univ. of Thessaloniki, Greece.

10. ________, 1966, "On the Existence of a New Integral in the Restricted Three-Body Problem", Astron. J., 71, 404-414.

11. Brouwer, D., 1959, "Solution of the Problem of Artificial Satellites without Drag", Astron. J., 64, 378-390.

12. __________ and Clemence, G. M., 1961, "Methods of Celestial Mechanics", Academic Press, New York.

13. Caley, A., 1848, Comb. Dublin Math. J., 3, 116.

14. Cesari, L., 1940, "Sulla Stabilità delle Soluzioni dei Sistemi di Equazioni Differenziali Lineari a Coefficienti Periodici", Atti Accad. Ital. Mem. Classe Fis. Mat. e Nat., 11, 633-692.

15. ________, 1963, "Asymptotic Behavior and Stability Problems in Ordinary Differential Equations", Springer-Verlag New York Inc., New York.

16. Choi, J. S. and Tapley, B. D., 1972, "An Extended Canonical Perturbation Method", Cel. Mech. (to appear).

17. Contopoulos, G., 1960, "A Third Integral of Motion in a Galaxy", Zeits. fur Astrophys., 49, 273-291.

18. ______________ and Barbanis, B., 1961, "An Application of the Third Integral of Motion", The Observatory, 82, 80-82.

19. ______________, 1962, "On the Existence of a Third Integral", Astron. J., 68, 1-14.

20. ______________, 1963, "A Classification of the Integrals of Motion", Astrophys. J., 138, 1297-1305.

21. ______________, 1963, "Resonances Cases and Small Divisors in a Third Integral of Motion. I", Astron. J., 68,

22. ______________ and Woltjer, L., 1964, "The Third Integral in Non-Smooth Potentials", Astrophys. J., 140, 1106-1119.

23. ______________, 1965, "The Third Integral in the Restricted Three-Body Problem", Astrophys. J., 142, 802-804.

24. ______________, 1966, "Adiabatic Invariants and the Third Integral", J. Math. Phys., 7, 788-797.

25. ______________ and Hadjidemetrious, J. P., 1968, "Characteristics of Invariant Curves of Plane Orbits", Astron. J., 73, 86-96.

26. ______________, 1970, "Resonance Phenomena in Spiral Galaxies", in "Periodic Orbits, Stability and Resonances" (Ed. G. E. O. Giacaglia), D. Reidel Publ. Co., Dordrecht, Holland.

27. ______________, "Orbits in Highly Perturbed Dynamical Systems", I (1970), Astron. J., 75, 96-107; II(1970), Astron. J., 75, 108-130; III(1971), Astron. J., 76, 147-156.

28. Deprit, A. et. al., 1969, "Birkhoff's Normalization", Cel. Mech., 1, 222-251.

29. __________ et. al., 1970, "Analytical Lunar Ephermeris: Brouwer's Suggestion", Astron. J., 75, 747-750.

30. __________ and Rom, A., 1970, "Characteristic Exponents of L_4 in the Elliptic Restricted Problem", Astron. Astrophys., 5, 416-428.

31. __________ and Rom, A., 1970, "The Main Problem of Artificial Sattelite Theory for Small and Moderate Eccentricities", Cel. Mech., 2, 166-206.

32. Diliberto, S. P., 1967, "New Results on Periodic Surfaces and the Averaging Principle", U.S.-Japanese Sem. on Diff. Funct. Eq., pp. 49-87, Benjamin, Philadelphia.

33. Dirac, P. A. M., 1958, "Generalized Hamiltonian Dynamics", Proc. Roy. Soc. London A246, 326-332.

34. Euler, L., 1753, "Theoria Motus Lunae" and 1772, "Theoria Motus Lunae, Novo Methodo", Petropoli, Typ. Acad. Imp. Scient.

35. Gelfand, I. M. and Lidskii, U. B., 1958, "On the Structure of the Regions of Stability of Linear Canonical Systems of Differential Equations with Periodic Coefficients", Am. Math. Soc. Transl. (2), 8, 143-182.

36. Giacaglia, G. E. O., 1964, "Notes on von Zeipel's Method", GSFC-NASA Publ. X-547-64-161.

37. ________, 1965, Evaluation of Methods of Integration by Series in Celestial Mechanics", Doctoral Thesis, Yale University, New Haven.

38. ________, 1967, "Nonintegrable Dynamical Systems", Chair Thesis, General Mechanics, Univ. of Sao Paulo, Sao Paulo.

39. ________, et al., 1970, "A Semi-Analytic Theory for the Motion of a Lunar Satellite", Cel. Mech., 3, 3-66.

40. ________ and Jefferys, W. H., 1971, "Motion of a Space Station", Cel. Mech. 4, 442-467.

41. ________, 1971, "Characteristic Exponents at L_4 and L_5 in the Elliptic Restricted Problem of Three Bodies", Cel. Mech., 4, 468-489.

42. ________, 1972, "Regularization of Conservative Central Fields", Public. Astron. Soc. Japan, 24, No. 3, July.

43. ________ and Nuotio, V. I., 1972, "Spinor Regularization of Conservative Central Fields", 3rd Annual Meeting, Div. Dynamical Astron., Am. Astron. Soc., Univ. of Maryland, In "Bull. Amer. Astron. Soc.", (to appear).

44. Goldstein, H., 1951, "Classical Mechanics", Addison-Wesley, Reading, Massachusetts.

45. Goursat, E., 1959, "Cours d'Analyse Mathematique", 7th. Ed., Vol. 2, Gauthier-Villars, Paris. (Reprinted by Dover Publ., New York).

46. Hale, J. K., 1954, "On the Boundedness of the Solution of Linear Differential Systems with Periodic Coefficients", Riv. Mat. Univ. Parma, 5, 137-167.

47. ________, 1961, "Integral Manifolds of Perturbed Differential Equations", Ann. Math., 73, 496-531.

48. ________, 1962, "On Differential Equations Containing a Small Parameter", Contrib. Diff. Eq., Vol. 1, J. Wiley, New York (Ed. J. P. LaSalle et. al).

49. Hale, J. K., 1963, "Oscillations in Nonlinear Systems", McGraw-Hill, New York.

50. __________, 1969, "Ordinary Differential Equations", (Chapt. 5), Wiley-Interscience, New York.

51. Henrand, J., 1970, "On a Perturbation Theory Using Lie Transforms", Cel. Mech., 3, 107-120.

52. Hènon, M. and Heiles, C., 1964, "The Applicability of the Third Integral of Motion; Some Numerical Experiments", Astron. J., 69, 73-79.

53. __________, 1965, "Exploration Numérique du Problème Restreint", Ann. Astrophys., 28, 499 and 992.

54. Hori, G., 1966, "Theory of General Perturbations with Unspecified Canonical Variables", Public. Astron. Soc. Japan, 18, 287-296.

55. __________, 1971, "Theory of General Perturbations for Noncanonical Systems", Publ. Astron. Soc. Japan, 23, 567-587.

56. Kamel, A. A., 1969, "Expansion Formulae in Canonical Transformations Depending on a Small Parameter", Cel. Mech., 1, 190-199.

57. __________, 1970, "Perturbation Method in the Theory of Nonlinear Oscillations", Cel. Mech., 3, 90-106.

58. Kevorkian, J., 1966, "The Two Variable Expansion Procedure for the Approximate Solution of Certain Nonlinear Differential Equations" in "Lectures in Applied Mathematics", vol. 7, p. 206, Amer. Math. Soc., Providence, R. I.

59. Kolmogorov, A. N., 1953, "On the Conservation of Quasiperiodic Motions for a Small change in the Hamiltonian Function", Dokl. Akad. Nauk USSR, 98, 527-530.

60. Kovalevsky, J., 1968, "Review of Some Methods of Programming of Literal Developments in Celestial Mechanics", Astron. J., 73, 203-209.

61. Krylov, N. and Bogoliubov, N. N., 1947, "Introduction to Nonlinear Mechanics", Ann. Math. Stud., 11, Princeton Univ. Press, Princeton, New Jersey.

62. Kustaanheimo, P., 1964, "Spinor Regularization of Kepler Motion", Ann. Univ. Turkuensis, A73, 3-7.

63. Kyner, W. T., 1961, "Invariant Manifolds", Rend. Circ. Mat. Palermo, 10, 98-110.

64. __________, 1963, "A Mathematical Theory of the Orbits about an Oblate Planet", Tech. Rep. to ONR, Det. Math., Univ. of Southern Calif., Los Angeles (51 pp.).

65. Kyner, W. T., 1968, "Rigorous and Formal Stability of Orbits about an Oblate Planet", Mem. Amer. Math. Soc., 81, 1-27.

66. Laricheva, V. V., 1966, "On Averaging a Certain Class of Systems of Nonlinear Differential Equations", Diff. Equa., 2, No. 1, 169-173.

67. Lefschetz, S., 1959, "Differential Equations. Geometric Theory", Interscience, New York.

68. Leimanis, E. and Minorsky, N., 1958, "Dynamics and Nonlinear Mechanics", (Ch. I), J. Wiley, New York.

69. Levi-Civita, T., 1903, "Traiettorie Singolari ed Urti nel Problema Ristretto dei Tre Carpi", Ann. Math., 9, 1-27.

70. Lindstedt, A., 1882, "Beitrag zur Integration der Differentialgleichungen der Störungtheorie", Abh. K. Akad. Wiss. St. Petersburg, 31, No. 4.

71. MacMillan, W. D., 1912, "A Method of Determining Solutions of a System of Analytic Functions in the Neighborhood of a Branch Point", Math. Annalen, 72, 180.

72. MacMillan, W. D., 1920, "Dynamics of Rigid Bodies", Dover Publ., New York (pp. 403-413).

73. Mersman, W. A., 1971, "Explicit Recursive Algorithms for the Construction of Equivalent Canonical Transformations", Cel. Mech., 3, 384-389.

74. Minorsky, N., 1962, "Nonlinear Oscillations", van Nostrand, Princeton, New Jersey.

75. Moser, J., 1955, "Nonexistence of Integrals for Canonical Systems of Differential Equations", Comm. Pure Appl. Math., 8, 409-436.

76. ________, 1958, "New Aspects in the Theory of Stability of Hamiltonian Systems", Comm. Pure Appl. Math., 2, 81-114.

77. ________, 1961, "A New Technique for the Construction of Solutions of Nonlinear Differential Equations", Proc. Nat. Acad. Sci., 47, 1824-1831.

78. ________, 1962, "On Invariant Curves of Area-Preserving Mappings of an Annulus", Nachr. Akad. Wiss. Göttingen, Math.-Phys. Kl. II, 1-20.

79. ________, 1963, "Perturbation Theory for Almost Periodic Solutions for Undamped Nonlinear Differential Equations", Int. Symp. on Nonlinear Diff. Eq. and Nonlinear Mech., Colorado Springs, 1961, Acad. Press, New York (pp. 71-79).

80. ________, 1964, "Hamiltonian Systems", Lecture Notes, New York Univ., New York.

81. Moser, J., 1966, "A Rapidly Convergent Iteration Method and Nonlinear Partial Differential Equations", I, Ann. Scu. Norm. Sup. Pisa, 20, 265-315; II, Ann. Scu. Norm. Sup. Pisa, 20, 499-533.

82. ________, 1966, "On the Theory of Quasi-Periodic Motions", SIAM Rev., 8, 145-172.

83. ________, 1967, "Convergent Series Expansions of Quasi-Periodic Motions", Math. Ann., 169, 136-176.

84. ________, 1970, "Regularization of Kepler's Problem and the Averaging Method on a Manifold", Comm. Pure Appl. Math., 23, 609-636.

85. Moulton, F. R., 1913, "Periodic Oscillating Satellites", Math. Ann., 73, 441-479.

86. ____________, 1920, "Periodic Orbits", Carnegie Inst. Washington Publ., 161, Washington, D. C.

87. Musen, P., 1965, "On the high order effects in the Methods of Krylov-Bogoliubov and Poincaré", J. Astronaut. Sci., 12, 129-134.

88. Nemitskii, V. V. and Stepanov, V. V., 1960, "Qualitative Theory of Differential Equations", Princeton Univ. Press, Princeton, New Jersey.

89. Pliss, V. A., 1966, "On the Theory of Invariant Surfaces", Diff. Eq., 2, 1139-1150.

90. __________, 1966, "Nonlocal Problems of the Theory of Oscillations", Acad. Press, New York.

91. Poincaré, H., 1893, "Les Methodes Nouvelles de la Mécanique Céleste", Vol. 2, Reprint by Dover Publ. New York (1957).

92. Poincaré, H., 1899, "Les Methodes Novelles de la Mécanique Céleste", (Vol. 3), Gauthier-Villars, Paris (Dover Publ. Reprint, New York).

93. Poincaré, H., 1909, "Leçons de Mécanique Céleste", (Vol. 2), Gauthier-Villars, Paris.

94. Roels, J. and Louterman, G., 1970, "Normalization des Systèmes Linèaires Canoniques et Application au Problème Restreinte des Trois Corps", Cel. Mech., 3, 129-140.

95. Sansone, G. and Conti, R., 1964, "Nonlinear Differential Equations", Pergamon Press, New York.

96. Siegel, C. L., 1941, "On the Integrals of Canonical Systems", Ann. Math. 42, 806-822.

97. Siegel, C. L., 1954, "Über die Existenz einer Normal form analytischer Hamiltonscher Differentialgleichungen in der Nähe einer Gleichgewichtslösung", Math. Am., 128, 144-170.

98. ____________, 1956, "Vorlesungen uber Himmelsmechanik", Springer-Verlag, Berlin.

99. Sternberg, S., 1969, "Celestial Mechanics", (2. Vols.), W. A. Benjamin, New York.

100. Szebehely, V. et al., 1970, "Mean Motions and Characteristic Exponents at the Libration Points", Astron. J., 75, 92-95.

101. Volosov, V. M., 1962, "Averaging in Systems of Ordinary Differential Equations", Russian Math. Surv., 17, 1-126.

102. Whittaker, E. T., 1961, "On the Adelphic Integrals of the Differential Equations of Dynamics", Proc. Roy. Soc. Edinburg, 37, 95.

103. ______________, 1937, "A Treatise on the Analytical Dynamics of Particles and Rigid Bodies", Cambridge, Univ, Press, London.

104. Wintner, A., 1947, "The Analytical Foundations of Celestial Mechanics", Princeton Univ. Press, Princeton, New Jersey.

105. Zeipel, H. von, 1916-17, "Recherches sur le Mouvement des Petits Planets", Arkiv. Astron. Mat. Phys., 11, 12, 13.

CHAPTER III

PERTURBATIONS OF INTEGRABLE SYSTEMS

1. Motion of an Integrable System.

In this section we give a more precise characterization of trajectories which are solutions of an integrable system. We start from Liouville's result. Given a Hamiltonian System

$$\begin{aligned} \dot{y}_k &= + H_{x_k}, \\ \dot{x}_k &= - H_{y_k}, \end{aligned} \qquad (k = 1,2,. .,n) \qquad (3.1.1)$$

let $H = H(y;x)$ be analytic in a given domain D of the phase space. If n uniform integrals $F_1, F_2,\ldots,F_n$, in involution, are known, in a domain $D' \subset D$, then in D' the system is integrable, i.e., reducible to quadratures. Let

$$F_i(y;x) = C_i = \text{const.} \qquad (3.1.2)$$

for $i = 1,2,\ldots,n$. In general one verifies that the closed manifolds generated by Eqs. (3.1.2) are tori and, on these, the motion is quasiperiodic. One can actually show that this is so, in general, for a Liouville system. More precisely, the following theorem (Arnol'd, 1963) can be stated. "Let the system (3.1.1), with n degrees of freedom, have n first uniform integrals $F_1,\ldots,F_n$ in involution. The equations $F_i = C_i$ define a compact manifold $M = M_C$ in every point of which the vectors grad F_i $(i = 1,2,\ldots,n)$ are linearly independent in the phase space of dimensions $2n$. Then M is a torus of dimension n and the point $(y(t); x(t))$, solution of (3.1.1) in D', has a quasiperiodic motion on M." This theorem justi fies the fact that we always consider integrable systems as given by a Hamiltonian $H = H_o(x)$, a function of the momenta only. The frequencies $\omega_k = H_{ox_k}$ $(k = 1,2,\ldots,n)$ will, in general, be rationally independent, so that the motion on the torus, defined by the parameters $y_1, y_2,\ldots,y_n$, is ergodic, in the sense that the trajectories cover such torus T^n densely everywhere. In other words, given any point $P^o \in T^n$, $P^o(y_1^o,y_2^o,\ldots,y_n^o)$, and an $\epsilon > 0$ arbitrary, it is possible to find a $T(\epsilon)$

such that for a given $\bar{t}$, $0 < \bar{t} \leq T(\epsilon)$, the relations $|y_k(\bar{t}) - y_k^o| < \epsilon$ are satisfies for all $k = 1,2,\ldots,n$.

2. Perturbations of an Integrable System.

Consider the system generated by the Hamiltonian

$$H = H_o(x) + \mu H_1(y;x;\mu) \qquad (3.2.1)$$

with $0 < \mu \leq 1$ and H_1 multiperiodic (period 2π) with respect to $y_1, y_2, \ldots, y_n$. We try to verify under what conditions the motion of the perturbed system described by (3.2.1) develops over tori which are "close" to the tori defined by $x_k = x_k^o =$ const. $(k = 1,2,\ldots,n)$. We initially give a brief description of the classical perturbations techniques. They consist in reducing, by a sequence of canonical transformations, the Hamiltonian H given by (3.2.1) to the successive forms

$$H = H_o^{(1)}(x') + \mu^2 H_1^{(1)}(y';x';\mu)$$

$$H = H_o^{(2)}(x'') + \mu^3 H_1^{(2)}(y'';x'';\mu)$$

$$\text{-------------------------------}$$

$$H = H_o^{(s)}(x^{(s)}) + \mu^s H_1^{(s)}(y^{(s)};x^{(s)};\mu)$$

$$\text{-------------------------------}$$

Such a method, as seen in the previous chapters, leads to equations of the type

$$\sum_{k=1}^{n} \omega_k \frac{\partial S}{\partial y_k} = F(x,y) = F(x,y + 2\pi)$$

whose solution contains small (if not zero) denominators and the resulting series are generally divergent, even if the frequencies ω_k are rationally independent. In this case the n-ple $\omega = (\omega_1, \omega_2, \ldots, \omega_n)$ of real numbers is supposed to satisfy the infinitely many inequalities

$$|k_1\omega_1 + k_2\omega_2 + \ldots + k_n\omega_n| \geq \{\sum_{i=1}^{n} |k_i|\}^{-p} \qquad (3.2.2)$$

with $p = n+1$ and all integers $k_i \neq 0$ and a conveniently chosen $K(\omega_1, \omega_2, \ldots, \omega_n) > 0$, (Koksma, 1936). Therefore, for a nonzero measure set of values of $\omega_1, \ldots, \omega_n$, the denominators of the classical perturbation theory are bounded below in absolute value. This is nevertheless insufficient to guarantee the convergence of the series in question. On the other hand, since we assume that the frequencies ω_k are continuous functions of the x_k^o, a continuous variation of these will necessarily introduce resonance values of ω_k and the above mentioned series cannot, in any event, be continuous functions of the x_k^o, that is, of the initial values of the problem. Under certain conditions on H_1 one can obtain conservation of quasiperiodic motions and the first theorem to be stated on this respect is due to Kolmogorov (1954). As Arnol'd says in his proof of Kolmogorov's Theorem (1963), such result is "A simple and novel idea, the combination of very classical and essentially modern methods, the solution of a 200 year old problem, a clear geometrical picture and great breadth of outlook..." This is actually so, since earlier results by Poincaré were considered in a too much general form and thought to prevent the lightest chance of integrability of dynamical systems. The only thing which one can conjecture is that non-integrable systems are dense, say in the space of all Hamiltonian functions. No statement is however available on the density of integrable system. If they are at least as dense as the rational numbers on any segment, we might still say there are quite a few integrable systems. In fact celebrated problems like the n-body gravitational problem, the restricted problem of three bodies, the asymmetric top and others have only been proved to be non-integrable in the sense that, in particular coordinates and cases, there are no uniform integrals or even more special cases like algebraic or analytic integrals (e.g., Whittaker, 1937; Siegel, 1954; Moser, 1961; Rüssman, 1959). The simplest statement one can make of Kolmogorov's Theorem is that, under the condition of non-degeneracy $|\partial^2 H_o/\partial x \partial x| \neq 0$, under a small analytic perturbation the majority of the invariant manifolds (tori) defined by H_o are not destroyed but simply deformed. For "majority" it is intended a nowhere dense set whose complement has measure small order ϵ. In fact, in any neighborhood of an invariant torus of the unperturbed system there is an invariant torus over which all trajectories are closed, that is, the frequencies ω_k are rationally dependent. Under however small perturbations these invariant tori collapse. In any event, for systems with more than one degree of freedom, essentially nothing is known about the long time (asymptotic) behavior of the trajectories. For conservative systems with two degrees of freedom, the manifold defined by the energy integral is three dimensional and contain the two-dimensional invariant tori. This is the maximum dimension for which the gaps between two such tori are finite and closed, so that, trajectories originating in these regions are confined to remain there for all times. For higher dimensions not even this is generally true.

As we have mentioned before, the formal series which produce the reduction of the Hamiltonian to a function of the actions only, have a questionable convergence mainly because of the appearance of small divisors. As Brouwer (1961) says, the convergence of the series depends on how fast the numerators decrease with the increasing order of approximation where more numerous and larger integers enter the combinations (3.2.2). This implies that a method should be devised so as to increase the rate of contraction of those numerators. This is probably one of the more important aspects present in Kolmogorov's suggested proof of his theorem. That method was obtained as a Newton's type method of approximation, which introduces quadratic convergence, in the sense that the error ϵ_n of the n-th approximation is order ϵ^2_{n-1}, for $n = 1,2,\ldots$ and $\epsilon_1 < 1$. More precisely, suppose that the perturbation μH_1 of (3.2.1) admits the bound $|\mu H_1| < M < 1$ for x,y in some domain D_1. If we write H_1 in the form

$$H_1(y;x) = \sum_k A_k(x)e^{ik\cdot y}$$

where $k\cdot y = k_1y_1 + k_2y_2 + \ldots + k_ny_n$, then for $|\text{Im } y| \leq \rho$, the coefficients $A_k(x)$ decrease as $M\, e^{-|k|\rho}$. Taking (3.1.2) into account, the bound which is obtained for $\mu^2 H_1^{(2)}$ is order $M^2\delta_1^{-q}$ for $|\text{Im } y'| \leq \rho - \delta_1$. The quantity δ_1 is related, in succession, with a convenient quantity δ_0 such that $M \leq \delta_0^N$ for $\delta_0 > 0$ sufficiently small and N sufficiently large, and, therefore, $H_1^{(s)} \leq \delta_s^N$ for $|\text{Im } y^{(s)}| \leq \rho - \delta_s = \rho_s < \rho_{s-1} < \ldots < \rho$. One may propose to fix the frequencies ω_k of the system and approximate the solution to the unknown torus defined in a space where the frequencies are exactly those. In the complete proof of the

Theorem, given by Arnol'd (1963), the frequencies ω_k are not fixed numbers but varying functions of the actions at every stage of approximation. A simplified version of Kolmogorov's Theorem was given by Barrar (1970). This version leads directly to some consequences in the problem of Poincaré's presentation of Lindstedt's Method. The original condition of analyticity for H is maintained, although Moser (1962) has shown that an H which is several times differentiable is sufficient. But this requires a method of smoothing to be discussed in the next chapter.

We begin considering the Hamiltonian in the form (all summations from 1 to n):

$$\begin{aligned} H(y;x) &= H_o + \sum_k \omega_k x_k + A(y) \\ &+ \sum_k B_k(y)x_k + \sum_{k,j} C_{kj}(y)x_k x_j + D(y;x) \end{aligned} \tag{3.2.3}$$

where H is analytic and multiperiodic of period 2π in each y_k, H_o is a constant, D contains terms of at least third degree in the x_k and the ω_k's satisfy the condition

$$|\sum_k j_k \omega_k| \geq \epsilon(\sum_k |j_k|)^{-s} \tag{3.2.4}$$

for $s = n+1$, all integers j_k and a conveniently chosen $\epsilon(\omega)$.

Next, let us consider a canonical transformation $(y;x) \to (y';x')$ defined by the Hamilton-Jacobi generator

$$S(y;x') = \sum_k (x'_k + \alpha_k)y_k + Y(y) + \sum_k x'_k Y_k(y) \tag{3.2.5}$$

where α_k are constants, so that

$$\begin{aligned} x_i &= S_{y_i} = x'_i + \alpha_i + Y_{y_i}(y) + \sum_k x'_k Y_{ky_i}(y) \\ y'_i &= S_{x'_i} = y_i + Y_i(y) \quad . \end{aligned} \tag{3.2.6}$$

From this last, assuming that the matrix

$$\beta = \frac{\partial y'}{\partial y} = I + \{\frac{\partial Y_i}{\partial y_j}\}$$

is not singular, say for y_i in a neighborhood of y'_i, one can write

$$y_i = F_i(y') = y'_i + \tilde{Y}_i(y') \tag{3.2.7}$$

and, therefore, (3.2.6) is invertible. By substituting x_i as given by (3.2.6) into (3.2.3) one finds

$$\begin{aligned} H = H_o &+ \sum_k \omega_k \alpha_k + \sum_k \omega_k x'_k + a_{(0)} + A^*(y) \\ &+ \sum_k B^*_k(y)x'_k + A^{(1)}(y') + \sum_k B^{(1)}_k(y')x'_k \\ &+ \sum_{k,j} C^{(1)}_{kj}(y')x'_k x'_j + D^{(1)}(y';x') \end{aligned} \tag{3.2.8}$$

where

$$\begin{aligned} A^*(y) &= \sum_k \omega_k \frac{\partial Y}{\partial y_k} - A(y) - a_{(0)} \quad , \\ B^*_k(y) &= \sum_j (\omega_j \frac{\partial Y_k}{\partial y_j} + C_{kj}(\alpha_j + \frac{\partial Y}{\partial y_j})) + B_k(y), \\ A^{(1)}(y') &= \sum_{k,j} C_{kj}(y)(\alpha_k + \frac{\partial Y}{\partial y_k})(\alpha_j + \frac{\partial Y}{\partial y_j}) + D_o(y) \ , \\ B^{(1)}_k(y') &= \sum_{i,j} C_{ij}(y)(\alpha_i + \frac{\partial Y}{\partial y_i})(\frac{\partial Y_k}{\partial y_j}) + D_k(y) \ , \\ C^{(1)}_{ij}(y') &= C_{ij}(y) + \sum_{k,\ell} C_{k\ell}(y) \frac{\partial Y_i}{\partial y_k} \frac{\partial Y_j}{\partial y_\ell} + D_{ij}(y) \ , \\ D^{(1)}(y';x') &= D(y;x' + \alpha + \frac{\partial Y}{\partial y^T} + \sum_j x'_j \frac{\partial Y_j}{\partial y^T}) - D_o(y) \\ &\qquad - \sum_i D_i(y)x'_i - \sum_{i,j} D_{ij}(y)x'_i x'_j, \\ D_o(y) &= D(y;\ \alpha + \frac{\partial Y}{\partial y^T}), \\ D_i(y) &= \sum_k \frac{\partial D}{\partial x_k}(y;\ \alpha + \frac{\partial Y}{\partial y^T})(\delta_{ik} + \frac{\partial Y_i}{\partial y_k}), \\ D_{ij}(y) &= \sum_{k,\ell} \frac{\partial^2 D}{\partial x_k \partial x_\ell}(y;\alpha + \frac{\partial Y}{\partial y^T})(\delta_{ij} + \frac{\partial Y_i}{\partial y_k})(\delta_{j\ell} + \frac{\partial Y_j}{\partial y_\ell}) \ , \end{aligned} \tag{3.2.9}$$

where y is substituted as a function of y' through (3.2.7). The constant $a_{(0)}$ is introduced for reasons to become clear in a moment. The point here is that the quantities $A^*(y)$, $B^*_k(y)$ are of the first order in α_k, Y, Y_k which are supposed to be small in a sense to be specified. On the other hand the quantities $A^{(1)}$, $B^{(1)}_k$ are of the second order with respect to the same variables. The aim of the method is to eliminate A^*, B^*_k by a proper choice of Y, α_k, Y_k $(k = 1,2,\ldots,n)$. This can in fact be easily accomplished as follows. If one defines

$$z_j = e^{iy_j},$$

by the hypotheses on H, one has, in particular,

$$A(y) = \sum_{(k)} a_{(k)} e^{i(k \cdot y)} = \sum_{(k)} a_{(k)} z^{(k)} \tag{3.2.10}$$

where $(k) = (k_1, k_2, \ldots, k_n)$, $(k \cdot y) = k_1 y_1 + k_2 y_2 + \ldots + k_n y_n$, and $z^{(k)} = z_1^{k_1} z_2^{k_2} \ldots z_n^{k_n}$. Also, by the same hypotheses,

$$\begin{aligned} Y(y) &= \sum_{(k)\neq 0} Y^{(k)} z^{(k)} , \\ Y_j(y) &= \sum_{(k)\neq 0} Y_j^{(k)} z^{(k)} . \end{aligned} \tag{3.2.11}$$

Setting $A^*(y) = 0$, from the first of (3.2.9), one finds

$$\begin{aligned} \sum_j \omega_j \frac{\partial Y}{\partial y_j} &= \sum_j i\omega_j z_j \frac{\partial Y}{\partial z_j} = \sum_j i\omega_j \sum_{(k)\neq 0} Y^{(k)} k_j z^{(k)} \\ &= \sum_{(k)\neq 0} i(\omega \cdot k) Y^{(k)} z^{(k)} = \sum_{(k)\neq 0} a_{(k)} z^{(k)} \end{aligned}$$

or

$$Y^{(k)} = -i(\omega \cdot k)^{-1} a_{(k)} \tag{3.2.12}$$

which defines Y and also $a_{(0)}$. Now, from the second of (3.2.9), defining

$$B_k = \sum_{(\ell)} B_k^{(\ell)} z^{(\ell)},$$

$$C_{kj} = \sum_{(\ell)} C_{kj}^{(\ell)} z^{(\ell)},$$

and

$$P_k(y) = \sum_j C_{kj} \frac{\partial Y}{\partial y_j} = \sum_{(\ell)} P_k^{(\ell)} z^{(\ell)},$$

one finds that α_j and $Y_j^{(k)}$ are defined by

$$\sum_k C_{jk}^{(0)} \alpha_k + P_j^{(0)} + B_j^{(0)} = 0 \tag{3.2.13}$$

and

$$i(\omega\cdot k)Y_j^{(k)} + P_j^{(k)} + B_j^{(k)} + \sum_\ell C_{j\ell}^{(k)}\alpha_\ell = 0 \tag{3.2.14}$$

respectively. We must, of course, satisfy certain conditions, viz.,

a) $(\omega\cdot k)$ should not be zero, for (3.2.12) to be meaningful.

b) The determinant of $\{C_{jk}^{(0)}\}$ should not be zero, for (3.2.13) to have a solution in the vector $\alpha = (\alpha_1,\ldots,\alpha_n)$.

c) Same as (a) for (3.2.14) to yield the constants $Y_j^{(k)}$.

It is also obvious that if $A(y)$, $B(y)$, $C_{kj}(y)$ have a finite trigonometric representation as Fourier's polynomials in y, the generator S defined by (3.2.5) is also a Fourier's polynomial in y. The process calls for a repeated application of successive canonical transformations and, in the limit, a reduction of the original Hamiltonian to

$$H = \sum_k \omega_k X_k + \sum_{k,j} K_{kj}(Y)X_k X_j + \Delta(Y;X) \tag{3.2.15}$$

where Δ is at least of third order in the components X_k of X. In this case, the system admits the solution

$$\begin{aligned} X_k &= 0, \\ Y_k &= \omega_k(t-t_k), \end{aligned} \tag{3.2.16}$$

for $k = 1,2,\ldots,n$.

With this in mind, one can now formulate Kolmogorov's Theorem in the following simplified form.

Theorem (Kolmogorov): "Consider the Hamiltonian H of a system in the form given in (3.2.3), analytic for $|x_k| \leq r_0$, $|\text{Im } y_k| \leq \rho_0$; let D_0 be this region and the following hypotheses be verified

a) $$\left| \sum_{k=1}^{n} j_k \omega_k \right| \geq \epsilon \left(\sum_{k=1}^{n} |j_k| \right)^{-s} \qquad (3.2.17)$$

for all integers j_k not all zero, $\epsilon = \epsilon(\omega)$ conveniently chosen and $s = n+1$.

b) The matrix $\{C_{kj}(0)\}$ is non-singular. (3.2.18)

Then for $|A(y)|$ and $|B_k(y)|$ sufficiently small in D_0, there exists a canonical transformation $(y;x) \rightarrow (Y;X)$ given by

$$\begin{aligned} x_k &= X_k + E_k(Y) + \sum_j E_{kj}(Y) X_j \\ y_k &= Y_k + N_k(Y) \end{aligned} \qquad (3.2.19)$$

such that E_k, E_{kj}, N_k are 2π-periodic in every Y_j and analytic in the region Δ_0 defined by $|X_k| < \frac{3}{4} r_0$, $|\text{Im } Y_k| < \frac{3}{4} \rho_0$. The transformation (3.2.19) maps Δ_0 into D_0 and, in Δ_0, the Hamiltonian H has the image (3.2.15)."

The proof of this theorem consists basically in showing that the application of successive canonical transformations of the type (3.2.6) is a convergent process of successive approximations from (3.2.3) to (3.2.15), and produces an analytical canonical transformation. The theorem follows from several basic Lemmas which were given by Arnol'd (1963). Here we limit ourselves to mention the two basic Lemmas.

Lemma 1. "If the n nonzero frequencies ω_k satisfy (3 2.17), if

$$F(z) = \sum_{(k)\neq 0} f_{(k)} z^{(k)} \tag{3.2.20}$$

and if S satisfies the equation

$$\sum_{j=1}^{n} \omega_j \frac{\partial S}{\partial y_j} \equiv i \sum_{j=1}^{n} \omega_j z_j \frac{\partial S}{\partial z_j} = F(z) \tag{3.2.21}$$

then the solution

$$S = - \sum_{(k)\neq 0} \frac{f_{(k)} z^{(k)}}{(k\cdot\omega)} \tag{3.2.22}$$

satisfies, for a conveniently chosen absolute constant C, the following relations

$$\begin{aligned} \|S\|_{\Gamma(\rho-h)} &\leq \frac{C}{\epsilon\, h^{s+n}} \|F\|_{\Gamma(\rho)}, \\ \left\|\frac{\partial S}{\partial y_k}\right\|_{\Gamma(\rho-h)} = \left\|z_k \frac{\partial S}{\partial z_k}\right\|_{\Gamma(\rho-h)} &\leq \frac{C}{\epsilon\, h^{s+m+1}} \|F\|_{\Gamma(\rho)}, \end{aligned} \tag{3.2.23}$$

where the norm $\| \ \|$ is the supremum of the absolute value in a ring domain

$$\Gamma(\rho) = \{e^{-\rho} \leq |z_k| \leq e^{+\rho}\} = \{|\text{Im } y_k| \leq \rho\}$$

for all $k = 1,2,\ldots,n$ and $0 < h < \rho$."

The proof of the Lemma depends on the irrationality hypothesis (3.2.17), which gives an upper bound for the divisors, i.e.,

$$|(\omega\cdot k)|^{-1} \leq \frac{\theta}{\epsilon} \sum_{j=1}^{n} |k_j|^{s} \tag{3.2.24}$$

where θ is a constant depending on n, s. The most tedious part of the proof depends on a correct estimate of $f_{(k)}/(k\cdot\omega)$ in the ring $\Gamma(\rho)$ and subsequent estimates of S and its derivatives. The necessary relations are obtained by Arnol'd in the Fundamental Lemma (1963; Section 2, 3).

Lemma 2. "Consider the quantities

$$\epsilon_o = \max(\|A\|_{\Gamma(\rho_o)};\ \|B_k\|_{\Gamma(\rho_o)};\ k = 1,2,\ldots,n)$$

$$\epsilon_1 = \max(\|A^{(1)}\|_{\Gamma(\rho_1)};\ \|B_k^{(1)}\|_{\Gamma(\rho_1)};\ k = 1,2,\ldots,n)$$

$$r_1 = r_o - 2h, \quad \rho_1 = \rho_o - 4h$$

and

$$0 < h < \frac{r_o}{2}, \quad 0 < h < \frac{\rho_o}{4} .$$

Let

$$\Gamma(r,\rho) = \{|x_k| \leq r;\ \mathrm{Im}(y_k) \leq \rho,\ k = 1,2,\ldots,n\}$$

$$\|F(y;x)\|_{\Gamma(r,\rho)} = \sup|F(y;x)|, \quad \text{for} \quad (y,x) \in \Gamma(r,\rho)$$

and consider the bounds:

$$\max |C_{kj}^*| \leq 2N$$

$$\|D(y;x)\|_{\Gamma(r_o,\rho_o)} \leq 2M$$

$$\|C_{kj}\|_{\Gamma(\rho_o)} \leq 2M,$$

where C_{kj}^* are the elements of the inverse of the matrix $\{C_{kj}^{(o)}\}$. We consider (3.2.4) and $h < \epsilon$, $t = s + n + 1$, $\frac{1}{2} \leq r_o \leq 1$.

Under these conditions there are constants C_k $(k = 1,2,\ldots,n)$ depending only on N, M, n such that, if $\epsilon_o \leq h^{2t+3}/C_o$, one has the following,

a) The transformation (3.2.6) can be written

$$\begin{aligned} y_i &= y_i' + f_i(y') = F_i(y') \\ x_i &= x_i' + g_i(y') + \sum_j g_{ij}(y')x_j' = G_i(y';x') \end{aligned} \tag{3.2.25}$$

where F_i, G_i are analytic for $y' \in \Gamma(\rho_1)$. Also

$$\|f_i\|_{\Gamma(\rho_1)} \leq 2h$$

$$\|g_i\|_{\Gamma(\rho_1)} \leq 2h/(n+2) \qquad (3.2.26)$$

$$\|g_{ij}\|_{\Gamma(\rho_1)} \leq h/(m+2)$$

<u>and it follows that the transformation</u> (3.2.25) <u>maps</u> $\Gamma(r_1,\rho_1)$ <u>into</u> $(r_0-h,\ \rho_0-2h)$. <u>Also</u> $\epsilon_1 \leq \frac{C_1}{h^{3t+3}}\epsilon_0^2$.

b) <u>The new Hamiltonian is defined for</u> $(y';x')$ <u>in</u> $\Gamma(r_1,\rho_1)$ <u>and one has</u>

$$\|C_{ij}^{(1)}\|_{\Gamma(\rho_1)} \leq \|C_{ij}\|_{\Gamma(\rho_0)} + hC_2 \qquad (3.2.27)$$

$$\|D^{(1)}(y';x')\|_{\Gamma(r_1,\rho_1)} \leq \|D(y;x)\|_{\Gamma(r_0,\rho_0)} + hC_3 \qquad (3.2.28)$$

$$|C_{ij}^{(1)}(0) - C_{ij}\ (0)| \leq (\frac{C_4}{\rho_1} + C_5)h." \qquad (3.2.29)$$

The proof of this Lemma is given in details by Barrar (1970). One uses Lemma 1 to estimate

$$\max\|\frac{\partial Y}{\partial y_k}\|_{\Gamma(\rho_0-h)} \leq \frac{C'\epsilon_0}{h^{t+1}}$$

$$\|P_k(y)\|_{\Gamma(\rho_0-h)} \leq \frac{C''\epsilon_0}{h^{t+1}} \qquad (3.2.30)$$

$$|\alpha_k| \leq \frac{C'''\epsilon_0}{h^{t+1}}$$

where C', C'', C''' are constants which depend only on N, M, n. After defining Y, P_k, α_k in $\Gamma(\rho_0-h)$ one solves for Y_k (from (3.2.14)) in $\Gamma(\rho_0-2h)$, giving, from Lemma 1,

$$\max\|Y_k\|_{\Gamma(\rho_0-2h)} \leq \frac{C^{iv}\epsilon_0}{h^{2t+1}}$$

$$\max\|\frac{\partial Y_k}{\partial y_j}\|_{\Gamma(\rho_0-2h)} \leq \frac{C^{v}\epsilon_0}{h^{2t+2}} \qquad (3.2.31)$$

where, again, C^{iv} and C^{v} are constants, depending only on M, N, n. Rouché's

Theorem is used to show that if

$$\max \|Y_k\|_{\Gamma(\rho_o-2h)} < 2h,$$

then the transformation (3.2.26) has the inverse

$$y_i = y_i' + f_i(y')$$

with

$$\|f_i\|_{\Gamma(\rho_o-4h)} < 2h$$

and maps $\Gamma(\rho_1)$ into $\Gamma(\rho_o-2h)$. Also, one easily sees that $f_i(y')$ is 2π-periodic in each y_k', so that part (a) of the Lemma follows by considering $C_o = \max(C', C'', C''', C^{iv}, C^{v}, n+2)$ and therefore the quantities $\max\|\frac{\partial Y}{\partial y_k}\|$, $\max\|\frac{\partial Y_i}{\partial y_k}\|$, α_k are all less than $h/(m+2)$ in $\Gamma(\rho_o-2h)$. Also, if $y' \in \Gamma(\rho_1)$, part (a) implies $y \in \Gamma(\rho_o-2h)$. All other estimates of part (b) of Lemma 2 also follow by simple applications of Cauchy's formula and Schwartz inequality.

Proof of Kolmogorov's Theorem.

The previous Lemmas lead directly to the proof of the Theorem. In fact, the Hamiltonian H in terms of $(y;x)$ in $\Gamma(r_o, \rho_o)$ has been transformed into one of the same form in $(y';x')$ in $\Gamma(r_o-2h, \rho_o-4h)$. The operation (defined in Lemma 1, with the estimates of Lemma 2) is then applied in succession. Let us define

$$r_{n+1} = r_n - 2h_{n+1}$$

and

$$\rho_{n+1} = \rho_n - 4h_{n+1}.$$

At the n-th stage, the Hamiltonian is defined in the space of $(y^{(n)};x^{(n)}) \in \Gamma(r_n, \rho_n)$ and, for conditions of Lemma 2, $\epsilon_n = \max(\|A\|^n_{\Gamma(\rho_n)}, \|B_k^{(n)}\|_{\Gamma(\rho_n)})$. The basic question is to verify the validity of Lemma 2 at every stage of repeated application of the transformation defined in Lemma 1. Again, for the details of the proof we refer to Barrar (1970) The basic assumptions are the estimates of Lemma 2, following the conditions

$$\epsilon_n \leq \frac{h_{n+1}^{2t+3}}{C_o} , \quad \epsilon_{n+1} \leq \frac{C_1}{h_{n+1}^{3t+3}} (\epsilon_n)^2$$

The choice of a convenient $L > 1$,

$$L^{n+1} \geq \frac{\max(C_o, C_1)}{h_{n+1}^{3t+3}} \tag{3.2.32}$$

$$L^2 \epsilon_o = a < 1$$

gives $|\epsilon_n| \leq a^{2n}$ and

$$\epsilon_n < \frac{1}{L^{n+2}} < \frac{1}{L^{n+1}} \leq \frac{h_{n+1}^{3t+3}}{C_o} ,$$

which substitutes the conditions $\epsilon_o \leq h^{2t+3}/C_o$ and $\epsilon_1 \leq \frac{C_1}{h^{3t+3}}$ of Lemma 2. The inequality (3.2.32) is satisfied by setting $h_n = \delta/2^n$ and $L \geq (2^{3t+3}/\delta^{3t+3})$ $\max(C_o, C_1)$ or $L > 1$ whichever is the case. For the original Hamiltonian (3.2.3) one assumes the bounds

$$|C_{kj}^*| \leq N$$

$$\|C_{kj}\|_{\Gamma(\rho_o)} \leq M$$

$$\|D(y;x)\|_{\Gamma(r_o, \rho_o)} \leq M$$

and by proper scaling, one can assume $r_o = 1$. With the choice of a $\delta > 0$ sufficiently small one shows that $\epsilon_n \to 0$ in $\Gamma(\frac{3}{4} r_o, \frac{3}{4} \rho_o)$ as $n \to \infty$. The remaining part is the proof that the limit of the transformation obtained by iteration of Lemma 1, satisfies the result announced in Kolmogorov's Theorem (Eqs. 3.2.17 and 3.2.15). The resulting transformation is obviously canonical since it is the composition of a sequence of canonical transformations. Also, the total transformation $T^{(n)}$, after n iterations, maps $\Gamma(r_{n+1}, \rho_{n+1})$ into $\Gamma(r_n, \rho_n)$ and its domain of validity contains $\Gamma(3r_o/4, 3\rho_o/4)$. In this, $T^{(n)}$ is uniformly bounded and the sequence converges to a T which is precisely (3.2.17).

Arnol'd's proof makes use of stronger methods of estimates, while the iterations on H, through the operation here defined in Lemma 1, follow along a line which is suggested by Newton's Method of approximation, and thus produce quadratic convergence.

The generalization of the Theorem by Arnol'd to degenerated systems would imply, in the foregoing approach, a starting Hamiltonian in the form

$$\begin{aligned} H = H_o + \sum_{k=1}^{m} \omega_k x_k + \sum_{j=m+1}^{n} \mu\omega_j x_j \\ + A(y) + \sum_{k=1}^{n} B_k(y) x_k + \sum_{k,j=1}^{n} C_{kj}(y) x_k x_j + D(y;x) \end{aligned} \tag{3.2.33}$$

in opposition to (3.2.3), and where μ is a "small" parameter, say, order ϵ (see Eq. (3.2.4)). Another case extremely difficult to treat corresponds to the case $\det\{C_{kj}\} = 0$. The first case can possibly be treated in a way analogous to the foregoing Lemmas and Theorems, but the speed of convergence is much slower than in the normal case. Moser's general theory (1967) assumes vast knowledge in measure theoretic results, algebra, and differential geometry. He is able to maintain the speed of convergence of the normal case. Moser's approach to the normal case is much simpler, especially if one considers analytic perturbations, which he does not. For this reason he is forced to use very sophisticated smoothing techniques, but, of course, gets a more general result. Arnol'd's presentation of Kolmogorov's Theorem is, eventually, more general than the form we have presented. He considers a Hamiltonian (3.2.1), analytic in a certain domain D of the phase space, say $D\{|\text{Im } y_k| \leq \rho;\ x_k \in G;\ k = 1,2,\ldots,n\}$, G an open set in R^n, and 2π-periodic in each y_k. The basic assumption is the non-vanishing of the determinant $\left|\frac{\partial^2 H_o}{\partial x_j \partial x_k}\right|$ in D. Then, he is able to show that, for every $K > 0$, there exists an $M(K,\rho,G,H_o) > 0$ such that if $|\mu H_1| \leq M$ in D, the trajectories defined by H are such that

1) There exists an invariant set D_1 (real), and if

$$\text{Re }(D) = D_1 + D_2, \text{ then } \text{mes } D_2 \leq K \text{ mes } D.$$

2) D_1 is composed of invariant, analytic, n-dimensional tori T_ω, defined by

$$x = x_\omega + f_\omega(\eta)$$

$$y = \eta + g_\omega(\eta)$$

where f_ω, g_ω are analytic functions of period 2π in each η_k $(k = 1,2,\ldots,n)$ and ω is a parameter specifying the torus T_ω.

3) The motion defined by H on T_ω is quasiperiodic with n frequencies $\omega = (\omega_1,\omega_2,\ldots,\omega_n)$:

$$\omega_k = \dot{\eta}_k = \partial H_0/\partial x_k\Big|_{x=x_\omega}$$

which satisfy

$$\Big|\sum_{k=1}^{n} m_k\omega_k\Big| \geq K\{\sum_{k=1}^{n} |m_k|\}^{-s},$$

$$s = n + 1,$$

for every set of integers m_k not all zero.

The conditions are basically those assumed before, while the condition on H_0 is transferred into a condition on the determinant of the quadratic part (in x_kx_j) of H as given by (3.2.3). Indeed, such form is quite similar to the one implicitly assumed by Arnol'd in the generalization of Kolmogorov's Theorem to degenerate cases, i.e., case where $|\partial^2H_0/\partial x_j\partial x_k| = 0$. Degeneracy, as far as the form (3.2.3) is concerned, would imply, for instance, that one of the x's is ignorable both in $\sum \omega_k x_k$ and in the quadratic part $\sum C_{kj}x_kx_j$. Obviously, the first case would make (3.2.4) an impossible condition, while the second case would produce a singular system for the definition of the α_k by (3.2.13). As we have seen in the theory of implicit functions, this leads to an expansion in fractional powers of the small parameter, in this case μ.

In simple words, however stated, Kolmogorov's theorem says that, if H_o is non-degenerate, under sufficiently small analytic perturbations, a set of non-zero measure of invariant tori defined by H_o is not destroyed, but simply slightly deformed. However, the passage from one torus to the other cannot be obtained by a continuous transformation since zones of rational dependent frequencies ω_k are necessarily crossed.

It should be noted, by other means, the condition on H_o can be made less stringent and one needs only to assume that

$$\det \begin{vmatrix} \frac{\partial^2 H_o}{\partial x_i \partial x_j} & \frac{\partial H_o}{\partial x_i} \\ \frac{\partial H_o}{\partial x_j} & 0 \end{vmatrix} \neq 0 \qquad (3.2.34)$$

3. Degenerate Systems.

In cases where the stated condition for $H_o(x)$ is not satisfied, or, in the simpler version given for Kolmogorov's Theorem, the matrix $\{C_{ij}\}$ is singular, still considering the irrational condition (3.2.4) satisfied, both Arnol'd's or Barrar's proofs cease to be valid. In an earlier version of the proof, Barrar (1966) indicated a possible way of handling the situation, giving to the Hamiltonian the form

$$H = A_o \sum_{k=1}^{m} \omega_k x_k + \sum_{j=m+1}^{n} \mu\omega_j x_j + A(x) + \mu H_1(y;x) \qquad (2.3.1)$$

where A_o is a constant and μ a constant parameter $0 \leq \mu \leq 1$. With respect to the original Hamiltonian written in the form

$$H = H_o(x) + \mu[H_{1s}(x)+H_{1p}(y;x)] \qquad (2.3.2)$$

and such that

$$\oint H_{1p} dy = 0,$$

the frequencies ω_k are defined by

$$\omega_k = \frac{\partial}{\partial x_k}(H_o + \mu H_{1s})\Big|_{x=x^o}, \quad k = 1,2,\dots,m$$

$$\omega_j = \frac{\partial}{\partial x_j} H_{1s}\Big|_{x=x^o}, \quad j = m+1, m+2,\dots,n$$

and are assumed to satisfy

$$\Big|\sum_{k=1}^{m} j_k\omega_k + \mu \sum_{i=m+1}^{n} j_i\omega_i\Big| \geq \mu\Big|\sum_{p=1}^{n} j_p\Big|^{-n-1}\delta$$

for all integers j_k not simultaneously zero. It is known that the exterior measure of all ω_k $(k = 1,2,\dots,n)$ not satisfying such condition is less than $K\delta\mu$, with K properly bounded and function of the ω_k's. Under the new assumption

$$\det\{\frac{\partial^2(H_o + \mu H_{1s})}{\partial x_i \partial x_j}\} = 2N\mu \neq 0$$

for $x = x^o \in D$, if H_{1p} is sufficiently small, then there are solutions of the system generated by H, which are quasi-periodic and take the form

$$y_k = \omega_k(t-\tau_k) + \phi_k(e^{i\lambda_1 t}, e^{i\lambda_2 t},\dots,e^{i\lambda_m t}, e^{i\mu\lambda_{m+1}t},\dots,e^{i\mu\lambda_n t})$$

$$y_j = \mu\omega_j(t-\tau_k) + \phi_j(\dots)$$

$$x_i = \alpha_i + \Psi_i(\dots)$$

for $k = 1,2,\dots,m$; $j = m+1, m+2,\dots,n$; $i = 1,2,\dots,n$. The functions ϕ_k, ϕ_j, Ψ_i have the same form. The proof actually is done by reducing (2.3.2) to the form (2.3.1) by essentially a Taylor expansion and applying an infinite sequence of canonical transformations which reduce the size of $H_1(y,x)$ at every stage. Such a method can be shown to converge for $n = 2$, $m = 1$. The convergence, however, cannot be uniform with respect to μ or x^o.

Arnol'd's Theorem (1963) is much more far reaching, especially insofar as convergence properties are concerned. He considers a function $H(y;x;\epsilon)$, where $y = (y_o;y_1)$, $x = (x_o;x_1)$, where y_o, x_o are n_o-vectors and y_1, x_1 are n_1-vectors and $n = n_0 + n_1$ is the number of degrees of freedom of the system. H is assumed to be 2π-periodic in each y_0 component, analytic in a domain $D\{x_0 \in G_0, |\text{Im } y_0| \leq \rho; |x_1, y_1| \leq R\}$ and ϵ is a real parameter $0 < \epsilon \leq \epsilon_0$. It is also assumed that the following expansion exists

$$H = H_0(x_0) + \epsilon H_1(y;x) + \epsilon^2 H_2(y;x;\epsilon)$$

where H_1 can be decomposed in a short periodic part $\tilde{H}_1$ and a long periodic part $\bar{H}_1$ as

$$H_1 = \bar{H}_1(x_0;x_1;y_1) + \tilde{H}_1(x_0;y_0;x_1;y_1)$$

and

$$\oint \tilde{H}_1 \, dy_0 = 0,$$

that is, the multiple Fourier series of $\tilde{H}_1$ has no constant term, or, better say, it has been thrown into $\bar{H}_1$. Also $\bar{H}_1$, the long periodic part, is composed by a secular part $\bar{H}_{1s}$ and a purely periodic (in each component of y_1) part $\bar{H}_{1p}$, i.e.,

$$\bar{H}_1 = \bar{H}_{1s}(x_0;\tau) + \bar{H}_{1p}(x_0;x_1;y_1)$$

where

$$\bar{H}_{1s} = \lambda_0 + \sum_{i=1}^{n_1} \lambda_i \tau_i + \sum_{i,j=1}^{n_1} \lambda_{ij} \tau_i \tau_j + \sum_{i,j,k=1}^{n_1} \lambda_{ijk} \tau_i \tau_j \tau_k .$$

The λ's are functions of x_0, $\lambda_{ij} = \lambda_{ji}$ and

$$2\tau_k = x_{1k}^2 + y_{1k}^2, \quad (k = 1,2,\ldots,n_1)$$

and also, in G_0,

$$\det\left\{\frac{\partial^2 H_0}{\partial x_{0i} \, \partial x_{0j}}\right\} \neq 0$$

$$\det\{\lambda_{ij}\} \neq 0.$$

Then, under proper bounds for H_2, $\tilde{H}_1$, $\bar{H}_1$, $\bar{H}_{1s}$, $\bar{H}_{1p}$, given an arbitrary $K > 0$ it is possible to obtain $E_o(K;H_o;H_1;G_o;\rho,R,C,\epsilon)$, such that, for $0 < E < E_o$, $0 < \epsilon < E_o^4$ and

$$|H_2| < C, \ |\tilde{H}_1| \leq C, \ |\bar{H}_{1s}| \leq C,$$

$$|\bar{H}_{1p}| \leq C|x_1,y_1|, \ |H_1| < C,$$

there are quasiperiodic solutions of the given Hamiltonian system covering invariant tori T_ω imbedded in a part D_1 of D, the complement of which, D_2, is small in the sense

$$\text{mes } D_2 < K \text{ mes } D_1.$$

The invariant sets T_ω are analytic and differ little (in some precise sense) from the tori defined by $x_{ok} = x_{ok\omega} = \text{const.}$, $\tau_k = \tau_{k\omega} = \text{const.}$ The frequencies of the quasiperiodic motions on such tori are

$$\omega_o = \frac{\partial H_o}{\partial x_{o\omega}}, \quad \omega_1 = \frac{\partial H_{1s}}{\partial \tau_\omega}$$

considered here, vector quantities.

The aspect of this theorem is basically geometric since, as for Komogorov's theorem, it shows the conservation of certain invariant sets under perturbations. It also implies that these invariant sets can be parametrized into tori, while the condition of non-degeneracy for H_o has been relaxed.

4. Perturbed Linear Oscillations.

Consider the Hamiltonian

$$H = H(x;y), \tag{3.4.1}$$

where x, y are n-dimensional canonical conjugate vectors, and suppose H to be real analytic in a domain D of the $2n$-dimensional phase space. We suppose $(x;y)$ to be real and $(0;0) \in D$ to be an isolated equilibrium point of the system

$$\dot{x}_k = -H_{y_k}, \quad \dot{y}_k = +H_{x_k}, \quad (k = 1,2,\ldots,n). \tag{3.4.2}$$

We also assume that

$$\det\left\{\frac{\partial^2 H}{\partial x_i \partial y_j}\right\} \neq 0$$

for $(x;y) = (0;0)$. By hypothesis, H is developable in Taylor series, in a conveniently restricted neighborhood of $(0;0)$, so that

$$H = H_2 + H_3 + H_4 + \cdots \tag{3.4.3}$$

where H_p is an homogeneous polynomial of degree p in $(x_1,\ldots,x_n; y_1,\ldots,y_n)$. We further suppose the eigenvalues associated with the quadratic form H_2 to be all distinct. Let these be $\lambda_1,\lambda_2,\ldots,\lambda_n; -\lambda_1,-\lambda_2,\ldots,-\lambda_n$. Then, there exists a linear symplectic transformation which reduces (3.4.3) to the form

$$H = \sum_{k=1}^{n} \lambda_k X_k Y_k + H_3(X;Y) + H_4(X;Y) + \cdots . \tag{3.4.4}$$

Let us write

$$\epsilon\xi_k = X_k, \quad \epsilon\eta_k = Y_k \quad (k = 1,2,\ldots,n) \tag{3.4.5}$$

with $0 \leq \epsilon \leq 1$. The equations (3.4.2) can now be written

$$\begin{aligned} \dot{\xi}_k &= -\frac{\partial F}{\partial \eta_k} \\ \dot{\eta}_k &= +\frac{\partial F}{\partial \xi_k} \end{aligned} \tag{3.4.6}$$

where

$$F = \sum_{k=1}^{n} \lambda_k \xi_k \eta_k + \epsilon F_1(\xi;\eta;\epsilon) \tag{3.4.7}$$

and F_1 is a power series in ϵ, ξ_k, η_k, beginning with terms of third degree in $(\xi;\eta)$. The eigenvalues may be real or complex and we shall suppose here that $(0;0)$ is an elliptic point in the sense that all λ_k are pure imaginary. We also define

$$\lambda_k = i\omega_k, \quad \omega_k \text{ real}, \quad k = 1,2,\ldots,n$$

the ω_k being positive or negative, none of them zero. In this case, the coefficients in F_1 are pure imaginary and we can write

$$F_1 = -i\, H_1$$

where H_1 is a power series in (ξ, η), with real coefficients and beginning with terms of the third degree.

By means of the canonical transformation

$$\begin{aligned} u_k &= \frac{1}{2} \xi_k \eta_k \\ v_k &= 2i\, \ell n(\xi_k/\eta_k) \end{aligned} \tag{3.4.8}$$

or

$$\begin{aligned} \eta_k &= (2u_k)^{1/2} e^{iv_k} \\ \xi_k &= (2u_k)^{1/2}\, e^{-iv_k} \end{aligned} \tag{3.4.9}$$

Equations (3.4.6) become

$$\begin{aligned} \dot{u}_k &= -\epsilon \frac{\partial H_1}{\partial v_k} \\ \dot{v}_k &= \omega_k + \epsilon \frac{\partial H_1}{\partial u_k} \end{aligned} \tag{3.4.10}$$

with a corresponding Hamiltonian

$$H = \sum_1^n \omega_k u_k + \epsilon H_1(u;v;\epsilon) ,$$

where H_1 is periodic in each v_k with period 2π and the coefficients of its Fourier series, polynomials in $u_1, u_2, \ldots, u_n$.

The problem of constructing solutions in the vicinity of an equilibrium point is therefore reduced to the study of the perturbation of the integrable system generated by

$$H_o = \sum_1^n \omega_k u_k .$$

The theorems stated in the previous sections do not apply to this problem since the determinant

$$\det\{\frac{\partial^2 H_o}{\partial u_i \partial u_j}\} \equiv 0.$$

On the other hand, it is also possible that

$$H_{1s} = \frac{1}{(2\pi)^n} \int_0^{2\pi} \cdots \int_0^{2\pi} H_1(u;v)dv_1\, dv_2 \ldots dv_n$$

is zero and moreover, the normal modes ω_k may be linearly dependent over the set of integers. This will not be the general case, however. We shall deal directly with this problem in the next chapter, where we construct formal series solution of system (3.4.10). It is important to note that the above mentioned problem is a particular case of the classical perturbation problem of a linear oscillator, that is

$$\ddot{x}_k + \omega_k^2 x_k = \epsilon f_k(x;\dot{x},\epsilon),$$
$$(k = 1,2,\ldots,n).$$

This is easily brought to the form

$$\dot{u}_k = \epsilon U_k$$
$$\dot{v}_k = \omega_k + \epsilon V_k$$

by means of the transformation

$$\dot{x}_k + i\omega_k x_k = u_k\, e^{iv_k}$$

that is,

$$\dot{x}_k = u_k \cos v_k$$
$$\omega_k x_k = u_k \sin v_k \qquad (3.4.11)$$

with x_k, $\dot{x}_k$ $(k = 1,2,\ldots,n)$ real quantities. This is the important problem of perturbed linear oscillations. Forerunners for such results are mainly Krylov and Bogoliukov (1943), Van der Pol (1926, 1927), Bogoliukov (1945, 1963), Mitropolski

(1964), Arnol'd (1961) and Moser (1956, 1965).

Consider first the autonomous case of the differential equation

$$\ddot{z} + \omega^2 z = \epsilon Z(z,\dot{z}) \tag{3.4.12}$$

where Z is supposed to be analytic in z, $\dot{z}$ in some domain of the jet space $(z,\dot{z})$ and for $0 \leq \epsilon \leq 1$. The frequency ω is real and constant. By means of the transformation

$$\begin{aligned} z &= \left(\frac{2x}{\omega}\right)^{1/2} \sin y \\ \dot{z} &= (2\omega x)^{1/2} \cos y \end{aligned} \tag{3.4.13}$$

the equation is transformed into

$$\begin{aligned} \dot{x} &= \epsilon \left(\frac{2x}{\omega}\right)^{1/2} \cos y \, Z^*(x,y) = \epsilon g(x,y) \\ \dot{y} &= \omega - \epsilon \frac{\sin y}{(2\omega x)^{1/2}} Z^*(x,y) = \omega + \epsilon f(x,y) \end{aligned} \tag{3.4.14}$$

and we assume g and f to be periodic in y with period 2π. We try to obtain a transformation (Moser, 1966):

$$\begin{aligned} x &= \xi + \epsilon v(\eta,\epsilon) + \epsilon V(\eta,\epsilon)\xi = x(\xi,\eta,\epsilon) \\ y &= \eta + \epsilon u(\eta,\epsilon) = y(\eta,\epsilon) \end{aligned} \tag{3.4.15}$$

such that (3.4.14) reduces to the form

$$\begin{aligned} \dot{\xi} &= O(\xi^2) \\ \dot{\eta} &= \tilde{\omega} + O(\xi) \end{aligned} \tag{3.4.16}$$

The choice $\xi = 0$, evidently, will reduce (3.4.16) to a linear integrable system.

Consider the more general system

$$\begin{aligned} \dot{x} &= \Omega + Mx + \epsilon g(x,y) \\ \dot{y} &= \omega + \lambda + \epsilon f(x,y) \end{aligned} \tag{3.4.17}$$

and we will show that there exist constants Ω, M, λ, which depend on ϵ, ω, such that (3.4.17) reduces to (3.4.16). The solution of the original problem will be possible if the system

$$\Omega(\epsilon,\omega) = M(\epsilon,\omega) = 0$$
$$\tilde{\omega} = \omega - \lambda(\epsilon,\omega) \tag{3.4.18}$$

can be solved for a given ω and for ϵ in a certain interval, $0 \leq \epsilon \leq \epsilon_0$. We suppose that the functions v, V, u are periodic of period 2π in η, analytic in ϵ and η, and zero mean with respect to η. Moreover, we assume the formal series

$$\lambda = \epsilon\lambda_1 + \epsilon^2\lambda_2 + \dots$$
$$M = \epsilon M_1 + \epsilon^2 M_2 + \dots \tag{3.4.19}$$
$$\Omega = \epsilon\Omega_1 + \epsilon^2\Omega_2 + \dots \quad .$$

In view of the hypotheses of analyticity,

$$v = v_0 + \epsilon v_1 + \epsilon^2 v_2 + \dots$$
$$V = V_0 + \epsilon V_1 + \epsilon^2 V_2 + \dots \tag{3.4.20}$$
$$u = u_0 + \epsilon u_1 + \epsilon^2 u_2 + \dots ,$$

with v_k, V_k, u_k $(k = 0,1,2,\dots)$ developable in Fourier series, as for example

$$v_k = \sum_j v_{kj}\, e^{ij\eta}, \quad v_{kj} = \text{const.}\ .$$

From the second of (3.4.15) it follows that

$$\dot{y} = \dot{\eta} + \epsilon \frac{\partial u}{\partial \eta}\dot{\eta}$$

and, comparing with the second of (3.4.16),

$$\dot{\eta} + \epsilon \frac{\partial u}{\partial \eta}\dot{\eta} = \omega + \lambda + \epsilon f(x(\xi,\eta,\epsilon),y(\eta,\epsilon)). \tag{3.4.21}$$

From the first of (3.4.15) and (3.4.16)

$$\dot{\xi} + \epsilon \frac{\partial v}{\partial \eta}\dot{\eta} + \epsilon \frac{\partial V}{\partial \eta}\xi\dot{\eta} + \epsilon V\dot{\xi} = \Omega + M(\xi + \epsilon v + \epsilon V\xi)$$
$$+ \epsilon g(x(\xi,\eta,\epsilon),y(\eta,\epsilon)). \tag{3.4.22}$$

The partial derivative of this, with respect to ξ, give

$$\frac{\partial \dot{\xi}}{\partial \xi} + \epsilon \frac{\partial v}{\partial \eta} \frac{\partial \dot{\eta}}{\partial \xi} + \epsilon \frac{\partial V}{\partial \eta} \dot{\eta} + \epsilon \frac{\partial V}{\partial \eta} \xi \frac{\partial \dot{\eta}}{\partial \xi} + \epsilon V \frac{\partial \dot{\xi}}{\partial \xi}$$

$$= M(1 + \epsilon V) + \epsilon \frac{\partial g}{\partial \xi}, \tag{3.4.23}$$

and supposing (3.4.16) to be verified for $\xi = 0$, we obtain from (3.4.21), (3.4.22) and (3.4.23),

$$\begin{aligned} \epsilon\omega \frac{\partial u}{\partial \eta} &= \lambda + \epsilon f(\epsilon v, \eta + \epsilon u) \\ \epsilon\omega \frac{\partial v}{\partial \eta} &= \Omega + \epsilon M v + \epsilon g(\epsilon v, \eta + \epsilon u) \\ \epsilon\omega \frac{\partial V}{\partial \eta} &= M(1+\epsilon V) + \epsilon \frac{\partial g}{\partial x}(\epsilon v, \eta + \epsilon u)(1 + \epsilon V). \end{aligned} \tag{3 4.24}$$

It remains to be shown that from (3.4.24) it is possible to obtain functions u, v, V conveniently defining λ, M and Ω. Making use of (3.4.19) and (3.4.20), the terms $O(\epsilon)$ give

$$\begin{aligned} \omega \frac{du_o}{d\eta} &= \lambda_1 + f(0,\eta) \\ \omega \frac{dv_o}{d\eta} &= \Omega_1 + g(0,\eta) \\ \omega \frac{dV_o}{d\eta} &= M_1 + \frac{\partial g}{\partial \xi}(0,\eta). \end{aligned} \tag{3.4.25}$$

In view of our hypotheses,

$$\begin{aligned} f(\xi,\eta) &= \sum_k f_k(\xi) e^{ik\eta} \quad (k \neq 0) \\ g(\xi,\eta) &= \sum_k g_k(\xi) e^{ik\eta} \quad (k \neq 0) \\ \frac{\partial g}{\partial \xi}(\xi,\eta) &= \sum_k g'_k(\xi) e^{ik\eta}. \end{aligned}$$

Setting

$$u_o = \sum_k u_{ok} e^{ik\eta} \quad (k \neq 0)$$

the first of (3.4.25) gives

$$\lambda_1 = -f_o(0) = \frac{1}{2\pi}\int_0^{2\pi} f(0,\eta)d\eta = \text{const.}$$

$$u_{ok} = \frac{f_k(0)}{i\omega k} = \text{const.}\ .$$

In a similar way, we obtain

$$\Omega_1 = -g_o(0) = -\frac{1}{2\pi}\int_0^{2\pi} g(0,\eta)d\eta = \text{const.}$$

$$v_{ok} = \frac{g_k(0)}{i\omega k} = \text{const.}$$

and

$$M_1 = \frac{dg_o(0)}{d\xi} = -\frac{1}{2\pi}\int_0^{2\pi}\frac{\partial g}{\partial \xi}(0,\eta)d\eta = \text{const.}$$

$$V_{ok} = \frac{g'_k(0)}{i\omega k} = \text{const.}\ .$$

In general, it is easily shown that for terms order ϵ^k $(k = 2,3,\ldots)$ the equations have the form (3.4.25), that is, it will result,

$$\begin{aligned}
\eta &= \omega t + \eta_o \\
x &= \epsilon v(\eta,\epsilon) = \epsilon \sum_{k=0}^{\infty}\sum_j \epsilon^k v_{kj} e^{ij(\omega t + \eta_o)} \\
y &= \omega t + \eta_o + \sum_{k=0}^{\infty}\sum_j \epsilon^k u_{kj}\, e^{ij(\omega t + \eta_o)}
\end{aligned} \tag{3.4.26}$$

and, therefore, in view of (3.4.13) we shall obtain a periodic solution $(\xi = 0)$ for (3.4.12), conveniently generalized to take into account the terms introduced in (3.4.17) from (3.4.14). The convergence of the method seems possible under the assumption that $|\omega|$ is bounded away from zero and we suppose the convergence of (3.4.20), (e.g. Moser, 1966).

A typical example is given by the autonomous equations

$$\dot{z}_1 = z_2$$

$$\dot{z}_2 = -\omega^2 z_1 + \epsilon Z(z_1,z_2)$$

where Z is a polynomial in z_1 and z_2 They can be written

$$\dot{z}_1 = z_2$$
$$\dot{z}_2 = (\epsilon\omega_1 - \omega^2)z_1 + \epsilon\omega_2 z_2 + \epsilon P_2(z_1, z_2),$$

where P_2 is a polynomial beginning with terms of at least the second degree.

Assume that the eigenvalues defined by

$$\begin{vmatrix} -\lambda & 1 \\ \epsilon\omega_1 - \omega^2 & \epsilon\omega_2 - \lambda \end{vmatrix} = 0$$

are distinct, that is $\omega^2 \neq \epsilon\omega_1 + \frac{1}{4}\epsilon^2\omega_2^2$. In this case, the linear part can be reduced to a diagonal form by means of a linear transformation

$$z = Bx, \quad |B| \neq 0,$$

provided

$$B^{-1}AB = \Lambda = \begin{pmatrix} \lambda_1 & 0 \\ 0 & \lambda_2 \end{pmatrix},$$

where A is the matrix of the linear part.

The result of the transformation is

$$\dot{x} = \Lambda x + \epsilon B^{-1}\begin{pmatrix} 0 \\ P_2 \end{pmatrix} = \Lambda x + g(x) \tag{3.4.27}$$

and, necessarily, g_k $(k = 1,2)$ are polynomials in (x_1, x_2) beginning with terms of at least the second degree.

We look for conditions under which there exists a transformation

$$x = \Phi(y) = y + \Phi^*(y) \tag{3.4.28}$$

such that (3.4.27) reduces to the linear form

$$\dot{y} = \Lambda y \tag{3.4.29}$$

that is $\dot{y}_k = \lambda_k y_k$, $(k = 1,2)$. We shall suppose Φ_k to be a power series in (y_1, y_2) and Φ_k^* to begin with terms of at least second degree.

We obtain, making use of the foregoing equations,

$$\frac{\partial\Phi}{\partial y}\Lambda y = \Lambda\Phi + g(\Phi) \tag{3.4.30}$$

which is an equation in Φ. In view of the form of g, it is easily verified that the equations

$$\sum_{k=1}^{2}\frac{\partial\Phi_n}{\partial y_k}\lambda_k y_k = \lambda_n\Phi_n = g_n(\Phi_1,\Phi_2) \tag{3.4.31}$$

for $n = 1,2$, give the coefficients of Φ if

$$\sum_{k=1}^{2}\lambda_k p_k - \lambda_j \neq 0 \quad (j = 1,2) \tag{3.4.32}$$

with $p_1 + p_2 \geq 2$, p_1 and p_2 non-negative integers.

If the real parts of λ_k have the same sign, the proof of the convergence is done in the same way as Siegel (1956) proves the convergence of Lyapunov's Theorem on Periodic Solutions, but is not known in general cases.

It is important to note that the above hypotheses are met by Van der Pol's equation

$$\ddot{z} + \omega^2 z = \epsilon(1-z^2)\dot{z}$$

provided

$$\omega > \frac{\epsilon}{2} .$$

5. Linear Periodic Perturbations.

We finally consider the example of a non-autonomous linear perturbation. In particular we deal with the equation

$$\ddot{z} + \omega^2 z = \epsilon F(t)z , \tag{3.5.1}$$

where $F(t) = F(t + 2\pi)$ for all t, and $F(t)$ is analytic for all finite t, with mean value zero. Problems related to the above equation have been studied in detail beginning with the work by Cesari (1940), then Gambill (1954) and Hale (1958), and several others. Here we shall deal with the problem from the point of view of the perturbation techniques of the kind discussed up to this point, but distinct from those of section (2.9).

We introduce the transformation

$$\begin{aligned} z &= \left(\frac{2x}{\omega}\right)^{1/2} \sin y \\ \dot{z} &= (2\omega x)^{1/2} \cos y \end{aligned} \tag{3.5.2}$$

and obtain the system

$$\begin{aligned} \dot{x} &= \frac{2\epsilon}{\omega} x \sin y \cos y\, F(t) \\ \dot{y} &= \omega - \frac{\epsilon}{\omega} \sin^2 y\, F(t) \end{aligned} \tag{3.5.3}$$

already considered in the autonomous case. Next, we consider the transformation

$$\begin{aligned} x &= \xi + \epsilon v(\eta,t,\epsilon) + \epsilon V(\eta,t,\epsilon)\xi \\ y &= \eta + \epsilon u(\eta,t,\epsilon) \end{aligned} \tag{3.5.4}$$

and we show that there exist functions v, V, u periodic of period 2π in η and t, such that (3.5.3) is reduced to the form

$$\dot{\eta} = \Omega(\epsilon,\omega) \tag{3.5.5}$$

for $\xi = 0$. This will allow the determination of Hill's characteristic exponent of the problem. Following the same procedure of the last section, we easily find

$$\epsilon\left(\Omega \frac{\partial u}{\partial \eta} + \frac{\partial u}{\partial t}\right) = \omega - \Omega - \frac{\epsilon}{\omega} \sin^2(\eta + \epsilon u)F(t),$$

$$\epsilon\left(\Omega \frac{\partial v}{\partial \eta} + \frac{\partial v}{\partial t}\right) = \frac{2\epsilon^2}{\omega} v \sin(\eta + \epsilon u) \cos(\eta + \epsilon u)F(t)$$

$$\epsilon\left(\Omega \frac{\partial V}{\partial \eta} + \frac{\partial V}{\partial t}\right) = \frac{2\epsilon}{\omega}(1 + \epsilon V)\sin(\eta + \epsilon u) \cos(\eta + \epsilon u)F(t).$$

From these, supposing,

$$\omega - \Omega = \epsilon\lambda$$

we obtain

$$\Omega \frac{\partial u}{\partial \eta} + \frac{\partial u}{\partial t} = \lambda - \frac{1}{\omega} \sin^2(\eta + \epsilon u)F(t), \tag{3.5.6}$$

$$\Omega \frac{\partial v}{\partial \eta} + \frac{\partial v}{\partial t} = \frac{\epsilon}{\omega} v \sin(2\eta + 2\epsilon u)F(t), \tag{3.5.7}$$

$$\Omega \frac{\partial(1 + \epsilon V)}{\partial \eta} + \frac{\partial(1 + \epsilon V)}{\partial \eta} = \frac{\epsilon}{\omega}(1 + \epsilon V)\sin(2\eta + 2\epsilon u)F(t). \tag{3.5.8}$$

Suppose that from (3.5.6) we can obtain λ and u, 2π periodic in t and η. Then, the right hand members of (3.5.7) and (3.5.8) divided by v and $(1 + \epsilon V)$ will be known functions of t and η, 2π periodic in these variables. If v and $1 + \epsilon V$ are subjected to the same initial (or boundary) conditions, it is evident that

$$v = 1 + \epsilon V. \tag{3.5.9}$$

On the other hand, by taking the partial derivative, with respect to η, of (3.5.6), we find

$$\Omega \frac{\partial}{\partial \eta} \left(\frac{\partial u}{\partial \eta}\right) + \frac{\partial}{\partial t} \left(\frac{\partial u}{\partial \eta}\right) = - \frac{1}{\omega} \sin(2\eta + 2\epsilon u)\left(1 + \epsilon \frac{\partial u}{\partial \eta}\right)F(t)$$

that is

$$\Omega \frac{\partial}{\partial \eta}\left(1 + \epsilon \frac{\partial u}{\partial \eta}\right) + \frac{\partial}{\partial t}\left(1 + \epsilon \frac{\partial u}{\partial \eta}\right) = - \frac{\epsilon}{\omega}\left(1 + \epsilon \frac{\partial u}{\partial \eta}\right)\sin(2\eta + 2\epsilon u)F(t)$$

which shows that, but by a minus sign, the equation for $1 + \epsilon\, \partial u/\partial \eta$ is of the same form that (3.5.7) or (3.5.8). In particular, it follows necessarily that

$$\ell n\left(1 + \epsilon \frac{\partial u}{\partial \eta}\right) + \ell n\, v = f(\eta - \Omega t)$$

where f is arbitrary. Let $f = \ell n\, g(\eta - \Omega t)$, so that

$$\left(1 + \epsilon \frac{\partial u}{\partial \eta}\right)v = g(\eta - \Omega t) \tag{3.5.10}$$

where g is an arbitrary function. If it is possible to find $u(\eta,t)$, then the functions v and $1 + \epsilon V$ are easily obtained, but by an arbitrary function, by purely algebraic operations. If u is 2π periodic in both η and t, and the properties are required for v and V, it will be necessary to choose the function $g(\eta - \Omega t)$ to be a constant, unless Ω happens to be a rational number. This condition , as we shall soon see, has nevertheless to be excluded if u has to be defined from (3.5.6). Therefore, it follows necessarily that

$$v(1 + \epsilon \frac{\partial u}{\partial \eta}) = k(\epsilon,\omega) = \text{const.} \tag{3.5.11}$$

With this, transformation (3.5.4), for $\xi = 0$, takes the form

$$x = \epsilon v(\eta,t,\epsilon)$$

$$y = \eta + \epsilon u(\eta,t,\epsilon)$$

or

$$\begin{aligned} x &= \epsilon v(\Omega t + \eta_0,t,\epsilon) \\ y &= \Omega t + \eta_o + \epsilon u(\Omega t + \eta_o,t,\epsilon) \end{aligned} \tag{3.5.12}$$

where u,v are 2π periodic in both $\eta = \Omega t + \eta_0$ and t. They will not be periodic in t, since it was excluded the case in which Ω is a rational number. Taking (3.5.2) and (3.5.12) into account, it is easily seen that

$$z = \alpha(t) \cos \Omega t + \beta(t) \sin \Omega t$$

so that Ω is, in fact, Hill's characteristic exponent. The application of this method to Hill's equation by Giacaglia (1968) has shown its convenience over the classical procedure of evaluating an infinite determinant.

It remains, therefore, to show that (3.5.6) defines $u(\eta,t)$, 2π periodic in both η and t, and the constant $\lambda = \lambda(\Omega,\omega,\epsilon)$. In fact, consider

$$u = u_0 + \epsilon u_1 + \epsilon^2 u_2 + \ldots$$

$$\lambda = \lambda_0 + \epsilon\lambda_1 + \epsilon^2\lambda_2 + \ldots,$$

and

$$u_k = u_k(\eta,t) = u_k(\eta + 2\pi,t) = u_k(\eta,t + 2\pi)$$

with

$$\int_0^{2\pi} dt \int_0^{2\pi} u_k d\eta = 0.$$

For the coefficient of ϵ^0 we find

$$\Omega \frac{\partial u_o}{\partial \eta} + \frac{\partial u_o}{\partial t} = \lambda_o - \frac{1}{\omega} \sin^2\eta \, F(t) \tag{3.5.13}$$

and, therefore, $\lambda_o = 0$. But, by hypothesis, we can write

$$F(t) = \sum_k F_k \, e^{ikt}, \quad F_o = 0$$

which defines u_o, provided

$$|k\Omega - j| \geq \epsilon |k|^{-p} \tag{3.5.14}$$

with k integer not zero and p a convenient positive number, for all integers j. This will be shown in the next chapter in connection with the results obtained by Moser on area preserving mappings (1962). The problem of convergence will also be dealt with in Moser's Theorem of the next chapter and at every stage k, the function u_k is defined by an equation of the type

$$\Omega \frac{\partial u_k}{\partial \eta} + \frac{\partial u_k}{\partial t} = \lambda_k + \sum_{p,q} A_{pq}^{(k)} e^{i(p\eta+qt)} \tag{3.5.15}$$

and we define

$$\lambda_k = -A_{00}^{(k)}$$

which is typical of an averaging method. It will result

$$\lambda = \epsilon\lambda_1 + \epsilon^2\lambda_2 + \ldots = \epsilon G(\epsilon, \Omega, \omega)$$

that is

$$\omega - \Omega = \epsilon^2 G(\epsilon, \Omega, \omega)$$

which gives

$$\Omega = \Omega(\epsilon, \omega) = \omega + \epsilon^2\Omega_2(\omega) + \epsilon^3\Omega_3(\omega) + \ldots$$

as the characteristic exponent.

If the equation (3.5.1) is taken in an n-dimensional space, that is

$$\ddot{z} + Az = \epsilon B(t) z$$

where $z = \mathrm{col}(z_1, z_2, \ldots, z_n)$, A a constant $n \times n$ matrix and $B(t)$ an $n \times n$ periodic matrix, the process above indicated can be applied as long as the eigenvalues of A are all distinct (and therefore none is zero). The essential difference, to be noted, is the fact that the equation corresponding to (3.5.6) will not be an equation for the unknown u_k only, but for all other components of u and also of the vector v. This fact, however, does not prevent the determination of u, v, simultaneously, by successive approximations and a generalized averaging procedure.

NOTES

The irrationality condition

$$|\omega\cdot k| \geq K\left(\sum_{j=1}^{n} |k_j|\right)^{-s} \tag{A}$$

where $s = n + 1$, can be stated in the following theorem due to Khinchin (1935) and also discussed by Koksma in 1936.

Theorem: "Almost every vector $\omega = (\omega_1,\omega_2,\ldots,\omega_n)$ satisfies the above inequalities, for all integers vectors $k = (k_1,k_2,\ldots,k_n)$ not zero and for a convenient $K(\omega) > 0$".

The proof is quite simple and depends on the fact that (A) does not hold only in resonance regions, of width less than $2K|k|^{-s}$, where $|k| = \Sigma\, |k_j|$, given k,K and ω in a bounded domain $\Omega(\omega)$. This fact contributes to the convergence of the iterations significantly, since, assuming $F(z)$ analytic [Eq. (3.2.20)], the coefficients $f_{(k)}$ decrease exponentially, that is, for some positive real numbers M,ρ

$$|f_{(k)}| \leq Me^{-|k|\rho}$$

and, therefore, from (3.2.22),

$$\left|\frac{f_{(k)}}{(k\cdot\omega)}\right| \leq \frac{ML}{K\delta^p}\, e^{-|k|(\rho-\delta)}$$

where $p = 2n + 3$ and for some convenient constants L,δ. This leads to the convergence of S, the generator of Poincare's Method, in a ring domain $\Gamma(\rho-\delta)$. Kolmogorov suggested in 1954 the following approach to the question of convergence of the successive applications of the canonical transformation defined by S. We look for an invariant torus $T(\omega^*)$ of the perturbed system, and the motion on this is quasi-periodic with frequencies $\omega^* = (\omega_1^*,\omega_2^*,\ldots,\omega_n^*)$, given a priori, and satisfying (A). $T(\omega^*)$ is in the neighborhood of the corresponding torus of the unperturbed system defined by $H_o(x)$, that is,

$$x^* = x + O(\mu),$$

$$\omega^* = \partial H_o/\partial x\Big|_{x=x^*} .$$

In that neighborhood, one can introduce new variables $(x';y')$ by means of an analytic canonical transformation generated by S, and the Hamiltonian takes the form

$$H(y;x) = H^{(1)}(y';x') = H_o^{(1)}(x') + H_1^{(1)}(y';x')$$

where $|H_1^{(1)}| \simeq |H_1|^2$, giving rise to quadratic convergence of Newton's type, as discussed by <u>Nash</u> and <u>Kantorovich.</u> This approach changes the convergence picture of the successive iterations indicated after Equation (3.2.1) and where only linear contraction is obtained, as in the classical approach to Poincaré's Method. The estimates of Lemma 1 can be obtained by a majorization process but, as Moser (1967) indicates, estimation by majorant series will not lead to convergence. His estimate leads to a majorant series $\sum_p (p!)^{2s}\mu^p$ which diverges for all $\mu > 0$. In the modified Newton method proposed by <u>Kolmogorov</u> the precision increases with an exponent 2 and the previous series are substituted by $\sum_p (p!)^{2s}\mu^{2^p}$ which converges for $\mu > 0$, sufficiently small. As indicated by <u>Kolmogorov</u> himself, the Theorem is directly applicable to some classical problems of dynamics, as for instance:

a) The motion of a point on an analytic surface which deviates little from a surface of revolution or from an ellipsoid.

b) The motion (planetary case) of an asterioid under the gravitational attraction of the sun and Jupiter, in two dimensions. Although for this problem the zero order Hamiltonian (in a uniformly rotating system) is of the form

$$H_o = 1/x_1^2 + \alpha x_2$$

and, therefore, formally degenerate, in the sense that $\det\{\partial^2 H_o/\partial x_i \partial x_j\} = 0$, the modified condition

$$\det\begin{Bmatrix} \frac{\partial^2 H_o}{\partial x_i \partial x_j} & \frac{\partial H_o}{\partial x_i} \\ \frac{\partial H_o}{\partial x_j} & 0 \end{Bmatrix} \neq 0$$

is satisfied, and Kolmogorov's Theorem applies. It is interesting to observe that if the above condition is verified there exists a function $F = \Phi(H) = \Phi(H_o) + \epsilon \cdots$ such that $\Phi(H_o)$ is not degenerate in the classical sense. In the current example, for instance, one can take

$$F = H^2 = (\frac{1}{x_1^2} + \alpha x_2)^2 + \cdots = F_o + \cdots$$

and it results $\det\{\partial^2 F_o/\partial x_i \partial x_j\} \neq 0$.

c) The stability of equilibrium and periodic solutions in systems with two degrees of freedom, in the elliptic case. An important application is to the circular restricted problem of three bodies (Arnol'd, 1961; Leontovich, 1962). Actually, a better solution to this last question was given by Deprit in 1967, following the important works of Moser (1958) and Gelfand-Lidskii (1958). In the vicinity of such an elliptic point, the Hamiltonian can be written

$$H = \sum_{k=1}^{2} \frac{1}{2} \omega_k (x_k^2 + y_k^2) + H_3 + \cdots .$$

In cases where $\omega_1\omega_2 > 0$, stability is guaranteed by the fact that the quadratic part H_2 of H is positive definite, although such a condition is only sufficient. To prove stability in the case $\omega_1\omega_2 < 0$, one reduces H to normal form up to the fourth degree, which, according to Birkhoff's results, can be done if $j_1\omega_1 + j_2\omega_2 \neq 0$ for $0 < |j_1| + |j_2| \leq 4$, or if $|\omega_1/\omega_2| \neq p/q$, with $p,q = 1,2,3,4$. The truncated resulting normal form is therefore

$$H = \sum_{j=1}^{2} \frac{1}{2} \omega_j (x_j^2 + y_j^2) + \sum_{k=1}^{2} \sum_{\ell=1}^{2} \frac{1}{4} \beta_{k\ell}(x_k^2 + y_k^2)(x_\ell^2 + y_\ell^2) + O(5).$$

Then, Arnol'd's theorem states that: "If for the system generated by H the determinant

$$\begin{vmatrix} \beta_{11} & \beta_{12} & \omega_1 \\ \beta_{21} & \beta_{22} & \omega_2 \\ \omega_1 & \omega_2 & 0 \end{vmatrix} \neq 0,$$

the equilibrium point $x_k = y_k = 0$ $(k = 1,2)$ is stable". Such a condition is strictly the general non-degeneracy condition put forth by Moser and Arnol'd, the first in the problem of existence of invariant curves in the perturbed twist mapping, the second in proving Kolmogorov's Theorem on the conservation of quasi-periodic motions. The extension of the above theorem to higher dimensions, in the same way, is not possible. The geometric reason underlying this fact is that the gaps between tori of dimensions larger than two are not generally bounded regions. In fact, one can give examples where an elliptic point of equilibrium is unstable.

The extension of Kolmogorov's results to degenerate systems was given in at least two simple examples by Arnol'd, in the classical problem of generation of quasi-periodic solutions from periodic unperturbed orbits (1961) and for the degenerate problem of interaction between two planets (1962). In the former, he considers the motion of a point (y_1, y_2) on a torus T_2. Such motion is quasi-periodic if

$$\frac{dy_1}{dy_2} = \lambda = \text{irrational.} \tag{B}$$

A neighboring system (perturbed) of differential equations on the torus can be written as

$$\frac{dy_1}{dy_2} = \lambda + \alpha + \epsilon f(y_1, y_2) \tag{C}$$

where α, ϵ are parameters, and $f(y_1, y_2)$ is supposed to be analytic. Kolmogorov's Theorem implies that if the perturbation $\epsilon f(y_1, y_2)$ is small enough, then one can find an $\alpha = \alpha_f(\epsilon)$ such that (C) is brought into form (B) by a proper variable transformation. This was shown by Arnol'd in 1961 (Izv. Akad. Nauk). The question of degeneracy reflects in this case when $\lambda = 0$ (or rational), so that the unperturbed motion is periodic and along the circles y_1 = constant. The reduction of (C) to form (B), in the irrational case, uses the fact that $n\lambda + m$ can be bounded away from zero by means of

$$|n\lambda+m| > K|\lambda|n^{-2} \tag{D}$$

for all integers m,n and $n \neq 0$. <u>Arnol'd</u> shows that if $\Lambda(K)$ is the set of all λ satisfying (D) and Λ the union of all $\Lambda(K)$ for $K > 0$, then the limit point of $\Lambda(K)$ for K in $(0,1/4)$ is zero and zero is also an accumulation point for Λ. The two basic theorems of Arnol'd are the following:

<u>Theorem 1:</u> "On the torus T_2, let there be given the differential equation

$$\frac{dy_1}{dy_2} = \epsilon f(y_1,y_2) \tag{E}$$

with ϵ a parameter and f analytic. Let $y_1 + 2\pi$ and $y_2 + 2\pi$ correspond to y_1 and y_2, and

$$\int_0^{2\pi} f(y_1,y_2)dy_2 > 0$$

for all y_1. Then for all sufficiently small $\lambda \in \Lambda(K)$ one can find an $\epsilon(\lambda)$ and a coordinates transformation $z = z_\lambda(y_1,y_2)$, analytic in y_1 and y_2, such that (E) takes the form

$$\frac{dz}{dy_2} = \lambda.$$

The set $\epsilon(\lambda)$, $\lambda \in \Lambda$, has positive measure and zero is its accumulation point".

One applies to system (E) the classical averaging procedure and reduces it to form

$$\frac{dy_1^{(1)}}{dy_2} = \epsilon c + \epsilon^2 F(y_1^{(1)},y_2,\epsilon). \tag{F}$$

After such reduction is performed, one considers the following:

Theorem 2: "Theorem 1 is valid for the equation (F) on T_2, provided c is constant and $F(y_1, y_2, \epsilon)$ is analytic".

It is seen that quadratic convergence is obtained. Eventually it will result that

$$\lambda = \epsilon c + \epsilon^2 c + \cdots$$

and after n applications of the averaging process one obtains

$$\frac{dy_1^{(n)}}{dy_2} = \epsilon c + \epsilon^2 c + \cdots + e^{2^n} F(y_1^{(n)}, y_2, \epsilon).$$

In general, if $|\epsilon^2 F| < M < |\epsilon c| \delta^4$, the new F, after one transformation, satisfies

$$|\epsilon^2 F_{new}| < M_1 = M^2/\delta^4 |\epsilon c|.$$

One can choose $\delta > 0$ such that

$$|\epsilon^2 F_{new}^{(n)}| < M_n = M^{4n/3}$$

which ensures convergence. As an example, consider the equation

$$\frac{dy}{dx} = \epsilon f(x,y)$$

and the average

$$f_s = \frac{1}{2\pi} \int_0^{2\pi} f(x,y) dx = f_s(y) > 0,$$

this condition being verified for $0 \leq y < 2\pi$. Also, define $f_p = f - f_s = f_p(x,y)$, and consider the change of variable

$$y_1 = y + \epsilon h(x,y).$$

It follows that

$$\frac{dy_1}{dx} = \epsilon f_s(y) + \epsilon f_p(x,y) + \epsilon \frac{\partial h}{\partial x} + \epsilon^2 \frac{\partial h}{\partial y} f(x,y)$$

and we choose

$$h(x,y) = -\int_0^x f_p(\xi,y)d\xi .$$

Therefore,

$$\frac{dy_1}{dx} = \epsilon f_s(y_1) + \epsilon^2\varphi(x,y_1,\epsilon).$$

Lastly, one defines the new variable

$$y_2(y_1) = \int_0^{y_1} \frac{cd\xi}{f_s(\xi)}$$

where the constant c is defined by the condition $y_2(2\pi) - y_2(0) = 2\pi$, that is,

$$\int_0^{2\pi} \frac{cd\xi}{f_s(\xi)} = 2\pi.$$

The condition $f_s(y) > 0$ is of evident significance. In terms of y_2, one finally obtains

$$\frac{dy_2}{dx} = \epsilon c + \epsilon^2\psi(x,y_2,\epsilon).$$

The process is repeated, and one finds

$$\frac{dy_3}{dx} = \epsilon c' + \epsilon^4\theta(x,y_3,\epsilon)$$

and so forth. Several more examples could be given which make use of the ideas contained in Kolmogorov's Theorem. The form we have assumed for H, given by (3.2.3), is not less general than that assumed by Kolmogorov, while the conditions of non-degeneracy can be made more evident and are reminiscent of the development of the Hamiltonian in the vicinity of an equilibrium or a periodic solution. Actually, the reduction to a system possessing invariant manifolds which are tori is obviously similar to the normalization of H in the sense of Birkhoff. That a majorization process might not be effective in the proof of convergence of the sequence of canonical transformations defined in Lemma 1, is not a major issue and, in fact, the majorization can be avoided as was done by Arnol'd and Moser. A much more important issue to be considered is that, normally, problems are degenerated and therefore neither Kolmogorov's nor Moser's Theorems, this to be discussed in the next chapter, can be applied. If n is the number of degrees of freedom, the absence of $m < n$ action variables in the unperturbed Hamiltonian is equivalent to the presence of m linear relations between the n non-zero frequencies defining a corresponding unperturbed system. The example just considered represents the first successful approach to the question, although previous studies by Arnol'd on the mapping of the circle onto itself, earlier in 1961, already contained the ideas necessary to solve the question in full generality, as he did in 1963.

For a system with n degrees of freedom, the existence of a single linear integral relation among the unperturbed frequencies, corresponds to the presence of a single type small divisor in the classical development of Poincaré's Method. The existence of $n - 1$ such relations correspond to a periodic unperturbed motion. In both cases one can show the existence of quasi-periodic motions under analytic and sufficiently small perturbations. The absence of a certain number of action variables in the unperturbed motion is called proper degeneracy by Arnol'd and such cases, except the linear one, is the most common in Physics. For cases of proper degeneracy one has the basic results obtained by Arnol'd, valid in general, that is, whatever the number of actions absent in H_o may be. In his work on the Classical Theory of Perturbations, in 1962, he produced an example where all the difficulties known to jeopardize the convergence of Classical Theories, like the Lindstedt-Poincaré Method, are encountered together, that is, the problem of mutual interaction of two planets with irrational mean motions (mean angular velocity about the sun). The explicit case of rational mean motions can be treated equivalently, and is actually found in the work by Moser in 1967. The problem has, in the plane case, four degrees of freedom, the Keplerian (zeroth order) Hamiltonian depends only on two momenta and the assumption of close to rational frequencies is also assumed, as close as allowed by condition (A). One has proper degeneracy and also that introduced by the transition from circular to eccentric orbits.

By defining Lagrangian Motion the one described by a slowing rotating ellipse in a plane, with short periodic small amplitude oscillations on the semi-major axis, accentricity and pericenter longitude, Arnol'd proves the following. Consider two planets moving in the same plane around the sun, and their common center of mass to be stationary, and let a_k and e_k $(k = 1,2)$ be the semi-major axes and eccentricities. Define a domain $D(\delta)$, in the eight dimensional phase-space, by $0 < c_k < a_k < C_k$; $e_k < \delta$; $k = 1,2$. And let $m_k = \mu \alpha_k$, $k = 1,2$, α_k = constant, be the masses.

Theorem: "For an arbitrary $\eta > 0$, there exists an $\epsilon > 0$ such that if $\mu, \delta < \epsilon$, the majority of points of $D(\delta)$, except a set of measure smaller than η mes $D(\delta)$, move in such a way that:

1) The point always remains in $D(\delta)$.
2) It has a quasi-periodic motion, on an analytic 4-dimensional torus in $D(\delta)$.
3) It always remains closer than η from a point of the phase-space which represents a certain Lagrangian Motion."

This is actually a stability result in the sense that, since the exceptional set above is connected, everywhere dense and not bounded, the motion is topologically unstable. Arnol'd also based such conclusion in the observational evidence of existence of gaps in the distribution of minor planets. In the absence of linear integer dependence between the frequencies of the slow varying variables (the planetary mean longitudes), an averaging method with respect to such variables is possible according to Kolmogorov's Theorem. The Hamiltonian is reduced to the form $F = F_0(L) + \mu F_1(L;\xi;\eta) + O(\mu^2)$ where $L = (L_1, L_2)$ are the average actions corresponding to the mean longitudes and

$$\xi_k + i\,\eta_k = e_k \exp(i\omega_k)$$

for $k = 1,2$, where ω_k is the longitude of perihelion of planet k. The Hamiltonian is then reduced to a (truncated) normal form in the vicinity of the stable equilibrium solution $\xi_k = \eta_k = 0$, that is,

$$F = F_2(r) + F_3(r,\theta)$$

where

$$r = (r_1, r_2),\ \theta = (\theta_1, \theta_2),\ F_3 = O(r^3)$$

and

$$\mu\, F_2 = \nu_1 r_1 + \nu_2 r_2 + c_{11} r_1^2 + 2c_{12} r_1 r_2 + c_{22} r_2^2 \ .$$

Therefore, in terms of the new canonical variables $(r;\theta)$,

$$F = F_0(L') + \mu F_2(L',r) + O(\mu^2, r^3)$$

while $r_k = O(e_k^2)$. The frequencies ν_1, ν_2 are both $O(\mu)$, that is, they correspond to slow varying angles θ_k. The quasi-periodic solutions are obtained using a Newton type iterative procedure, where quadratic convergence is achieved. The original form of Arnol'd's Theorem for degenerate systems, a generalization of the above result, can be given as follows. Consider the Hamiltonian to have the form

$$H = H_0(x_1, x_2, \ldots, x_k) + \epsilon H_1(y_1, y_2, \ldots, y_2;\ x_1, x_2, \ldots, x_n),$$

where $k < n$, H_1 is 2π-periodic with respect to each y_i and analytic for $x \in D$ and $|\mathrm{Im}\ y| < \zeta$. For $\epsilon = 0$, we assume the motion to be quasi-periodic and defined by

$$\dot{y}_i = \frac{\partial H_0}{\partial x_i} = \omega_i,\quad \dot{x}_i = 0 \quad (i = 1,2,\ldots,k)$$

and

$$\dot{y}_j = \dot{x}_j = 0 \qquad (j = k+1, k+2, \ldots, n).$$

Under the condition that the average of H_1 with respect to $y_1, y_2, \ldots, y_k$ does not depend on $y_{k+1}, y_{k+2}, \ldots, y_n$, that is,

$$\int_0^{2\pi} \cdots \int_0^{2\pi} H_1(y;x)dy_1 dy_2 \cdots dy_k = \overline{H}_1(x),$$

it is possible to show that for ϵ sufficiently small, for the majority of initial conditions (as in the previous theorem, the exceptional cases form a set connected, everywhere dense and unbounded), the motion generated by H differs little, for all time, from the quasi-periodic motion defined by the frequencies $\dot{y}_j = \partial\overline{H}/\partial x_j = \omega_j$, where $x_j = \text{const.}$, $j = 1,2,\ldots,n$ and $\overline{H}(x) = H_0(x) + \epsilon\overline{H}_1(x)$. The initial conditions x must be such that

$$|j_1\omega_1 + j_2\omega_2 + \cdots + j_k\omega_k| > K(\omega)\left(\sum_{\ell=1}^{k} |j_\ell|\right)^{-k-1},$$

for a convenient constant K. More precisely one has the following theorem.

Theorem: "Let $x \in D$ where the matrices $\{\partial^2 H_0/\partial x_i \partial x_j\}$ $(i,j = 1,2,\ldots,k)$ and $\{\partial^2 H_1/\partial x_i \partial x_j\}$ $(i,j = k+1,k+2,\ldots,n)$, are not singular. Let T be the toroidal domain $\{\text{Im } x = \text{Im } y = 0,\ x \in D,\ 0 \leq y_i < 2\pi,\ i = 1,2,\ldots,n\}$. Given $\eta > 0$ arbitrary, there exists an $\epsilon_0 > 0$ such that if $|\epsilon| < \epsilon_0$, then in T there are analytic invariant n-dimensional tori and the motion of these is quasi-periodic. The tori form a nowhere dense set in T and its complement has measure $< \eta$ mes T." This is a simpler though equivalent form of Arnol'd's complex theorem presented in section 3 of this chapter. It is worth mentioning that the preliminary theorem on planetary motion given previously is important since it produces the solution to a case where two different kinds of degeneracy are present. The limiting degeneracy $r_k = 0$ (or $e_k = 0$), corresponding to circular orbits and the proper degeneracy $\mu = 0$, introducing fewer frequencies necessary to describe the unperturbed system than there are for the perturbed one.

For the general linear systems considered in section 4 (Eq. 3.4.17) and 5 (Eq. 3.5.3), early results were obtained by Bogoliubov in 1963 ("On Quasi-Periodic Solutions in Nonlinear Problems of Mechanics", Publ. Akad. Nauk Ukrain. SSR, 1963) and reported by Mitropolski in 1964. New results were obtained by Moser in 1966 and augmented in a beautiful work by the same author in 1967, on "Convergent Series Expansions for Quasi-Periodic Motions". A theorem by Bogoliubov (1945) stated that "Under the hypothesis $\text{Re}\ \Omega_k \leq -\gamma < 0$ and for sufficiently small ϵ, the system

$$\dot{y} = \Omega y + \epsilon g(t;y;\epsilon) \tag{G}$$

where $\Omega = \text{diag}(\Omega_1, \Omega_2, \ldots, \Omega_n)$, possesses an almost periodic solution if $g(t;y;\epsilon)$ is almost periodic. If $g(t;y;\epsilon)$ is quasi-periodic with basic frequencies $\omega_1, \omega_2, \ldots, \omega_n$ and analytic, so is the solution."

Also, as we have seen earlier in these Notes (Equation C, Theorem 1), the system

$$\dot{x} = \omega + \lambda + \epsilon f(x,\epsilon,\lambda) \tag{H}$$

admits solutions, for $\lambda = \lambda(\epsilon)$, in the form

$$x = \omega t + c + \epsilon u(\omega t + c, \epsilon)$$

where c is a convenient constant vector $c = (c_1, c_2, \ldots, c_n)$, as shown by Arnol'd.

The main theorem of Bogoliubov, puts together these two kinds of systems and can be stated as follows.

Theorem (Bogoliubov): "If in the family of differential equations

$$\begin{aligned} \dot{x} &= \omega + \lambda + \epsilon f(x,y,\lambda,\epsilon) \\ \dot{y} &= \Omega y + \epsilon g(x,y,\lambda,\epsilon) \end{aligned} \tag{I}$$

we assume that

$$\begin{aligned} |j \cdot \omega| &\geq \gamma |j|^{-\tau} \\ \text{Re}\ \Omega_k &\leq -\gamma < 0, \end{aligned} \tag{J}$$

then there exists a $\lambda = \lambda(\epsilon)$ such that the system possesses an n-parameter family of quasi-periodic solutions

$$x = \omega t + c + \epsilon u(\omega t+c, \epsilon)$$
$$y = \epsilon v(\omega t+c, \epsilon),$$

where all the functions involved are analytic in all arguments". The vector x is supposed n-dimensional and y is m-dimensional. Moser, (1966) has shown the validity of such results in cases f and g are differentiable, transforming the problem to that of a flow on a torus. But he goes further and proves the following theorem.

Theorem: "Under the conditions

$$|i(j\cdot\omega) - \Omega_k| \geq r|j|^{-\tau}$$
$$|i(j\cdot\omega) - \Omega_k + \Omega_\ell| \geq r|j|^{-\tau}$$
$$k, \ell = 1,2,\ldots,n; \; j = (j_1, j_2, \ldots, j_n),$$

there exist $\lambda = \lambda(\epsilon)$, $\mu = \mu(\epsilon)$, $M = M(\epsilon)$ analytic in ϵ, vanishing at $\epsilon = 0$ and satisfying

$$\Omega^*\mu = 0, \quad M\Omega^* - \Omega^*M = 0,$$

where $*$ indicates adjoint, such that the system

$$\begin{aligned} \dot{x} &= \omega + \lambda + \epsilon f(x,y,\epsilon,\lambda,\mu) \\ \dot{y} &= \Omega y + \mu + My + \epsilon g(x,y,\epsilon,\lambda,\mu), \end{aligned} \qquad (K)$$

where x is n-dimensional, y is m-dimensional, f,g are analytic in all arguments, and 2π-periodic in $x_1, x_2, \ldots, x_n$, possesses a quasi-periodic solution with the characteristic numbers ω, Ω."

That is, there exists an analytic transformation

$$\begin{aligned} x &= \xi + \epsilon u(\xi,\epsilon) \\ y &= \eta + \epsilon v(\xi,\epsilon) + \epsilon V(\xi,\epsilon)\eta \end{aligned} \qquad (L)$$

such that the above system (K) becomes

$$\dot{\xi} = \omega + O(\eta)$$

$$\dot{\eta} = \Omega\eta + O(\eta^2).$$

Hill's Equation.

As an example, for the specific case we have dealt with in section 5, consider Hill's equation

$$\ddot{z} + \omega^2 z = 2\epsilon^2 z \sum_{k\geq 1} \alpha_k \cos 2k\tau \tag{M}$$

where

$$\omega = 1 + \epsilon - \frac{3}{4}\epsilon^2 + \frac{3}{4}\epsilon^3 + O(\epsilon^4).$$

We introduce the change of variables

$$z = \sqrt{2x/\omega}\ \sin y$$

$$\dot{z} = \sqrt{2\omega x}\ \cos y$$

and obtain

$$\begin{aligned} \dot{x} &= \epsilon^2 \frac{x}{\omega} \sum_{k\geq 1} \alpha_k[\sin 2(y+k\tau) + \sin 2(y-k\tau)] \\ \dot{y} &= \omega - \epsilon^2 \frac{1}{2\omega} \sum_{k\geq 1} \alpha_k[2 \cos 2k\tau - \cos 2(y+k\tau) - \cos 2(y-k\tau)]. \end{aligned} \tag{N}$$

We now look for a coordinate transformation of type (L), that is,

$$x = \xi + \epsilon X(\eta,\tau,\epsilon) + \epsilon\tilde{X}(\eta,\tau,\epsilon)\xi$$

$$y = \eta + \epsilon\ (\eta,\tau,\epsilon).$$

and reduce (M) to the form

$$\dot{\xi} = O(\xi^2)$$

$$\dot{\eta} = \Omega + O(\xi),$$

where, for some λ,

$$\Omega = \omega - \epsilon\lambda(\epsilon,\omega)$$

and

$$|j\Omega - k| \geq \epsilon' |j|^{-\alpha}$$

for ϵ', α conveniently chosen.

In fact, Siegel (1956) has shown that for $0 < \Omega < 1$, the set of Ω's which do not satisfy the above conditions $(j \neq 0,\ k,j$ integers) has measure less than $2\pi\epsilon'(1+\epsilon')/3$.

We assume

$$\lambda = \sum_{k \geq 0} \lambda_k \epsilon^k$$

$$Y = \sum_{k \geq 0} Y_k \epsilon^k$$

$$Y_k = \sum_{p,q} Y_k^{(p,q)} \exp i(p\eta + q\tau)$$

$$X = \left(1 + \epsilon \frac{\partial Y}{\partial \eta}\right)^{-1},$$

according to equations (3.5.4) through the end of section 5. The equation defining Y_k is, for $k = 0,1,2,\ldots,$

$$\Omega \frac{\partial Y_k}{\partial \eta} + \frac{\partial Y_k}{\partial \tau} = \lambda_k + G_k\ (\lambda_0,\ldots,\lambda_{k-1},\eta,\tau) \qquad \text{(P)}$$

and we define λ_k as the average of $-G_k$ with respect to η, τ. In fact, it is easily seen that G_k in (P) assumes the form

$$G_k = \sum_{p,q} G_k^{(p,q)} \exp[i(p\eta + q\tau)]$$

so that, since by hypothesis, Ω is not rational, $\lambda_k = G_k^{(0,0)}$ and

$$Y_k^{(p,q)} = -i\ G_k^{(p,q)}/(p\Omega + q)$$

for all p,q not simultaneously zero. In case of Hill's equation (M) one finds (Giacaglia, 1968)

$$\lambda_0 = \lambda_1 = \lambda_2 = 0$$

$$\lambda_3 = \frac{225\ \Omega}{16\omega^2(\Omega^2 - 1)} \quad .$$

Hence

$$\omega - \Omega = \epsilon\lambda = \frac{225\ \epsilon^4 \Omega}{16\omega^2(\Omega^2-1)} + O(\epsilon^4)$$

or, introducing the value of ω,

$$\Omega = 1 + \epsilon - \frac{3}{4}\epsilon^2 - \frac{201}{32}\epsilon^3 + O(\epsilon^4)$$

which coincides with the expression for c_0 (Brown, 1896, p. 276) obtained via computation of an infinite determinant, by successive approximations, from the dominant (main) diagonal. Here we have used, as in our original paper,

$$\tau = t - t_0$$

$$\epsilon \equiv m = n - n'$$

where t, t_0, n, n' are defined by Brown.

For a discussion of general methods of development of quasi-periodic solutions, by means of convergent series expansions, of nonlinear equations of the type considered above, the work by Moser (1967) is highly suggested. This work, in our opinion, contains most of what has been done in perturbation theory. Systems of variational equations, in the vicinity of equilibrium or periodic solutions, fall in this category and find many interesting answers. As far as cases of resonance, or rationality among the frequencies ω, Ω, they will be dealt with at the end of chapter 5, from this same point of view.

More Applications of Kolmogorov's Theorem.

Besides the examples already mentioned, we end these Notes by indicating the fact that Barrar (1966) made use of Kolmogorov's Theorem to show the existence of quasi-periodic orbits in the case of artificial satellites of the oblate earth. However, he cannot allow the eccentricity of the orbit to go to zero for reasons of limiting degeneracy. Such is possible by the use of Arnol'd's modified approach, as mentioned before. Also, Moser in 1965 applied Kolmogorov's Theorem to the construction of quasi-periodic solutions of Duffing's Equation without damping

$$\ddot{x} + ax + bx^3 = \epsilon f(t,x,\dot{x}) \tag{Q}$$

where f is quasi-periodic in t, with basic frequencies $\omega_1,\omega_2,\ldots,\omega_n$, and real analytic in $x,\dot{x}$. As usual one considers the fact that for $\tau > n - 1$ there exists a constant $K > 0$ such that the conditions

$$|j\cdot\omega + j_0| \geq K|j|^{-\tau}, \qquad j = (j_1,j_2,\ldots,j_n),$$

are satisfied for almost all ω's. Under these conditions, Moser concludes that for f satisfying the foregoing properties and, in addition,

$$f(t,x,\dot{x}) = f(-t,x,-\dot{x}),$$

there exists a real analytic function $\tilde{a}(\epsilon)$ and a quasi-periodic solution $x = \varphi(t,\epsilon)$ with basic frequencies $\omega_1,\omega_2,\ldots,\omega_n$, such that $\tilde{a}(0) = a$, $\varphi(t,0) = 0$. This, eventually, will not produce a solution for equation (Q), since the coefficient of x shall have to be modified to a certain $\tilde{a}(\epsilon) \neq a$. The method developed by Moser is not a mere application of Kolmogorov's Theorem but a novel approach to the specific problem in view, also because it does not require the equations to have a Hamiltonian character. The term bx^3 is considered a perturbation. The equation is written in complex form

$$\dot{z} = i\tilde{a}z + bx^3 + \epsilon f(t,x,\dot{x})$$

where $z = x + i\dot{x}$, and $\tilde{a}/a$ is supposed to vary in $[1 \pm \mu]$ with μ small. One writes

$$y = \tilde{a} - a = O(\mu)$$

so that the equation is written

$$\dot{z} = i(a+y)z + g(\theta,z,\bar{z})$$

where

$$\theta = (\theta_1,\theta_2,\ldots,\theta_n),\ \theta_k = \omega_k t.$$

There exists a coordinate transformation

$$y = \eta + u(\eta)$$

$$z = \zeta + v(\theta,\zeta,\bar{\zeta},\eta)$$

such that the equation is reduced, by successive approximations, to

$$\dot{\zeta} = i(a+\eta)\zeta,$$

which is used to prove the announced result.

REFERENCES

1. Arnol'd, V. I., 1961, "On the Stability of Positions of Equilibrium of a Hamiltonian System in the General Elliptic Case", Dokl. Akad. Nauk 137, 255-257.

2. _______, 1961, "On the Generation of a Quasi-periodic Motion from a Set of Periodic Motions", Dokl. Akad. Nauk 138, 13-15.

3. _______, 1961, "Small Denominators I. Mapping the Circle onto Itself", Izv. Akad. Nauk USSR, Ser. Mat. 25, 21-86.

4. _______, 1962, "On the Classical Theory of Perturbations and the Problem of Stability of Planetary Systems", Dokl. Akad. Nauk 145, 487-490.

5. _______, 1963, "A Theorem of Liouville Concerning Integrable Problems of Dynamics", Sibirsk. Mat. Zh. 4, 471-474.

6. _______, 1963, "Proof of A. N. Kolmogorov's Theorem on the Preservation of Quasi-Periodic Motions under Small Perturbations of the Hamiltonian", Uspekhi Mat Nauk USSR, 18, 13-40.

7. _______, 1963, "Small Denominators and Problems of Stability of Motion in Classical and Celestial Mechanics", Uspekhi Mat. Nauk USSR 18, 91-192.

8. Barrar, R. B., 1966, "Existence of Conditionally Periodic Orbits for the Motion of a Satellite Around an Oblate Planet", Q. of Appl. Math., 24, 47-55.

9. _______, 1966, "A Proof of the Convergence of the Poincaré-von Zeipel Procedure in Celestial Mechanics", Am. J. Math., 88, 206-220.

10. _______, 1970, "Convergence of the von Zeipel Procedure", Cel. Mech., 2, 494-504.

11. Birkhoff, G. D., 1927, "Dynamical Systems", Am. Math. Soc. Colloq. Publ. IX, Providence, R. I.

12. Bogoliubov, N. N., 1945, "On Some Statistical Methods in Mathematical Physics", Izv. Akad. Nauk USSR Publ., Moscow.

13. _______, 1963, "On Quasi-Periodic Solutions in Nonlinear Problems of Mechanics", Publ. Akad. Nauk Ukrain. SSR, Karev.

14. Brouwer, D. and Clemence, G. M., 1961, "Methods of Celestial Mechanics", Acad. Press, New York.

15. Brown, E. W., 1896, "An Introductory Treatise on the Lunar Theory", Cambridge Univ. Press, London (Dover re-public., 1960).

16. Cesari, L., 1940, "Sulla Stabilitá delle Soluzioni dei Sistemi di Equazioni Differenziali Lineari a Coefficienti Periodici", Atti Accad. Ital. Mem. Classe Fis. Mat. Nat. 11, 633-692.

17. Deprit, A. and Deprit-Bartholomé, A., 1967, "Stability of the Triangular Lagrangian Points", Astron. J., 72, 173-179.

18. Gambill, R. A., "Criteria for Parametric Instability for Linear Differential Systems with Periodic Coefficients", Riv. Mat. Univ. Parma, 5, 169-181 (1954).

19. Gelfand, I. M. and Lidskii, V. B., 1958, "On the Structure of the Regions of Stability of Linear Canonical Systems of Differential Equations with Periodic Coefficients", Am. Math. Soc. Transl. (2), 8, 143-182.

20. Giacaglia, G. E. O., 1968, "Transformation of Hill's Equation - A Method of Solution", Astron. J., 72, 998-1001.

21. Hale, J. K., 1958, "Sufficient Conditions for the Existence of Periodic Solutions of First and Second Order Differential Equations", J. Math. Mech., 7, 163-172.

22. Kantorovich, L. V., 1948, "Functional Analysis and Applied Mathematics", Uspekhi Mat. Nauk USSR, 3, 89-185.

23. Khinchin, A. Ya., 1935, "Continued Fractions", O.N.T.I., Moscow.

24. Koksma, J. F., 1936, "Diophantische Approximationen", Springer-Verlag, Berlin.

25. Kolmogorov, A. N., 1954, "General Theory of Dynamical Systems and Classical Mechanics", Proc. Int. Cong. Math. Amsterdam, 1, 315-333 (Noordhoff, 1957).

26. Krylov, N. and Bogoliubov, N. N., 1947, "Introduction to Nonlinear Mechanics", Ann. Math. Stud., 11, Princeton Univ. Press, Princeton.

27. Leontovich, A. M., 1962, "On the Stability of the Lagrangian Periodic Solutions of the Restricted Three-Body Problem", Dokl. Akad. Nauk USSR 143, 525-528.

28. Mitropolski, J. A., 1966, "Problems in the Asymptotic Theory of Non-Stationary Oscillations", Gauthier-Villars, Paris.

29. Moser, J., 1956, "The Resonance Lines for the Synchroton", Proceed. CERN Symp. Geneva, 290-292.

30. _______, 1965, "Combination Tones for Duffing's Equation", Comm. Pure Appl. Math., 43, 167-181.

31. _______, 1965, "The Stability Behavior of the Solution of Hamiltonian Systems", Lecture Notes, S.I.D.A.

32. _______, 1958, "New Aspects in the Theory of Stability of Hamiltonian Systems", Comm. Pure Appl. Math., 2, 81-114.

33. _______, 1962, "On Invariant Curves of Area-Preserving Mappings of an Annulus", Nachr. Akad. Wiss. Göttingen Math.-Phys. Kl. II, 1-20.

34. _______, 1966, "On the Theory of Quasi-Periodic Motions", SIAM Rev., 8, 145-172.

35. _______, 1967, "Convergent Series Expansions of Quasi-Periodic Motions", Math. Ann., 169, 136-176.

36. _______, 1961, "A New Technique for the Construction of Solutions of Nonlinear Differential Equations", Proc. Nat. Acad. Sci., 47, 1824-1831.

37. Nash, J., 1956, "The Imbedding of Riemmanian Manifolds", Ann. Math., 63, 20-63.

38. Rüsman, H., 1959, "Über die Existenz einer Normalform inhaltstreuer elliptischer Transformationen", Math. Annalen, 137, 64-77.

39. Siegel, C. L., 1941, "On the Integrals of Canonical Systems", Ann. Math., 42, 806-822.

40. _______, 1954, "Über die Existenz einer Normalform analytischer Hamiltonscher Differentialgleichungen in der Nähe einer Gleichgewichtslösung", Math. Ann., 128, 144-170.

41. _______, 1956, "Vorlesungen über Himmelsmechanik", Springer-Verlag, Berlin.

42. Pol, B. van der, 1926-27, "Forced Oscillations in a Circuit with Nonlinear Resistance", Phil. Mag., 2, 21-27; 3, 65-80.

43. Whittaker, E. T., 1937, "A Treatise on the Analytical Dynamics of Particles and Rigid Bodies", Cambridge, Univ. Press, London.

CHAPTER IV

PERTURBATIONS OF AREA PRESERVING MAPPINGS

1. Preliminary Considerations.

According to Liouville's Theorem, flow of a conservative system in phase space is measure preserving, that is, the volume is conserved.

Considering in general a system of n ordinary differential equations

$$\dot{x} = f(x,t) \qquad (4.1.1)$$

with $x = (x_1,x_2,\ldots,x_n)$, $f = (f_1,\ldots,f_n)$ we suppose f to be analytic in a given region D of R^n and periodic, period 2π, in t. We also suppose f to be at least C^2 with respect to t, for all t. If $x_o \in D$, exists and is unique the solution

$$x = g(x_o,t) \qquad (4.1.2)$$

such that

$$g(x_o,0) = x_o \ ,$$

$$\frac{\partial g}{\partial t}(x_o,t) = f(g(x_o,t),t).$$

Consider the transformation

$$T:x \rightarrow g(x,2\pi) \qquad (4.1.3)$$

which is certainly 1:1 in the neighborhood of $x = x_o$. We indicate by T^q the transformation obtained by applying T, q times in succession, as for example

$$T^2:x \rightarrow T:g(x,2\pi) \rightarrow g[g(x,2\pi),2\pi].$$

Let $\bar{x} = 0$ be a solution of (4.1.1). Then, evidently, $\bar{x} = 0$ is a fixed point of T^q, q positive integer, that is

$$T^q : \bar{x} = \bar{x}. \tag{4.1.4}$$

The following definitions are worth mentioning here:

a) The point $\bar{x}$ is singular if there exists a neighborhood in D which does not contain any other fixed point of T. Note that, in general, a fixed point of T^q, q positive integer, corresponds to a periodic solution of (4.1.1) with period $2\pi q$. If the point is singular, the periodic solution is singular. The concept given by Whittaker (1960) represents, obviously, a particular case of the above definition.

b) The point $\bar{x}$ is ordinary if in every neighborhood of $\bar{x}$, in D, there exists at least another fixed point. The corresponding periodic solution is then called ordinary.

We will mainly be concerned with area preserving transformations, i.e., those for which, given

$$|J|\, dx_1^* \, dx_2^* \, \ldots \, dx_n^* = dx_1 \, dx_2 \, \ldots \, dx_n$$

where J is the Jacobian

$$|J| = \left| \frac{\partial(x_1, x_2, \ldots, x_n)}{\partial(x_1^*, x_2^*, \ldots, x_n^*)} \right| = \left| \frac{\partial x}{\partial x^*} \right| ,$$

we have $|J| = 1$.

Consider now a Hamiltonian system

$$\begin{aligned} \dot{x}_k &= -H_{y_k} \\ & \qquad\qquad (k = 1, 2, \ldots, n) \\ \dot{y}_k &= +H_{y_k} \end{aligned} \tag{4.1.5}$$

where

$$H = H(x;y;t) = H(x;y;t+2\pi)$$

is analytic in a certain domain D of the phase space, containing the origin, and is at least C^2 in t for all t finite. Suppose $(x = 0, y = 0)$ is an equilibrium point of (4.1.5). Let $(x_o;y_o) \in D$ and let

$$x = f(x_o;y_o;t)$$
$$y = g(x_o;y_o;t)$$

be the corresponding solution, analytic in D. Consider the transformation

$$T\begin{cases} x^* = f(x;y;2\pi) \\ y^* = g(x,y;2\pi) \end{cases} \tag{4.1.6}$$

in the neighborhood of the origin. Obviously, the point $(0;0)$ is a fixed point of T. Since f and g are analytic in a neighborhood of $(0;0)$ they admit the Taylor series

$$x^* = f(0;0;2\pi) + \left.\frac{\partial f}{\partial x}\right|_o x + \left.\frac{\partial f}{\partial y}\right|_o y + \cdots$$

$$y^* = g(0;0;2\pi) + \left.\frac{\partial g}{\partial x}\right|_o x + \left.\frac{\partial g}{\partial y}\right|_o y + \cdots \quad .$$

Taking into account the fact that

$$f_o = g_o = 0$$

and defining

$$z^* = \binom{x^*}{y^*}, \qquad z = \binom{x}{y},$$

$$J = \begin{pmatrix} \frac{\partial f}{\partial x} & \frac{\partial f}{\partial y} \\ \frac{\partial g}{\partial x} & \frac{\partial g}{\partial y} \end{pmatrix} = J(x;y), \quad J_o = J(0;0),$$

we have

$$z^* = J_o z + \cdots \quad . \tag{4.1.7}$$

Evidently, the transformation T is canonical, that is, $|J| = 1$ and moreover J has to be symplectic (with unit multiplier):

$$J^T E J = E$$

where

$$E = \begin{pmatrix} 0_n & -I_n \\ I_n & 0_n \end{pmatrix} .$$

Such properties are evidently valid for J_o and the transformation defined by (4.1.7) is area preserving. Consider the linear mapping

$$\zeta^* = J_o \zeta \tag{4.1.8}$$

and the eigenvalues problem

$$|J_o - \lambda I| = 0,$$

corresponding to an algebraic equation of degree $2n$ in λ. As the independent term is $|J_o|$, no root is zero. Let u_k be an eigenvector corresponding to λ_k, that is

$$J_o\, u_k = \lambda_k\, u_k \ .$$

From this, we obtain successively

$$\begin{aligned} EJ_o\, u_k &= \lambda_k\, Eu_k \ , \\ J_o^T EJ_o\, u_k &= \lambda_k\, J_o^T E\, u_k \\ E\, u_k &= \lambda_k\, J_o^T E\, u_k \\ J_o^T E\, u_k &= \frac{1}{\lambda_k}\, E\, u_k \end{aligned}$$

so that λ_k is also a solution of the characteristic equation

$$|J_o^T - \frac{1}{\lambda_k} I| = 0$$

and since $|J_o^T - \lambda I| = |J_o - \lambda I| = 0$, if λ_k is an eigenvalue, $\frac{1}{\lambda_k}$ also is. Thus

$$\lambda_1 \lambda_2 \dots \lambda_{2n} = 1.$$

Consider all λ_k distinct (real or complex). If there exists a linear transformation

$$\zeta = L\omega$$

such that

$$\omega^* = L^{-1} J_o L\omega = \Lambda\omega$$

where

$$L^{-1} J_o L = \Lambda = \text{diag}(\lambda_1, \lambda_2, \dots, \lambda_{2n}) ,$$

then

$$\omega_k^* = \lambda_k \omega_k$$

and the transformation (4.1.7) can be written

$$Z^* = \Lambda Z + \psi(Z). \qquad (4.1.9)$$

Since (4.1.9) is analytic in the neighborhood of the origin $Z = 0$, we can write

$$X_k^* = \lambda_k X_k + \sum_{m=2}^{\infty} p_m^{(k)}(X;Y)$$

$$Y_k^* = \frac{1}{\lambda_k} Y_k + \sum_{m=2}^{\infty} q_m^{(k)}(X;Y)$$

for $k = 1,2,\dots,n$, where p_m^k and q_m^k are homogeneous polynomials of degree m in $X_1, X_2, \dots, X_n$, $Y_1, Y_2, \dots, Y_n$. In vector form we can also write

$$\begin{aligned} X^* &= \lambda X + \phi(X;Y) \\ Y^* &= (\tfrac{1}{\lambda}) Y + \psi(X;Y) \end{aligned} \qquad (4.1.10)$$

where, necessarily

$$|J| = |\frac{\partial(X^*,Y^*)}{\partial(X,Y)}| = 1$$

and therefore (4.1.10) is area preserving. Let us now write (4.1.10) as

$$\begin{aligned} X_k^* &= X_k U_k \\ & \qquad\qquad (k = 1,2,\ldots,n) \\ Y_k^* &= Y_k V_k \end{aligned} \tag{4.1.11}$$

where

$$\begin{aligned} U_k &= \sum_{m=0}^{\infty} P_m^{(k)}(X;Y) \\ V_k &= \sum_{m=0}^{\infty} Q_m^{k}\ (X;Y) \end{aligned} \tag{4.1.12}$$

and $P_m^{(k)}$, $Q_m^{(k)}$ are homogeneous polynomials of degree m. Of course, we have

$$P_o^{(k)} = \lambda_k \ ,$$

$$Q_o^{(k)} = \frac{1}{\lambda_k} \ .$$

The discussion of fixed points of mappings of the type (4.1.10), with an area preserving property, is of paramount importance in the theory of dynamical systems, but it is out of the scope of these notes. Nevertheless, for reasons of introduction to the following sessions, we wish to mention some facts about properties of transformation in the vicinity of a fixed point (Birkhoff, 1927; Siegel, 1956). The basic theorem states that given an autonomous system

$$\dot{x} = f(x), \qquad x = (x_1,x_2,\ldots,x_n)$$

corresponding to the initial condition $t = 0$, $x = \xi \in D$, $f(x)$ analytic in D, exists and is unique the solution $x = x(\xi,t)$, $x(\xi,0) = \xi$ and $x(\xi,t)$ is analytic in t, for $0 \leq t < T$, the maximal interval. If, in correspondence to $\bar{\xi} \in D$, the solution $x(\xi,t)$ is periodic of period 2π and is contained in D, then the transformation

$$x = x(\xi, 2\pi)$$

has a fixed point $\bar{\xi}$ in D. In general, given a transformation defined in some domain D, the problem is to determine the existence and obtain fixed points of the transformation. Brouwer's and Poincaré-Birkhoff's fixed point theorems are typical examples of such problems. The second has immediate applications to several problems of dynamics since the proper domain is typical of invariant manifolds of dynamics. The theory of perturbations, in this respect, deals with the properties of mappings in the vicinity of fixed points or, in a more general sense, with the behavior of invariant manifolds, corresponding to area preserving maps, under perturbations. The historically crucial problem is the normalization of a mapping in the vicinity of a fixed point, the original work being attributed to Birkhoff (1922) and recent developments to Gustavson (1966). It is well known that the area preserving mapping

$$\begin{aligned} x'_k &= \lambda_k x_k + f_k(x;y) \\ y'_k &= \mu_k y_k + g_k(x;y) \end{aligned} \tag{4.1.13}$$

with $\mu_k = 1/\lambda_k$, cannot be reduced, but for exceptional cases, to the normal form

$$\begin{aligned} \xi'_k &= \lambda_k \xi_k \\ \eta'_k &= \mu_k \eta_k \end{aligned} \tag{4.1.14}$$

by a transformation

$$U = \begin{cases} x_k = \xi_k + \phi_k(\xi;\eta) \\ y_k = \eta_k + \psi_k(\xi;\eta). \end{cases} \tag{4.1.15}$$

Birkhoff's normal form

$$T\begin{cases}\xi_k' = \xi_k u_k(\xi;\eta)\\ \eta_k' = \eta_k v_k(\xi;\eta)\end{cases} \tag{4.1.16}$$

where

$$u_k = \lambda_k + \sum_{j=1}^{n} a_{2,j}^{k}\, \xi_j\eta_j + \sum_{j=1}^{n}\sum_{\ell=1}^{n} a_{4,j,\ell}^{k}\, \xi_j\eta_j\, \xi_\ell\eta_\ell + \cdots$$

$$v_k = \mu_k + \sum_{j=1}^{n} b_{2,j}^{k}\, \xi_j\eta_j + \sum_{j=1}^{n}\sum_{\ell=1}^{n} b_{4,j,\ell}^{k}\, \xi_j\eta_j\, \xi_\ell\eta_\ell + \cdots ,$$

by means of a transformation (4.1.15) again can only be achieved in a purely formal series, generally divergent. It is nevertheless possible to obtain a transformation which reduces (4.1.13) to a form coincident with Birkhoff's normal form up to any degree of approximation, although the remainder does not go to zero as the approximation is extended to infinity. This fact, for systems with one degree of freedom, leads naturally to Birkhoff's Fixed Point Theorem.

Consider now, the origin to be an elliptic fixed point of a transformation, and let the transformation be (4.1.13), and area preserving. The Jacobian matrix of this, J, is sympletic, that is

$$J^T E J = E.$$

If, in particular, we consider the variables transformation (4.1.15), for $k = 1,2,\ldots,$, the derivatives $\partial x_k/\partial\xi_k$, $\partial y_k/\partial\eta_k$ are equal to 1 at the origin $\xi = \eta = 0$, and, therefore, there is a generating function $W(y;\xi)$ such that

$$x_k = W_{y_k} , \qquad \eta_k = W_{\xi_k} . \tag{4.1.17}$$

W is analytic in the neighborhood of $(y = 0;\ \xi = 0)$, and the series

$$W = \sum_k y_k\xi_k + \cdots \tag{4.1.18}$$

converges in that neighborhood. The generator W will represent every transformation of type U. For the reduction to Birkhoff's Normal Form, W is nevertheless generally divergent. In this case, let us consider the function W^* which is obtained by eliminating all terms of degree larger than $2m + 2$ in W, that is

$$W^* = \sum_{k=1}^{n} y_k \xi_k + \sum_{p,q} A_{pq} y_1^{p_1} \cdots y_n^{p_n} \xi_1^{q_1} \cdots \xi_n^{q_n} \tag{4.1.19}$$

where the summation in p,q extends for all p_i, q_i satisfying

$$3 \leq \sum (p_i + q_i) \leq 2m + 2.$$

We have

$$\begin{aligned} x_k &= W^*_{y_k} = \xi_k + \tilde{X}_k(y;\xi) \\ \eta_k &= W^*_{\xi_k} = y_k + N_k(y;\xi) \end{aligned} \tag{4.1.20}$$

and, by hypothesis,

$$\frac{\partial^2 W^*}{\partial \xi_k \partial y_k} = 1$$

at the origin $\xi = y = 0$. Therefore, one can write in a neighborhood of this,

$$\begin{aligned} y_k &= \eta_k + Y_k(\xi;\eta) \\ x_k &= \xi_k + X_k(\xi;\eta) \ . \end{aligned} \tag{4.1.21}$$

Series (4.1.21) are certainly convergent in a neighborhood of the origin and reduce (4.1.13) to a form T^* which coincides with Birkhoff's normal form (4.1.16) up to terms of order $2m + 1$. In order to obtain T^* the sole condition is that $\lambda_j^s \neq 1$ for $j = 1,2,\ldots,n$ and $s = 1,2,\ldots,2m + 2$. If such a condition is not verified, zero divisors are introduced in W^*, precisely at the order $2 \leq k \leq 2m + 2$, for which $\lambda_j^k = 1$, for some j.

Let $m = 2$. Since we have assumed the origin to be an elliptic point,

T^* takes the form

$$\xi'_k = e^{i\Omega_k}\,\xi_k + \phi_k^{(4)}(\xi;\eta)$$

$$\eta'_k = e^{-i\Omega_k}\,\eta_k + \psi_k^{(4)}(\xi;\eta)$$

where $\phi_k^{(4)}$ and $\psi_k^{(4)}$ are terms of order ≥ 4, so that it is sufficient to take

$$\Omega_k = \alpha_k + \sum_{j=1}^{n} \beta_k^j\, \xi_j \eta_j \,.$$

An important condition is that not all β_k^j can be zero. If this happens, the approximation has to be extended to a higher order. If, to any order, this phenomenon is verified, the system is highly singular. This situation can be associated with the problem of resonance in a conservative system (self excitation). Evidently α_k and β_k^j are real and moreover $\eta_k = \bar{\xi}_k$ (the complex conjugate of ξ_k). It follows that $\eta'_k = \bar{\xi}'_k$ so that one needs only to consider the relation

$$\xi'_k = e^{i\Omega_k}\,\xi_k + \phi_k^{(4)}(\xi;\bar{\xi})$$

$$\Omega_k = \alpha_k + \sum_{j=1}^{n} \beta_k^j\, \xi_j \bar{\xi}_j \,, \qquad \lambda_k = e^{i\alpha_k} \,.$$

On the other hand, expanding the exponential

$$e^{i\Omega_k} = \lambda_k(1 + i\delta_k)$$

where terms of higher order are thrown into $\phi_k^{(4)}$, and

$$\delta_k = \sum_{j=1}^{n} \beta_k^j\, |\xi_j|^2 \,.$$

Therefore, in vector form

$$\xi' = \Lambda\xi + \Phi^{(4)} \tag{4.1.22}$$

$$\Lambda = \text{diag}\{\lambda_1(1 + i\delta_1), \ldots, \lambda_n(1 + i\delta_n)\}.$$

For $n = 1$, the transformation satisfies Birkhoff's Fixed Point Theorem. For $n > 1$, an extension of such a theorem is not available. When the number of degrees of freedom is $n = 2$, there are situations where, the system being conservative, the theorem can still be applied. For instance, when the Hamiltonian is periodic in one of the variables.

2. Regions of Motion. Perturbation of a Truncated Birkhoff's Normal Form.

The asymptotic behavior of solutions of Hamiltonian systems consists, in part, in the characterization of the possible regions of motion, invariant sets, corresponding to a certain set of initial conditions. This is also part of the problem of the stability of such solutions. Solutions which are well characterized are, in this case, periodic or quasiperiodic. A restrictive definition, useful to our purposes, is that a function $Z(t) = \Phi(\theta_1, \theta_2, \ldots, \theta_n)$ is quasiperiodic if

a) Φ is periodic in each $\theta_k (k = 1, 2, \ldots, n)$.

b) Φ has certain properties of regularity, e.g., $\in C^p$ or $\in C^\infty$ or is analytic.

c) $\theta_k = \omega_k t + \alpha_k$, $k = 1, 2, \ldots, n$; $\alpha_k, \omega_k = \text{const}$.

d) $\sum_{k=1}^{n} p_k \omega_k = 0$, with $p_1, p_2, \ldots, p_n$ integers, if, and only if, $p_1 = p_2 = \cdots = p_n = 0$.

As we have seen, if $H(x;y)$ is integrable in the Liouville sense then, in general, the motion is quasiperiodic and the pertinent invariant set is an n-dimensional torus parametrized by angle variables $\eta_k = \omega_k t + \alpha_k$, $k = 1, 2, \ldots, n$, with the ω_k linearly independent over the set of integers. It is equivalent to say that there exists a canonical transformation $(x;y) \to (\xi;\eta)$ such that the image of H in the $(\xi;\eta)$ phase space, is only a function of ξ (i.e., $\xi_1, \xi_2, \ldots, \xi_n$). Such a transformation can be produced by a Hamilton-Jacobi generator $S(\xi;y)$ such that

$$\eta_k = S_{\xi_k} , \qquad x_k = S_{y_k}$$

and therefore

$$\xi_k(x;y) = \text{const.} \qquad (k = 1,2,\ldots,n)$$

and

$$\dot{\eta}_k(x;y) = H_{\xi_k}(x(\xi;\eta);y(\xi;\eta)) = \text{const} = \omega_k(\xi)$$

or

$$\eta_k = \omega_k t + \alpha_k .$$

The trajectories remain on an n-dimensional torus. Changing the initial conditions y_k we will only be changing the initial phases $\eta_k(0)$ and the torus is unchanged. Changing the initial values of x_k we will be changing the frequencies ω_k, or the "radii" of the torus, i.e., $\xi_1,\ldots,\xi_n$. In fact, W has to satisfy

$$H(\frac{\partial W}{\partial y_1},\ldots,\frac{\partial W}{\partial y_2}; y_1,\ldots,y_n) = \alpha(\xi_1,\ldots,\xi_n)$$

so that the energy depends only on $\xi_1,\ldots,\xi_n$, and any change in the initial phases α_k does not affect such energy.

Quasiperiodic motions satisfy the following important properties:

1) The trajectory is dense everywhere on the torus. That is, given a domain D, the point $(x(t);y(t))$ is such that there exists a finite $t = \tau$ for which $x(\tau),y(\tau) \in D$.

2) The trajectory is uniformly distributed, that is, the time Δt during which the point stays in a domain D is proportional to the measure of D, for Δt sufficiently large. In another way, for every function $F(y_1,y_2,\ldots,y_n)$ with $y_k = \omega_k t + \alpha_k$, R-integrable on the torus, the average with respect to time is

equal to spacial average:

$$\lim_{T \to \infty} \int_0^T F(\omega_1 t + \alpha_1, \ldots, \omega_n t + \alpha_n)dt =$$

$$= \frac{1}{(2\pi)^n} \int_0^{2\pi} \cdots \int_0^{2\pi} F(y_1, \ldots, y_n) dy_1 dy_2 \cdots dy_n .$$

3) There exist values $\xi_1^0, \ldots, \xi_n^0$ such that for some integers $p_1, \ldots, p_n$ not all zero, one verifies

$$\sum_{k=1}^{n} p_k \left.\frac{\partial H}{\partial \xi_k}\right|_{\xi=\xi^0} = 0.$$

In such cases the solution of the system is periodic. Now, a periodic solution is associated with the existence of a fixed point of an area preserving mapping, so that, the existence of a periodic solution corresponds to the research of such fixed points. Moreover, the same is true with respect to the stability.

As we have seen, in the vicinity of an elliptic point, one can define invariant tori by a possible truncated normal form of Birkhoff. In case of one degree of freedom, the torus can be represented by a circle. The normal form shows that in the vicinity of an elliptic point every such invariant circle of radius r and center at the origin (which is a fixed point), is mapped onto itself by means of a rotation

$$\alpha(r) = \alpha_0 + \alpha_1 r + \alpha_2 r^2 + \cdots .$$

If one considers the perturbation due to terms which where neglected in the reduction, the basic problem is to find out what happens to the invariant circles (or tori). As a consequence of Kolomogorov's Theorem (1954) the majority of the circles are not destroyed but simply deformed. The fixed point is, therefore, enclosed by close, analytic and invariant curves, arbitrarily small, and is therefore, stable. Such curves cover a region of nonzero measure, and have the origin as an accumulation point. But they don't cover completely any arbitrary neighborhood of the origin. In between them, there are zones of instability resulting from circles such that $\alpha(r)$ is commensurable with π (Arnol'd, 1963). What actually happens is that, subjected to the perturbations of higher order, such circles brake into an even number of fixed points, half elliptic and half hyperbolic (Birkhoff, 1927). From the hyperbolic point emanate orbits which are known as wild motions, where expected by Poincaré and have recently been actually generated by Danby (1970).

3. Moser's Theorem.

Consider initially a system with 2 degrees of freedom $(n = 2)$. The quasiperiodic orbits are described by

$$\begin{aligned} y_k &= \omega_k(x_1^o, x_2^o)t + y_k^o \\ x_k &= x_k^o \\ (k &= 1,2). \end{aligned} \tag{4.3.1}$$

The torus T^2 can be visualized in two ways: either as the product of two circles of radii x_1, x_2 or by a square in the plane (y_1, y_2). The second way is simpler and easier to visualize in higher dimensions. As the measure of y_1 and y_2 can be made modulus 2π, we can identify points of opposite sides of a square of side 2π, that is, $(0, y_2)$ with $(2\pi, y_2)$ and $(y_1, 0)$ with $(y_1, 2\pi)$. This way the torus has the classical definition through a relation of equivalence. The trajectories are line segments in the square, and the slope to the y_1 axis is ω_2/ω_1. It is evident that if ω_2/ω_1 irrational the trajectory cover the square and is quasiperiodic. (Ergodic flow). In any event, the solutions remain in the torus, which is, therefore, said invariant with respect to the flow.

If we assume Kolmogorov's hypothesis that

$$\det \left| \frac{\partial^2 H_o}{\partial x_j \partial x_j} \right| \neq 0,$$

then it follows that with $H = H_o + \mu H_1$ analytic and μ sufficiently small, and for all ω_1, ω_2 rationally independent and satisfying the inequality $|m_1\omega_1 + m_2\omega_2| \geq K$ for all integers m_1, m_2 not simultaneously zero, there exist invariant tori

$$x_k = A_k(Y_1, Y_2) \qquad\qquad (k = 1,2)$$
$$y_k = Y_k + \phi_k(Y_1, Y_2)$$

and the solutions on each torus are given by $Y_k = \omega_k t + Y_k^o$. We can express this fact by saying that the ergodic flow can be "continued" under perturbations.

The reduction of this problem to an area preserving mapping of a set into itself, presents advantages which follow from the geometric structure of such transformation. Forerunners along geometric lines are certainly Poincaré (1912), Birkhoff (1927) and Moser (1955, 1962, 1964).

Consider Moser's problem on the area preserving transformation of a circular ring into itself. Given a circular ring $A\{0 \leq a \leq r \leq b\}$, consider the transformation

$$U_o \begin{cases} \theta^* = \theta + \omega(r) \\ r^* = r \end{cases} \qquad (4.3.2)$$

with

$$\frac{d\omega}{dr} > 0, \qquad r \in A.$$

Such transformation does not change circles but simply introduces a rotation of an angle $\omega(r)$, increasing with r. Consider now the perturbed transformation

$$U \begin{cases} \theta^* = \theta + \omega(r) + F(r,\theta) \\ r^* = r + G(r,\theta) \end{cases} \qquad (4.3.3)$$

with

$$|F(r,\theta)| < \omega(r) \quad \text{for all} \quad \theta$$

$$|G(r,\theta)| < r$$

for $r \in A$, and F,G periodic (or period 2π) in θ. Then it can be shown that, under certain conditions for $\omega(r)$, F, G and with U_o and U area preserving, the transformation U (twist mapping) has closed invariant curves which are near the invariant circles of U_o. One of the essential conditions is the necessity of ω to be irrational with respect to π. The connection of this result with conservative systems of one degree of freedom is evident. In fact, the invariant sets of an integrable one dimensional Hamiltonian system are circles.

Let us now consider a system with two degrees of freedom, generated by the Hamiltonian

$$H = H_o(x_1,x_2) + \mu\{H_{1s}(x_1,x_2) + H_{1p}(x_1,x_2,y_1,y_2)\}$$

analytic in a domain D of the phase space and periodic (period 2π) with respect to both y_1 and y_2, and $|\mu| < \mu_o$, $0 < \mu_o \leq 1$. Suppose the conditions of Kolmogorov's Theorem to be satisfied, and, in D,

$$\frac{\partial H_o}{\partial x_2} \neq 0.$$

We can, therefore, solve the equation

$$H_o(x_1,x_2) + \mu H_1(x_1,x_2,y_1,y_2) = h = \text{const.}$$

with respect to x_2 and find

$$x_2 = \tilde{K}(x_1,y_1,y_2,h,\mu) \qquad (4.3.4)$$

with $\tilde{K}$ analytic in a certain domain D' (x_1,h), periodic in y_1,y_2 with period 2π, and for $|\mu| < \mu_o$. It is, therefore, possible to develop $\tilde{K}$ in a power series in μ. In fact, it is sufficient to recognize that, under the assumed hypotheses, we can write

$$\tilde{K} = \tilde{K}_0(x_1,h) + \mu K_1(x_1,y_1,y_2,h,\mu). \tag{4.3.5}$$

We can now eliminate the time from the system, together with the variable x_2. In fact,

$$\begin{aligned} \frac{dx_1}{dy_2} &= -\frac{\partial H}{\partial y_1}\Big/\frac{\partial H}{\partial x_2} = \frac{\partial \tilde{K}}{\partial y_1} \\ \frac{dy_1}{dy_2} &= \frac{\partial H}{\partial x_1}\Big/\frac{\partial H}{\partial x_2} = -\frac{\partial \tilde{K}}{\partial x_1} \end{aligned} \tag{4.3.6}$$

with $\tilde{K}$ given by (4.3.5). Let us define

$$y_2 = \tau, \qquad x_1 = x, \qquad y_1 = y$$

so that, indicating with primes derivatives with respect to τ,

$$x' = \frac{\partial K}{\partial y}, \qquad y' = -\frac{\partial K}{\partial x} \tag{4.3.7}$$

with

$$K = K_0(x) + \mu K_1(x,y,\tau,\mu).$$

By considering the space (x,y,τ), the study of solutions of (4.3.7) can be reduced to the study of a transformation T_μ of the plane $\tau = 0$ into the plane $\tau = 2\pi$. Such transformation is completely defined by (4.3.7). Let the initial conditions be $\tau = 0$, $x = x_0$, $y = y_0 \in D$ and the corresponding solution of (4.3.7)

$$\begin{aligned} x &= \phi(\tau,x_0,y_0,\mu) \\ y &= \psi(\tau,x_0,y_o,\mu). \end{aligned} \tag{4.3.8}$$

Then T_μ is given by

$$x^* = \phi(2\pi,x,y,\mu)$$
$$y^* = \psi(2\pi,x,y,\mu). \qquad (4.3.9)$$

For $\mu = 0$, the transformation T_o is

$$x^* = \phi_o(2\pi,x,y) = \phi(2\pi,x,y,0)$$
$$y^* = \psi_o(2\pi,x,y) = \psi(2\pi,x,y,0)$$

where ϕ_o and ψ_o can be written immediately. In fact, for $\mu = 0$, $K = K_o(x)$ and therefore

$$y(\tau,x_o,y_o) = y_o + \frac{\omega_1}{\omega_2}\tau$$
$$x(\tau,x_o,y_o) = x_o$$

where

$$\omega_i = \frac{\partial H_o}{\partial x_i}(x_1,x_2(y_1,y_2,x_1)) = \omega_i(x)$$

and therefore

$$\frac{\omega_1}{\omega_2} = \omega(x).$$

Letting $\alpha(x) = 2\pi\omega(x)$, the transformation T_o is,

$$x^* = x$$
$$y^* = y + \alpha(x)$$

<u>which is precisely of the form studied by Moser</u>. The transformation T_μ can evidently be written as

$$x^* = x + \mu G(x,y,\mu)$$
$$y^* = y + \alpha(x) + \mu F(x,y,\mu) \qquad (4.3.10)$$

and x will be defined in the ring $0 < a \leq x \leq b$. Evidently, the transformation T_μ is area preserving, that is, $dx\, dy$ is invariant with T_μ. On the other hand, it is easily seen that, from the hypotheses, G and F are periodic in y, period 2π. If in the ring $\alpha'(x) > 0$ and $\alpha(x)$ satisfy the condition

$$|\alpha - \frac{m}{n} 2\pi| > \mu n^{\frac{3}{2} - \sigma}$$

with σ integer ≥ 4, m, n integers, $n \neq 0$, then Moser's Theorem guarantees that T_μ has invariant curves which, for μ sufficiently small, are close to the circles $x^* =$ const. which are the invariant curves of T_0. In other words, the invariant curves of T_0 are slightly distorted under small perturbations.

On the other hand, if there exists an integral $J(x,y) =$ const., then, obviously

$$J(x^*,y^*) = J(x,y)$$

and therefore, the curves $J(x,y) =$ const. are invariant. This gives a necessary condition for the existence of integrals of (4.3.7) close to the integrals $x =$ const. for $\mu = 0$.

Finally, let us see how the condition $\alpha'(x) \neq 0$ is expressed in terms of the original Hamiltonian. By definition

$$\alpha = 2\pi \frac{\frac{\partial H_0}{\partial x_1}(x_1, x_2(x_1))}{\frac{\partial H_0}{\partial x_2}(x_1, x_2(x_1))} = \alpha(x_1)$$

Thus

$$\frac{d\alpha}{dx_1} = \frac{2\pi}{\left(\frac{\partial H_0}{\partial x_2}\right)^2} \frac{\partial H_0}{\partial x_2}\left(\frac{\partial^2 H_0}{\partial x_1^2} + \frac{\partial^2 H_0}{\partial x_1 \partial x_2} \frac{dx_2}{dx_1}\right) -$$

$$- \frac{\partial H_o}{\partial x_1}\left(\frac{\partial^2 H_o}{\partial x_1 \partial x_2} + \frac{\partial^2 H_o}{\partial x_2^2}\frac{dx_2}{dx_1}\right) .$$

But from $H_o(x_1,x_2) = h$, it follows that

$$\frac{\partial H_o}{\partial x_1} + \frac{\partial H_o}{\partial x_2}\frac{dx_2}{dx_1} = 0$$

or

$$\frac{dx_2}{dx_1} = - \frac{\partial H_o}{\partial x_1} \Big/ \frac{\partial H_o}{\partial x_2}$$

and, therefore

$$\frac{d\alpha}{dx_1} = - \frac{2\pi}{\left(\frac{\partial H_o}{\partial x_2}\right)^3} \Delta(x_1,x_2)$$

where Δ is the determinant

$$\Delta = \begin{vmatrix} \frac{\partial^2 H_o}{\partial x_1^2} & \frac{\partial^2 H_o}{\partial x_1 \partial x_2} & \frac{\partial H_o}{\partial x_1} \\ \frac{\partial^2 H_o}{\partial x_1 \partial x_2} & \frac{\partial^2 H_o}{\partial x_2^2} & \frac{\partial H_o}{\partial x_2} \\ \frac{\partial H_o}{\partial x_1} & \frac{\partial H_o}{\partial x_2} & 0 \end{vmatrix} \tag{4.3.11}$$

which has to be different from zero. <u>This condition is less restrictive than the one considered by Kolmogorov (1954) and, in fact, it is the one assumed by Arnold (1963), in his theorem. Here, it is a consequence of Moser (1962) hypotheses.</u>

4. Systems with n Degree of Freedom

Consider the analytic (in some domain) Hamiltonian

$$H = H_o(x) + \mu H_1(x;y) \tag{4.4.1}$$

2π-periodic in every component $y_1, y_2, \ldots, y_n$ of y. For $\mu = 0$, the quasi-periodic solutions

$$\begin{aligned} y_k &= \Omega_k(x^o)t + y_k^o \\ x_k &= x_k^o \end{aligned} \qquad (k = 1,2,\ldots,n) \tag{4.4.2}$$

are on a torus T_n and the flow is ergodic. Along every solution of the system produced by (4.4.1) we have

$$H(x(t);y(t)) = h = \text{const.}$$

and let us suppose that $\partial H/\partial x_n \neq 0$. Then we can solve $H = h$ for x_n and find

$$x_n = f(x_1, \ldots, x_{n-1}, y_1, \ldots, y_n, h, \mu) \tag{4.4.3}$$

By calling $y_n = \tau$ and eliminating t from the equations, we get

$$\begin{aligned} x_k' &= \frac{\partial f}{\partial y_k} \\ y_k' &= -\frac{\partial f}{\partial x_k} \end{aligned} \qquad (k = 1,2,\ldots,n-1) \tag{4.4.4}$$

with

$$f = f(x_1, \ldots, x_{n-1}, y_1, \ldots, y_{n-1}, \tau, h, \mu).$$

If H is analytic, so is f. Again, the study of the solutions of (4.4.4) can be reduced to the study of a transformation T_μ of the plane $\tau = 0$ into the plane $\tau = 2\pi$, T_μ generated by (4.4.4). In fact, let (x_k^o, y_k^o), $k = 1,2,\ldots,n-1$ be chosen in the plane $y_n = \tau = 0$. The solution of (4.4.4) going through this point at $\tau = 0$ will be

$$x_k = x_k(\tau, x^0; y^0; \mu) \qquad (k = 1,2,\ldots,n-1) \qquad (4.4.5)$$
$$y_k = y_k(\tau, x^0; y^0; \mu)$$

and therefore

$$T_\mu \begin{cases} x^* = x_k(2\pi, x; y; \mu) \\ y^* = y_k(2\pi, x; y; \mu) \end{cases} \qquad (k = 1,2,\ldots,n-1). \qquad (4.4.6)$$

On the other hand

$$\frac{\partial f_0}{\partial x_k} = - \frac{\partial H_0}{\partial x_k} \Big/ \frac{\partial H_0}{\partial x_n} = - \frac{\Omega_k}{\Omega_n}(x_1, x_2, \ldots, x_{n-1}) \qquad (4.4.7)$$

so that, for $\mu = 0$,

$$x_k = x_k^o$$
$$y_k = \frac{\Omega_k}{\Omega_n}(x_1^0, \ldots, x_{n-1}^0)\tau + y_k^o \qquad (4.4.8)$$

for $k = 1,2,\ldots,n-1$. Let $\alpha_k = 2\pi\Omega_k/\Omega_n$, it follows that

$$T_0 \begin{cases} x_k^* = x_k \\ y_k^* = \alpha_k(x_1, \ldots, x_{n-1}) + y_k \end{cases} \qquad (k = 1,2,\ldots,n-1) \qquad (4.4.9)$$

and, for $\mu \neq 0$,

$$T_\mu \begin{cases} x_k^* = x_k + G_k(x_1, \ldots, x_{n-1}, y_1, \ldots, y_{n-1}) \\ y_k^* = y_k + \alpha_k(x_1, \ldots, x_{n-1}) + F_k(x_1, \ldots, x_{n-1}, y_1, \ldots, y_{n-1}) \end{cases} \qquad (4.4.10)$$

for $k = 1,2,\ldots,n-1$, and x_k in the ring $0 < a_k \leq x_k \leq b_k$. Clearly G_k and F_k

depend on μ and should go to zero with μ. Since (4.4.4) is Hamiltonian T_μ is area preserving. If there exists a first integral of (4.4.4) it will be invariant with respect to T_μ. Moreover, a periodic solution of period $2\pi p$, p integer > 0, will be such that

$$T_\mu^p(x^o,y^o) = (x^o,y^o)$$

so that (x^o,y^o) is a fixed point of T_μ.

In geometric terms now, Kolmogorov's Theorem can be put in the following form. We wish to determine under what conditions the transformation T_μ has invariant sets which are "close" to the invariant sets (tori) of T_o.

Consider T_o defined in $0 < a_k \leq x_k \leq b_k$ $(k = 1,2,\ldots,N)$ with the condition

$$\det\left\{\frac{\partial\alpha_k}{\partial x_j}\right\} \neq 0.$$

The torus T^N, product of the N circles $x_k = \text{const.}$ is invariant with respect to T_o. Consider now T_μ, with F_k, G_k bounded and 2π-periodic in $y_1,y_2,\ldots,y_N$. Let us assume that "Every closed and limited set which is "near" T^N and represented by

$$x_k = f_k(y_1,\ldots,y_N) = f_k(y_1 + 2\pi,\ldots,y_N + 2\pi) \qquad (4.4.11)$$

with $\partial f_k/\partial y_j$ conveniently bounded, intercepts its image set through T_μ". It can now be shown that, under the above hypotheses, T_μ has limited and closed invariant sets, for values of F_k and G_k sufficiently small, provided these functions are differentiable up to a certain order.

Here we only present a sketch of the proof of the main theorem, beginning with some fundamental lemmas. For $N = 1$, the proof has been given in details by Moser (1962).

Theorem. "Given an $\epsilon > 0$ and an integer $s \geq 1$, the transformation T_μ has a limited and closed invariant set

$$y = y' + p(y')$$
$$x = x^o + q(y') \qquad (4.4.12)$$

where p, q are N-vectors, 2π-periodic in every component $y_1, y_2, \ldots, y_N$ of y, $p, q \in C^s$ and $|p_k|_s + |q_k|_s < \epsilon$, provided:

a) For T_μ, every closed set (4.4.10) and its image, have at least a common point.

b) In the ring $0 < a_k \leq x_k \leq b_k$, $b_k - a_k \geq 1$, one can find a $C_o > 1$ such that

$$\frac{1}{N! C_o^N} \leq \left| \det \left\{ \frac{\partial \alpha_k(x)}{\partial x_j} \right\} \right| < N! C_o^N .$$

It will then be possible to find an $f_o(\epsilon, s, C_o) > 0$ and an integer $m(s)$ such that $F_k, G_k \in C^m$ and

$$|F_k|_o + |G_k|_o < \delta_o$$
$$|\alpha_k|_o + |F_k|_m + |G_k|_m < C_o . \qquad (4.4.13)$$

Moreover, we state that the transformation induced over the set (4.4.11) is given by

$$\tilde{y}_k = y'_k + \alpha_k(x^o) = y'_k + \alpha_k \qquad (4.4.14)$$

and, more precisely, given α_k satisfying the relations

$$\alpha_k(a) + \epsilon < \alpha_k < \alpha_k(b) - \epsilon \qquad (4.4.15)$$

for which

$$\left| \sum_{j=1}^{N} j_k \alpha_k + j_{N+1} 2\pi \right| \geq \epsilon \left\{ \sum_{k=1}^{N} |j_k| \right\}^{-N-1/2} \qquad (4.4.16)$$

for all integers j_k not all zero, there exists an invariant set (4.4.11) with $\alpha_k(x^0) = \alpha_k$ (rotation numbers)." We have used the notation

$$|f|_s = \sup \left| \left(\frac{\partial}{\partial x_1}\right)^{s_1} \left(\frac{\partial}{\partial x_2}\right)^{s_2} \cdots \left(\frac{\partial}{\partial x_n}\right)^{s_n} f(x_1,\ldots,x_n) \right|$$

with $s_1 + s_2 + \cdots + s_n = s$ and $x \in D$, the domain of definition of $f(x)$.

Lemma. Initially one can prove that "given the transformation

$$Q \begin{cases} x_k^* = x_k + g_k(x;y) \\ y_k^* = y_k + x_k + f_k(x;y) \end{cases} \qquad (k = 1,2,\ldots,N)$$

defined in $0 < a_k \leq x_k \leq b_k$, where f_k, g_k are 2π-periodic in every component $y_1, y_2, \ldots, y_N$ of y and $b_k - a_k \geq 1/C_0$ and

a) $|f|_1 + |g|_1 < \delta_0$ (4.4.17)

b) $|D^m f_k| + |D^m g_k| < C_0$ (4.4.18)

where $D^m = \left(\frac{\partial}{\partial x_1}\right)^{p_1} \cdots \left(\frac{\partial}{\partial x_N}\right)^{p_N} \left(\frac{\partial}{\partial y_1}\right)^{q_1} \cdots \left(\frac{\partial}{\partial y_N}\right)^{q_N}$

with $p_1 + p_2 + \cdots + p_N + q_1 + q_2 + \cdots + q_N = m$

c) $a_k + \epsilon < \alpha_k < b_k - \epsilon$

d) $\left| \sum_1^N j_k \alpha_k + 2\pi j_{N+1} \right| \geq \epsilon \left\{ \sum_1^N j_k \right\}^{-\sigma+N+3/2}$

$\sigma \geq 2(N+1)$, integer,

it is possible to reduce Q to

$$Q_0 \begin{cases} x_k^* = x_k \\ y_k^* = y_k + x_k \end{cases} .$$

This is done by an iteration procedure, much like the one used in Kolmogorov's Theorem. However, in this case, the iteration converges in such a way that the deviation from Q_0, decreases with $\delta_0^{q^n}$ where n is the stage of the iteration and q may be taken equal to $4/3$.

In fact, let Q, M, δ be three parameters satisfying the relations

$$M = Q^{\nu} > Q > 1$$

$$\delta = M^{-2q}$$

with

$$q = \frac{4}{3}, \qquad \nu = 6(\sigma+1), \qquad m = 3 + 11\nu. \tag{4.4.19}$$

The parameters Q_{n-1}, M_{n-1}, δ_{n-1} referring to the stage $n - 1$ of the iteration, are related with n, M, δ by

$$Q = Q_{n-1}^{q}, \qquad Q_{n-1} = Q_0^{q^{n-1}}$$

$$M = M_{n-1}^{q}, \qquad M_{n-1} = M_0^{q^{n-1}}$$

$$\delta = \delta_{n-1}^{q}, \qquad \delta_{n-1} = \delta_0^{q^{n-1}},$$

supposing that at every stage

$$|f_k|_1 + |g_k|_1 < \delta_{n-1} \quad \text{in} \quad |x_k - \alpha_k| < \frac{1}{M_{n-1}} \tag{4.4.20}$$

$$|D^m Q_{n-1} f_k| + |D^m M_{n-1} g_k| \leq Q_{n-1}^{q_1+\cdots+q_N+1} M_{n-1}^{p_1+p_2+\cdots+p_N} \tag{4.4.21}$$

with $q_1 + q_2 + \cdots + q_N + p_1 + p_2 + \cdots + p_N = m$.

For $n = 1$, (4.4.20) coincides with (4.4.17) for δ sufficiently small. Moreover, taking $M_0 > 1/\epsilon$ and $|x_k - \alpha_k| < 1/M_0 < \epsilon$, we certainly have $a_k \leq x_k \leq b_k$.

Also (4 4.21), for $n = 1$, coincides with (4.4.18). In fact, (4.4.18) implies

$$|D^m Q_o f_k| + |D^m M_o g_k| \leq C_o M_o$$

since $M > Q$. Thus, being

$$q_1 + \cdots + q_N + p_1 + \cdots + p_N = m \geq \nu$$

$$Q_o^{q_1+\cdots+q_N} M_o^{p_1+\cdots+p_N} \geq Q_o^n > Q_o^\nu \geq M_o$$

it follows that

$$C_o M_o \leq C_o Q_o^{q_1+\cdots+q_N} M_o^{p_1+\cdots+p_N} \leq Q_o^{q_1+\cdots+q_N+1} M_o^{p_1+\cdots+p_N+1}$$

because $Q_o > C_o$. Therefore, for $n = 1$, we ought to take $M_o > 1/\epsilon$ and $Q_o > C_o$, that is

$$\delta_o < \epsilon^{2q}, \qquad C_o^{-2q\nu}$$

and eventually δ_o might have to be taken "much smaller than" the above quantities.

Now we proceed to prove a Basic Lemma which, nevertheless needs the necessary background of two well known propositions

A) Consider the difference equation

$$S(y+\alpha) - S(y) = F(y) \tag{4.4.22}$$

where S and F are to be periodic of period 2π in each component $y_1, y_2, \ldots, y_N$ of y, with zero mean. Let the vector ω satisfy

$$\left| \sum_1^N j_k \alpha_k + 2\pi j_{N+1} \right| \geq \epsilon \left\{ \sum_1^N |j_k| \right\}^{-\tau+N+1/2} \tag{4.4.23}$$

where $\tau \geq 2N + 1$, integer. Then if $F(y)$ has $\tau \geq 2N + 1$ continuous derivative with respect to every y_j, the equation (4.4.22) has a continuous solution, such that

$$|S(F)|_0 \leq \frac{C}{\epsilon} |F|_\tau \qquad (4.4.24)$$

with C an arbitrary constant independent of any parameter. Moreover, it is possible to take $S(0) = 0$. In fact using for F the Fourier's series

$$F = \sum_{j \neq 0} F_j e^{ij \cdot y}$$

the solution is

$$S = \sum_{j \neq 0} \frac{F_j}{e^{ij \cdot \alpha} - 1} e^{ij \cdot y} \qquad (4.4.25)$$

where $j = (j_1, j_2, \ldots, j_N)$. Since F has τ continuous derivatives with respect to each y_j,

$$|F_j| \leq \left(\sum_{k=1}^{N} |j_k|\right)^{-\tau} |F|_\tau$$

with $\sum |j_k| \neq 0$. On the other hand $|e^{ij \cdot \alpha} - 1| = |2 \sin \frac{1}{2} (j \cdot \alpha)| \geq \frac{2}{\pi} \epsilon \left\{ \sum_1^N |j_k| \right\}^{-\tau+N+1/2}$ for ϵ sufficiently small, and therefore

$$\left| \frac{F_j}{e^{ij \cdot \alpha} - 1} \right| \leq \frac{\pi}{2\epsilon} \left(\sum_{k=1}^{N} |j_k|\right)^{-N-1/2} |F|_\tau$$

and the series (4.4.25) is absolutely convergent. To show (4.4.24) it is sufficient to take

$$C = \sum_{j \neq 0} \left(\sum_{k=1}^{N} |j_k|\right)^{-N-1/2}$$

It follows as a corollary that if $F(x;y)$ is 2π-periodic in $y_1, y_2, \ldots, y_N$ with zero mean, the difference equation

$$S(x;y+\alpha) - S(x;y) = F(x;y)$$

can be solved as before and

$$|S|_0 \leq \frac{C}{\epsilon} |D_y^\tau F|_0.$$

B) We can define a smoothing operation (Moser, 1962; Arnol'd, 1963) on a function of 2N variables. It will approximate, in a sense to be specified, functions $F(x;y)$ with functions $\Phi(x;y)$ which have smoother behavior than $F(x;y)$. It is basically an interpolation through high derivatives of F, which leaves invariant (interpolates exactly) a polynomial of a specified degree.

Let $F(x;y)$ be a continuous function in the domain

$$a_k < x_k < b_k$$

$$-\infty < y_k < +\infty$$

for $k = 1,2,\ldots,N$. The degree of approximation with respect to the two vector variables x,y can be set independently (in fact, it could be set independently for every one of the 2N components) and measured by two parameters $M, Q > 1$. The smoothed function $\Phi_{M,Q}(F)$ is defined in the smaller domain

$$\alpha_k = a_k + \frac{1}{M} < x_k < b_k - \frac{1}{M} = \beta_k$$

and we suppose $2/M < b_k - a_k$. It is defined by

$$\Phi_{M,Q}(F(x;y)) = \underset{a_k < \xi_k < b_k}{\int \cdots \int} \; \underset{-\infty < \eta_k < \infty}{\int \cdots \int} K_{M,Q}(x-\xi, y-\eta) F(\xi,\eta) d\xi \, d\eta \tag{4.4.27}$$

The kernel $K_{M,Q}$ is taken to be

$$K_{M,Q}(x;y) = MK(Mx) \cdot QK(Qy)$$

where $K(x)$ is C^∞ and satisfies

$$K(x) = 0 \quad \text{for} \quad |x_k| > 1 \tag{4.4.28}$$

and

$$\int \overset{-\infty}{\underset{-\infty}{\cdots}} \int x_1^{k_1} x_2^{k_2} \cdots x_N^{k_N} K(x)dx = \begin{cases} 1 & \text{for } \Sigma k_i = 0 \\ 0 & \text{for } 0 < \Sigma k_i < m \end{cases} \tag{4.4.29}$$

where $k_i \geq 0$ and m is a fixed number. Thus, $\Phi_{M,Q}(F)$ depends on m which is chosen to satisfy (4.4.13).

Condition (4.4.28) shows that $K_{M,Q}$ is zero for $|y_k| > 1/Q$ and $|x_k| > 1/M$. Hence the integration interval in (4.4.29) can be restricted to an interval contained in $a_k < \xi_k < b_k$ since $\alpha_k < x_k < \beta_k$. Therefore, (4.4.27) can be written, in terms of new variables of obvious significance,

$$\Phi_{M,Q}(F(x;y)) = \underset{|u_k| < 1}{\int \cdots \int} \underset{|v_k| < 1}{\int \cdots \int} K_{1,1}(u;v)\, F(x - \frac{u}{M}\, ;\, y - \frac{v}{Q})du\, dv \tag{4.4.30}$$

Finally, condition (4.4.29) shows that polynomials $P_m(x;y)$ of maximum degree m are invariant under Φ, that is

$$\Phi_{M,Q}(P_m(x;y)) = P_m(x;y).$$

Now, "if $F(x,y)$ is continuous in $a_k < x_k < b_k$, in $\alpha_k < x_k < \beta_k$ one verifies

$$|D_x^q\, D_y^p\, \Phi_{Q,M}(F)| \leq \tilde{C}\, Q^{p_1+\cdots+p_n} M^{q_1+\cdots+q_N} \cdot \sup_{a_k \leq x_k \leq b_k} |F| \tag{4.4.31}$$

for all integers $p_1,\ldots,p_N$, $q_1,\ldots,q_N$ and $\tilde{C}$ a constant which depends on the kernel $K(x)$, p and q. Also if $F(x,y) \in C^m$, then, in $\alpha_k < x_k < \beta_k$ one verifies

$$|F - \Phi_{M,Q}(F)| \leq CQ^{-p_1-p_2-\cdots-p_N} M^{-q_1-q_2-\cdots-q_N} \cdot \sup |D_x^q\, D_y^p\, F|$$

for

$$p_1 + p_2 + \cdots + q_N = m \text{ and } a_k < x_k < b_k \text{".} \tag{4.4.32}$$

This can be shown by observing that from (4.4.27) it follows that

$$|D_x^q D_y^p \Phi_{M,Q}(F)| \leq \sup_{a_k<x_k<b_k} |F| \int \cdots \int |D_x^q D_y^p K_{M,Q}(x-\xi;y-\eta)| d\xi \, d\eta \leq$$

$$\leq \sup_{a_k<x_k<b_k} |F| \, Q^{p_1+\cdots+p_N} M^{q_1+\cdots+q_N} \int \cdots\cdots \int |D_x^q K(x)| dx \int \cdots \int |D_y^p K(y)| dy$$

which proves (4.4.31). Next, we develop $F(x-u/M;\ y-v/Q)$ in Taylor series in the neighborhood of $u = v = 0$, up to terms order $< m$, with rest. Since $\Phi_{M,Q}$ conserves polynomials of degree $< m$, only the rest $R_m(x;y;u;v)$ of the series will contribute to $F - \Phi_{M,Q}(F) = (I - \Phi_{M,Q})F$. This rest can be estimated by

$$|R_m| \leq \sup_{\substack{(u;v) \\ p_1+\cdots+q_N=m}} N \frac{|u_1^{q_1} \cdots u_N^{q_N} v_1^{p_1} \cdots v_N^{p_N}|}{Q^{p_1+\cdots+p_N} M^{q_1+\cdots+q_N}} |D_x^p D_y^p F(\xi;\eta)|$$

where

$$|\xi_k - x_k| < 1/M \ , \quad |\eta_k - y_k| < 1/Q \ ,$$

$$|u_k| < 1 \ , \quad |v_k| < 1.$$

Hence, from (4.4.30) we obtain

$$|(I - \Phi_{M,Q})F(x;y)| \leq C \sup_{(u;v)} Q^{-p_1-\cdots-p_N} M^{-q_1-\cdots-q_N} \cdot |D_x^q D_y^p F(\xi;\eta)|$$

where $p_1 + p_2 + \cdots + p_N + q_1 + q_2 + \cdots + q_N = m$ and

$$C = N \sup_{p_1+\cdots+q_N=m} \int \cdots \int |u_1^{p_1} \cdots u_N^{q_N} v_1^{p_1} \cdots v_N^{p_N} K_{1,1}(u;v)| du \, dv$$

which proves (4.4.32).

<u>Note</u>. For $a_k < x_k < b_k$ we have, with $p_1 + \cdots p_N = \tau$, $q_1 + \cdots + q_N = 0$,

$$S^* = |S(\Phi_{M,Q}(F(x;y))| \leq \frac{C}{\epsilon} |D_y^\tau \Phi_{M,Q}(F)|_o \leq \tilde{C}\, C\, \frac{Q^\tau}{\epsilon} |F|_o$$

and, choosing $Q > 1/\epsilon$, $\sigma = \tau + 1$, it follows that

$$S^* \leq \tilde{C}\, CQ^{\tau+1}|F|_o = \tilde{C}\, CQ^\sigma\, |F|_o \qquad (4.4.33)$$

and similarly,

$$S^{*2} = |S(S(\Phi_{M,Q}(F(x,y))))| \leq \tilde{C}\, CQ^{2\sigma}\, |F|_o\, . \qquad (4.4.34)$$

It is also possible to show that one can construct a function $K(x)$ with the mentioned properties (Moser, 1962).

Lemma. We now proceed to the evaluation of an iteration stage in the process of reducing a mapping to the linear form of Moser's twist mapping.

"Consider the transformation

$$T \begin{cases} x_k^* = x_k + g_k(x;y) \\ y_k^* = y_k + x_k + f_k(x;y) \end{cases} \qquad (4.4.35)$$

in $|x_k - \alpha_k| < 1/M_{n-1}$, satisfying conditions already mentioned

$$|f|_1 + |g|_1 < \delta_o$$

$$|D^m f_k| + |D^m g_k| < c_o$$

and such that every closed finite regular surface near the surface $x_k = \text{const.}$ $(k = 1,2,\ldots,N)$ and its image have at least one point in common. Then, for δ_o sufficiently small there exists a transformation

$$\begin{aligned} x_k &= \xi_k + u_k(\xi;\eta) \\ y_k &= \eta_k + v_k(\xi;\eta) \end{aligned} \qquad (4.4.36)$$

for

$$|\xi_k - \alpha_k| < \frac{1}{M_{n-1}} - \frac{1}{M} > \frac{1}{M}$$

such that

$$|v_k|_1 + |u_k|_1 < \frac{1}{Q} \tag{4.4.37}$$

and the transformation T takes the form

$$\begin{aligned} \xi_k^* &= \xi_k + \phi_k(\xi;\eta) \\ \eta_k^* &= \eta_k + \xi_k + \psi_k(\xi;n) \end{aligned} \tag{4.4.38}$$

where

$$|\phi_k|_0 + |\psi_k|_0 < \delta = \delta_{n-1}^q \tag{4.4.39}$$

in

$$|\xi_k - \alpha_k| < \frac{1}{Q}$$

and, moreover

$$|D_\eta^p D_\xi^q Q\phi_k| + |D_\eta^p D_\xi^q M\psi_k| < Q^{p_1+\cdots+p_N+1} M^{q_1+\cdots+q_N} \tag{4.4.40}$$

for $p_1 + \cdots + p_N + q_1 + \cdots + q_N = m$."

In fact, from the above relations, it follows that

$$\xi_k + \phi_k(\xi;\eta) + u_k(\xi^*;\eta^*) = \xi_k + u_k(\xi;\eta) + g_k(\xi+u,\eta+v),$$

$$\eta_k + \xi_k + \psi_k(\xi;\eta) + v_k(\xi^*;\eta^*) = \eta_k + v_k(\xi;\eta) + \xi_k + u_k(\xi;\eta) + f_k(\xi+u;\eta+v)$$

or

$$\begin{aligned} \phi_k(\xi;\eta) + u_k(\xi^*;\eta^*) &= u_k(\xi;\eta) + g_k[\xi + u(\xi;\eta);\ \eta + v(\xi;\eta)] \\ \psi_k(\xi;\eta) + v_k(\xi^*;\eta^*) &= v_k(\xi;\eta) + u_k(\xi;\eta) + f_k[\xi + u(\xi;\eta);\ \eta + v(\xi;\eta)]. \end{aligned} \tag{4.4.41}$$

We now linearize the equations which result from $\phi_k = \psi_k = 0$ by considering f_k, g_k, u_k, v_k and $|\xi_k - \alpha_k|$ to be of the same order of magnitude, that is,

$O(\mu)$. We shall neglect terms order $\geq \mu^2$. This way one finds

$$\begin{aligned} v_k(\xi;\eta+\omega) - v_k(\xi;\eta) &= u_k(\xi;\eta) + f_k(\xi;\eta) \\ u_k(\xi;\eta+\omega) - u_k(\xi;\eta) &= g_k(\xi;\eta). \end{aligned} \tag{4.4.42}$$

The second of (4.4.42) can only be solved to our satisfaction, according to the previous lemmas, if $g_k(\xi;\eta)$ has zero mean with respect to η. And, in any event, the functions f_k and g_k shall need a smoothing operation in order to take care of the loss of derivatives.

We shall define u_k, v_k as the solutions of the system

$$\begin{aligned} u_k(\xi;\eta+\alpha) - u_k(\xi;\eta) &= \Phi(g_k - \overline{g}_k) \\ v_k(\xi;\eta+\alpha) - v_k(\xi;\eta) &= u_k(\xi;\eta) + \Phi(f_k) \end{aligned} \tag{4.4.43}$$

with $\overline{v}_k = 0$, a bar indicating of the average with respect to the η-vector and Φ is the smoothing operator previously defined.

The average of the second of the above equations gives

$$\overline{u}_k + \overline{\Phi}(f_k) = 0$$

so that, from the first,

$$u_k = S(\Phi(g_k)) + \overline{u}_k = S(\Phi(g_k)) - \overline{\Phi}(f_k) \tag{4.4.44}$$

where S is defined by (4.4.22). From the second of (4.4.43), it follows

$$v_k = S(u_k + \Phi(f_k)) = S(S(\Phi(g_k))) + S(\Phi(f_k)). \tag{4.4.45}$$

The Equations (4.4.44) and (4.4.45) define u_k, v_k in the ring

$$|\xi_k - \alpha_k| < \frac{1}{M_{n-1}} - \frac{1}{M} = \mu.$$

It remains to verify (4.4.37), (4.4.39) and (4.4.40).

a) Making use of (4.4.33) and (4.4.34) and considering (4.4.20),

$$|u_k| + |v_k| \leq c_1 Q^{2\sigma}(|f_k|_o + |g_k|_o) <$$
$$< c_1 Q^{2\sigma} \delta_{n-1} = c_1 Q^{2\sigma} M^{-2} \tag{4.4.46}$$

and, in like manner

$$|D_\xi u_k| + |D_\xi v_k| \leq c_2 Q^{2\sigma} M^{-1}$$
$$|D_\eta u_k| + |D_\eta v_k| \leq c_2 Q^{2\sigma} M^{-1}. \tag{4.4.47}$$

As we have taken $\nu > 2\sigma + 1$,

$$|u_k| + |v_k| < \frac{1}{QM},$$

and $|u_k|_1 + |v_k|_1 < \frac{1}{Q}$, which proves (4.4.37). It also follows that u_k, v_k are defined in the ring

$$|\xi_k - \alpha_k| < \frac{1}{M_{n-1}} - \frac{1}{M} > \frac{3}{M}$$

provided

$$M_o > 4^{1/1+q}$$

and there (4.4.46) is verified. By the Implicit Functions Theorem, the image of $|\xi_k - \alpha_k| < 3/M$ under the mapping (4.4.36), will cover at least the ring $|x_k - \alpha_k| < 2/M$ and, therefore, the inverse transformation is uniquely defined in $|x_k - \alpha_k| < 2/M$ and, in here, it is continuously differentiable if $M(\geq M_o)$ is chosen sufficiently large. It follows that (4.4.38) is defined and differentiable in $|x_k - \alpha_k| < 1/M$ since (4.4.36) maps this region into

$$|x_k - \alpha_k| \leq |\xi_k - \alpha_k| + |u_k| < \frac{1}{M} + \frac{1}{QM}$$

and (4.4.35) into

$$|x_k^* - \alpha_k| < |x_k - \alpha_k| + \delta_{n+1} < \frac{1}{M} + \frac{1}{QM} + \frac{1}{M^2} < \frac{2}{M}$$

where the inverse of (4.4.38) is uniquely defined.

b) The estimation of $|\phi_k|$, $|\psi_k|$ in $|\xi_k - \alpha_k| < 1/M$ needs the hypothesis that every closed surface near the invariant manifold $x_k = \text{const.}$ $(k = 1,2,\ldots,N)$, that is, $\xi_k = \xi_k^o$, has an image

$$\xi_k^* = \xi_k^o + \phi_k(\xi^o;\eta)$$

which intercepts $\xi_k = \xi_k^o$ and, therefore, it is equivalent to suppose that $\phi_k(\xi^o;\eta)$ has at least one zero with respect to η. Hence

$$\sup_\eta |\phi_k(\xi^o;\eta)| \leq \underset{\eta}{\text{amplitude}}\ \phi_k(\xi^o;\eta) \leq 2 \sup_\eta |\phi_k(\xi^o;\eta) + Z_k(\xi^o)|$$

where $Z_k(\xi)$ are convenient functions of ξ. We shall take

$$Z_k(\xi) = -\overline{\Phi}(g_k(\xi;\eta)).$$

From (4.4.41),

$$\frac{1}{2} |\phi_k(\xi;\eta)|_o \leq |\phi_k(\xi;\eta) - \overline{\Phi}(g_k(\xi;\eta))|_o =$$

$$= |u_k(\xi;\eta) - u_k(\xi^*;\eta^*) + g_k(\xi+u;\ \eta+v) - \overline{\Phi}(g_k(\xi;\eta))|_o$$

and using (4.4.43), (4.4.47) we find

$$\frac{1}{2} |\phi_k(\xi;\eta)|_o \leq c_2 Q^{2\sigma+1} M^{-2} |\xi_k - \alpha_k| +$$

$$+ |u_k|_1 (|\phi_k|_o + |\psi_k|_o) + |\Phi(g_k(x;y))|_1 (|u_k|_o +$$

$$+ |v_k|_o) + |(I - \Phi)(g_k(x;y))|_o$$

where the norm of the terms in $(x;y)$ stands for the fact that these functions are defined in the ring $|\xi_k - \alpha_k| < 1/M_{n-1} - 1/M$.

Similarly, by subtracting $(4.4.42)_1$ from $(4.4.41)_2$,

$$|\psi_k(\xi;\eta)|_o \leq c_2 Q^{2\sigma+1} M^{-2} |\xi_k - \alpha_k| + |v_k|_1 (|\phi_k|_o + |\psi_k|_o)$$

$$+ |\Phi(f_k(x;y))|_1 (|u_k|_o + |v_k|_o) + |(I - \Phi)(f_k(x;y))|_o .$$

By adding the last two estimates and considering the already established conditions

$$|\Phi(f_k)|_1 \le C_3 \, M|f_k|_0 < \frac{C_3}{M}$$

$$|\Phi(g_k)|_1 \le \frac{C_3}{M}$$

together with the hypothesis (4.4.21), i.e., $|D^m Q_{n-1} f_k| + |D^m M_{n-1} g_k| \le$ $Q_{n-1}^{q_1+\cdots+q_N+1} M_{n-1}^{p_1+\cdots+p_N} (q_1 + \cdots + q_N + p_1 + \cdots + p_N = m)$, we find

$$|\phi_k|_0 + |\psi_k|_1 \le C_4 (Q^{2\sigma+1} M^{-3} + Q_{n-1}^{1+(1-q)m}). \tag{4.4.48}$$

On the other hand

$$C_4 Q^{2\sigma+1} M^{-3} = C_4 \, M^{\frac{2\sigma+1}{\nu} - 3} < \frac{1}{2} M^{-2q}$$

for M sufficiently large and

$$\nu > \frac{2\sigma+1}{3-2q} = 6\sigma + 3.$$

Also, since it is true that

$$m(q-1) > 1 + 2q^2\nu,$$

we have

$$C_4 \, Q_{n-1}^{1+(1-q)m} < \frac{1}{2} M^{-2q} = \frac{1}{2} Q_{n-1}^{-2q^2\nu}$$

and from (4.4.48) it follows that

$$|\phi_k|_0 + |\psi_k|_0 < M^{-2q} = \delta$$

which proves (4.4.39)

c) In order to prove the last part of the Lemma, let us introduce

$$x_k' = Mx_k\,,\quad \xi_k' = M\xi_k$$

$$y_k' = Qy_k\,,\quad \eta_k' = Q\eta_k$$

so that the transformation (4.4.35) can be written

$$T' \left\{ \begin{aligned} x_k^{*\prime} &= x_k' + g_k'(x';y'), \quad g_k' = Mg_k \\ y_k^{*\prime} &= y_k' + \frac{Q}{M}x_k' + f_k'(x';y'), \quad f_k' = Qf_k\,. \end{aligned} \right. \tag{4.4.49}$$

From (4.4.21) it follows that

$$|D_{x'}^p\, D_{y'}^q\, Qf_k| + |D_{x'}^p\, D_{y'}^q\, Mg_k| \leq Q_{n-1}\frac{M}{M_{n-1}}\left(\frac{Q_{n-1}}{Q}\right)^{q_1+\cdots+q_N} \cdot$$

$$\cdot \left(\frac{M_{n-1}}{M}\right)^{q_1+\cdots+q_N} \leq Q_{n-1}\frac{M}{M_{n-1}}\left(\frac{Q_{n-1}}{Q}\right)^m =$$

$$= Q_{n-1}^{1+(1-q)(m-\nu)} \leq 1.$$

On the other hand

$$|Qf_k| + |Mg_k| \leq M\delta_{n-1} = \frac{1}{M} \leq 1$$

for $|x_k - \alpha_k| < 1/M_{n-1}$. Hence, f_k' and g_k' are defined for every real y' and in the region

$$|x_k' - M\alpha_k| < \frac{M}{M_{n-1}} > 1$$

are bounded by 1 as their derivatives of order m. From the Theorem of the Mean, applied in succession to every one of these derivatives up to that order, it follows that

$$|f_k'(x';y')|_m + |g_k'(x';y')|_m < C_5$$

where C_5 is a constant depending only on m. Therefore, the transformation T' (4.4.49) is such that

$$|x_k' + g_k'|_m + |y_k' + \frac{Q}{M} x_k' + f_k'|_m < C_6$$

which we write

$$|T'|_m < C_6 .$$

For the coordinates transformation (4.4.36) written in the form

$$W' \begin{cases} x_k' = \xi_k' + u_k'(\xi';\eta') \\ y_k' = \eta_k' + v_k'(\xi';\eta') \end{cases}$$

we find, analogously, for $p_1 + p_2 + \cdots + p_N + q_1 + q_2 + \cdots + q_N \leq m + 2\sigma$,

$$|u_k'|_m + |v_k'|_m < C_7 M Q^{2\sigma} \delta_{n-1} = C_7 M^{-1} Q^{2\sigma} < \frac{1}{Q} < \frac{1}{4}$$

and, now, the Implicit Functions Theorem implies the existence of the inverse transformation with all derivatives up to the mth order bounded by constants which do not depend on Q,M in $|x_k - \alpha_k| < \frac{2}{M}$. Therefore,

$$|W'|_m < 2, \qquad |W'^{-1}|_m < 2$$

and the transformation $W'^{-1}T'W$ is given in its components by functions whose derivatives up to order m are bounded by a constant C_8 independent of M and Q. Hence

$$|D_\eta^p D_\xi^q Q\psi_k| + |D_\eta^p D_\xi^q M\phi_k| < C_8 Q^{p_1+\cdots+p_N} M^{q_1+\cdots+q_N}$$

for $p_1 + p_2 + \cdots + p_N + q_1 + q_2 + \cdots + q_N \leq m$ in $|\xi_k - \alpha_k| < 1/M$, provided $Q > C_8$, which proves (4.4.40).

The proof of Lemma at p. 233 now follows from the previous Lemma. In fact, let us define

$$T^\circ \begin{cases} x_k^* = x_k + g_k(x;y) \\ y_k^* = y_k + x_k + f_k(x;y) \end{cases}$$

in $a_k \leq x_k \leq b_k$, and

$$T^\infty \begin{cases} x_k^* = x_k \\ y_k^* = y_k + x_k \end{cases}$$

with $|f_k| + |g_k| < \delta_o$ for $|T^o - T^\infty| < \delta_o$. The second Lemma states the existence of a transformation

$$W_1 \begin{cases} x_k = \xi_k + u_k(\xi;\eta) \\ y_k = \eta_k + v_k(\xi;\eta) \end{cases}$$

which maps T^o into T^1 with

$$T^1 = W_1^{-1} T^o W_1$$

where, formally,

$$|T^1 - T^\infty| < \delta_o^q = \delta_1 .$$

By repeating the reduction process of T^o defined in an ever smaller ring, we find

$$T^n = W_n^{-1} T^{n-1} W_n \qquad (n = 1,2,3,\ldots) \tag{4.4.50}$$

with

$$|T^n - T^\infty| < \delta_{n-1}^q = \delta_n .$$

If $(\xi;\eta)$ are the coordinates relative to T^n, then this is defined in

$$|\xi_k - \alpha_k| < \frac{1}{M_n} = \frac{1}{M_{n-1}^q} . \tag{4.4.51}$$

In view of the second Lemma, the transformation W_n of coordinates from the stage $n - 1$ to the stage n, is such that

$$|W_n - I|_1 < \frac{1}{Q_n} = \frac{1}{Q_{n-1}^q} \tag{4.4.52}$$

where I is the identity transformation. Therefore, on the torus $\xi_k = \alpha_k$, T^n converges to T^∞ which is a "rotation" $\eta_k^* = \eta_k + \alpha_k$. The relation between $(\xi;\eta)$ of T^n and $(x;y)$ of T^0 is

$$S_n = W_1 W_2 \cdots W_n$$

$$T^n = S_n^{-1} T^0 S_n$$

as a consequence of (4.4.50). The transformation S_n is defined in $|\xi_k - \alpha_k| < \frac{1}{M_n}$ and maps this ring into the smaller ring $|x_k - \alpha_k| < \frac{1}{M_o}$. If we write S_n in the form

$$x_k = \xi_k + p_k^n(\xi;\eta)$$

$$y_k = \eta_k + q_k^n(\xi;\eta)$$

it will suffice to show that p_k^n, q_k^n and their derivatives converge to functions $p_k(\eta)$ and $q_k(\eta)$, uniformly. If this is so, the invariant manifold of the Theorem at page 231 (for the specific case $\alpha_k = x_k$ and $\delta = 1$) is precisely

$$x_k = \alpha_k + p_k(\eta)$$

$$y_k = \eta_k + q_k(\eta).$$

This is readily verified, by initially observing that

$$|p_k^n| + |q_k^n| < \sum_{t=1}^{n} (|u_t| + |v_t|) < \sum_{t=1}^{n} \frac{1}{Q_t}$$

and, for Q_o sufficiently large, the last term can be made $< \epsilon$. For the convergence of the derivatives of p_k^n, q_k^n consider the Jacobian matrix J_n of W_n. Since

$$\frac{\partial x_k}{\partial \xi_j} = \delta_{kj} + \frac{\partial u_k}{\partial \xi_j}, \quad \frac{\partial x_k}{\partial \eta_j} = \frac{\partial u_k}{\partial \eta_j}$$

$$\frac{\partial y_k}{\partial \xi_j} = \frac{\partial v_k}{\partial \xi_j}, \quad \frac{\partial y_k}{\partial \eta_j} = \delta_{kj} + \frac{\partial v_k}{\partial \eta_j}$$

it follows that

$$J_n - I = \begin{pmatrix} \frac{\partial u}{\partial \xi} & \frac{\partial u}{\partial \eta} \\ \frac{\partial v}{\partial \xi} & \frac{\partial v}{\partial \eta} \end{pmatrix} .$$

The maximum absolute value of the elements of this matrix, $|J_n - I|$, is given, in view of (4.4.52), for $|u_k|_1 + |v_k|_1 < 1/Q$, by

$$|J_n - I| < \frac{1}{Q_n} .$$

The Jacobian matrix G_n of S_n is now, obviously, the product of the n matrices $J_1, J_2, \ldots, J_n$, i.e.,

$$G_n = J_1 \, J_2 \cdots J_n$$

and, also, the convergence of the derivatives of p_k^n, q_k^n is equivalent to the convergence of the product G_n. But J_t is majorized by

$$I + \frac{1}{Q_t} U, \qquad U = \begin{pmatrix} 1 & 1 & \ldots. & 1 \\ 1 & 1 & \ldots. & 1 \\ \cdot & \cdot & \cdot & \cdot \\ 1 & 1 & \ldots. & 1 \end{pmatrix} .$$

Hence, it is sufficient to show the convergence of the product

$$\prod_{t=1}^{\infty} (I + \frac{1}{Q_t} U)$$

which, being commutative, is less or equal to

$$\prod_{t=1}^{\infty} \exp \frac{U}{Q_t} = \exp \sum_{t=1}^{\infty} \frac{U}{Q_t}$$

which is obviously convergent. Also,

$$|G_n - I| \leq |\prod_{t=1}^{\infty} (I + \frac{1}{Q_t} U) - I| \leq$$

$$\leq |\exp(\sum_{t=1}^{\infty} \frac{1}{Q_t} U) - I| \leq \exp \frac{C_9}{Q_o} - 1 \leq \frac{C_{10}}{Q_o} .$$

By choosing Q_o sufficiently large,

$$|G_n - I| < \epsilon$$

and therefore $|p_k|_s + |q_k|_s < \epsilon$, for $s = 1$, as stated in the Theorem.

In order to prove such a Theorem we now have to eliminate the restriction $\alpha_k(x) = x_k$. It is sufficient to introduce a change of variable and the proof of the Theorem follows again from the second Lemma. In fact, let

$$T_\mu \begin{cases} x_k^* = x_k + G_k(x;y) \\ y_k^* = y_k + \alpha_k(x) + F_k(x;y) \end{cases}$$

be defined in $a_k \leq x_k \leq b_k$.

Let

$$\xi_k = a_k(x), \qquad \eta_k = y_k \, . \tag{4.4.53}$$

T_μ becomes now

$$T \begin{cases} \xi_k^* = \xi_k + g_k(\xi;\eta) \\ \eta_k^* = \eta_k + \xi_k + f_k(\xi;\eta) \end{cases}$$

since the transformation $(x;y) \to (\xi;\eta)$ is 1:1 with bounded m derivatives. Also

$$f_k(\xi;\eta) = F_k(x(\xi);\eta)$$

$$g_k(\xi;\eta) = \alpha_k(x + G(x;y)) - \alpha_k(x) =$$

$$= \alpha_k(x(\xi) + G(x(\xi);\eta)) - \alpha_k(x(\xi))$$

and therefore, (4.4.17) and (4.4.18) are satisfied for a C_o conveniently chosen. The ring where T is defined contains the ring of width C_o^{-1}, i.e.,

$$\alpha_k(b) - \alpha_k(a) \geq C_o^{-1}$$

with

$$\alpha_k(a) \leq \xi_k \leq \alpha_k(b)$$

which results from (4.4.53).

The proof of the Theorem for $s > 1$ is done in a similar way by establishing new appropriate relations among the parameters $M, Q, \delta, q, s, \nu, \sigma, m$ (e.g., Moser, 1962, 1966).

It is important to note that

I) With respect to Kolmogorov's Theorem, Moser's Theorem is equivalent but does not require analyticity of the Hamiltonian (in the case analyticity of mapping T_μ). The existence of derivatives up to a certain order only is required. This was possible by smoothing out the function involved by a convenient operation.

II) The presence of small divisors in the series, of the form $\exp[i(j\cdot\alpha)] - 1$, is controlled by estimate of type (4.4.23).

Classical results of diophantine approximation show that such an estimate, for all integers j_k, will not be satisfied only by a set of values of $\alpha_1,\alpha_2,\ldots,\alpha_n$ whose measure, with respect to the unit cube $0 < \alpha_k < 1$ $(k = 1,2,\ldots,n)$ is $O(\epsilon)$ and goes to zero with ϵ. The possible small values of such denominators are balanced by the fact that the convergence to the transformation T^∞ is as fast as $\delta_o^{q^n}$.

III) The nondegeneracy condition required by Moser's Theorem is, as mentioned before, the same given by Arnol'd, a less stringent condition than required by Kolmogorov.

In fact, since

$$\alpha_k = 2\pi\omega_k = 2\pi\Omega_k/\Omega_n , \qquad \Omega_n \neq 0$$

condition (4.4.16) can be written

$$\left|\sum_{k=1}^{n} j_k \Omega_k\right| \geq \epsilon' \left(\sum_{k=1}^{n} |j_k|\right)^{-n+1/2}$$

which is the generalized irrationality condition of Siegel. The condition

$$\det \left(\frac{\partial\alpha_k}{\partial x_j}\right) \neq 0,$$

making use of the definition

$$\alpha_k = 2\pi \frac{\dfrac{\partial H_o}{\partial x_k}}{\dfrac{\partial H_o}{\partial x_n}} (x_1,\ldots,x_{n-1}, x_n(x_1,\ldots,x_{n-1},h))$$

and the relations

$$\frac{\partial x_n}{\partial x_j} = - \frac{\partial H_o}{\partial x_j} \Big/ \frac{\partial H_o}{\partial x_n} ,$$

lead to the generalization of (4.3.11), that is,

$$\begin{vmatrix} \dfrac{\partial^2 H_o}{\partial x_1^2} & \dfrac{\partial^2 H_o}{\partial x_1 \partial x_2} & \cdots & \dfrac{\partial^2 H_o}{\partial x_1 \partial x_n} & \dfrac{\partial H_o}{\partial x_1} \\ \dfrac{\partial^2 H_o}{\partial x_2 \partial x_1} & \dfrac{\partial^2 H_o}{\partial x_2^2} & \cdots & \dfrac{\partial^2 H_o}{\partial x_2 \partial x_n} & \dfrac{\partial H_o}{\partial x_2} \\ \hdashline \dfrac{\partial^2 H_o}{\partial x_n \partial x_1} & \dfrac{\partial^2 H_o}{\partial x_n \partial x_2} & \cdots & \dfrac{\partial^2 H_o}{\partial x_n^2} & \dfrac{\partial H_o}{\partial x_n} \\ \dfrac{\partial H_o}{\partial x_1} & \dfrac{\partial H_o}{\partial x_2} & \cdots & \dfrac{\partial H_o}{\partial x_n} & 0 \end{vmatrix} \neq 0. \qquad (4.4.54)$$

It is also obvious that if $\Omega_k \neq 0$ $(k = 1,2,\ldots,u)$ and if the usual non-degeneracy condition of Kolmogorov's Theorem is satisfied, so will (4.4.54). The reverse is not necessarily true.

5. Degenerate Systems

Here we limit ourselves to some remarks which should allow the proof of a theorem equivalent to Arnold's Theorem in the degenerate case.

Let μ be the small parameter entering H, the Hamiltonian of the system with $0 < \mu \leq 1$. In general, however, μ is much less than one. We shall also admit that some $\Omega_k = 0$ $(k = 1,2,\ldots,p < n)$. We shall define, as in Arnold's Theorem the "secular" part H_{1s} of H and, from this, being

$$H = H_o + \mu(H_{1s} + H_{1p}),$$

we define

$$\Omega_k = \frac{\partial H_{1s}}{\partial x_k} \qquad k = 1,2,\ldots,p,$$

and

$$\Omega_j = \frac{\partial H_o}{\partial x_j} \qquad j = p+1,p+2,\ldots,n.$$

Hence, the transformation T_μ will be written in the form

$$\begin{aligned} x_k^* &= x_k + \mu\, G_k(x;y) && (k = 1,2,\ldots,n) \\ y_k^* &= y_k + \mu\{\alpha_k(x) + F_k(x;y)\} + \beta_k && (k = 1,2,\ldots,p) \qquad (4.5.1) \\ y_j^* &= y_j + \alpha_j(x) + \mu\, F_j(x;y) && (j = p+1,p+2,\ldots,n) \end{aligned}$$

defined in $0 < a_k \leq x_k \leq b_k$, $b_k - a_k \geq 1$, for $k = 1,2,\ldots,n$. Under such conditions, Moser's Theorem should still be valid with the following modifications:

a) The transformation (4.4.14) has to be substituted by

$$\tilde{y}_k = y'_k + \mu\,\alpha_k(x^0) = y'_k + \mu\,\alpha_k \qquad (k = 1,2,\ldots,p)$$

$$\tilde{y}_k = y'_j + \alpha_j(x^0) = y'_j + \alpha_k \qquad (j = p+1,p+2,\ldots,n)$$

and δ_0 can again be chosen independently of μ. This is quite important, since is for $\mu \to 0$ we would have $\delta_0 \to 0$ and the proof would not hold.

b) The transformation $T_0[T_\mu(\mu = 0)]$ has to be substituted by

$$x^*_k = x_k \qquad (k = 1,2,\ldots,n)$$

$$y^*_k = \mu\,\alpha_k(x) + y_k + \beta_k \qquad (k = 1,2,\ldots,p)$$

$$y^*_j = \alpha_j(x) + y_j \qquad (j = p+1,p+2,\ldots,n)$$

with

$$\mu\,\alpha_k(a) < \mu\,\alpha_k(x) < \mu\,\alpha_k(b).$$

c) The interval defined by (4.4.15) must be modified to

$$\alpha_k(a) + \epsilon < \frac{\alpha_k - \beta_k}{\mu} < \alpha_k(b) - \epsilon$$

and, finally, the condition (4.4.16) has to be written

$$\left|\sum_{k=1}^{p} m_k \mu_k \alpha_k + \sum_{j=p+1}^{N} m_j \alpha_j + 2\pi j_{N+1}\right| \geq \mu\epsilon \left\{\sum_{1}^{N} |m_k|\right\}^{-N-1/2}$$

for all integers m_k, m_j not all zero.

NOTES

The problem of reducing a Hamiltonian system of differential equations or a measure preserving mapping in the vicinity of an equilibrium or a fixed point respectively, are analogous and present the same kind of difficulties, leading in general to divergence of the asymptotic series which represent the transformation. This does not, however, indicate that a normal form does not exist, but simply, at most, that such form cannot be obtained by a power series. The basic idea behind Birkhoff's Normalization of a Hamiltonian System is to describe all motions in the neighborhood of a point of equilibrium, as a natural extension of Lyapunov's construction of periodic solutions in that neighborhood.

Birkhoff's basic theorem states that if the characteristic exponents (the normal modes at the equilibrium point) are rationally independent, a normal form exists in the formal sense and the Hamiltonian can be written as a formal power series in $\zeta_1, \zeta_2, \ldots, \zeta_n$ where $\zeta_k = x_k^2 + y_k^2$, in some convenient coordinates and momenta y and x. Obviously, as noted before, the ζ's become automatically n integrals of motion. In general, they are only formal integrals since the transformation from the original form to the normal form is done by a formal series, in general divergent. In cases of resonance, that is, when rational dependence exists among the normal modes, the situation is not changed and, generally, the normal form obtained by Gustavson in 1966 is divergent, as his numerical experiments also indicate when compared with similar experiments by Hènon and Heiles in 1964. About the denseness of Hamiltonians for which the reduction to Normal Form can be obtained by convergent series, a precise result was obtained by Siegel in 1941. For a system with two degrees of freedom, the Taylor expansion of H about an elliptic equilibrium point is

$$H = \sum_{k=1}^{2} \frac{1}{2} \nu_k (y_k^2 + x_k^2) + H_3(y;x) + \cdots \tag{A}$$

where

$$H_n = \sum_{k,\ell} c_{k\ell} x^k y^\ell = \sum_{k_1} \sum_{k_2} \sum_{\ell_1} \sum_{\ell_2} c_{k_1 k_2 \ell_1 \ell_2} x_1^{k_1} x_2^{k_2} y_1^{\ell_1} y_2^{\ell_2}$$

and, from the convergence of H, one can assume by proper normalization of time and space that

$$|c_{k\ell}| \leq 1. \tag{B}$$

Let S be the set of all H of form (A) that satisfy (B). Siegel has shown that the Hamiltonians with irrational ν_1/ν_2 for which the Birkhoff transformation to Normal Form diverges are dense in S with the topology defined by choosing the neighborhoods of prescribed $c^o_{k\ell}$ as satisfying $|c_{k\ell} - c^o_{k\ell}| < \epsilon_{k\ell}$, for all k,ℓ and a given sequence of $\epsilon_{k\ell} > 0$. The rate of decrease of $\epsilon_{k\ell}$ with k,ℓ is arbitrary. Therefore a complete knowledge of H is necessary to decide on convergence or divergence, a fact which makes the reduction to normal form of very little physical significance when one considers error in estimating parameters. A stronger result by Siegel in 1954 indicates that cases in which convergence of Birkhoff's Transformation occurs are exceptional, in the sense that they form a set which is the union of at most denumerably many nowhere dense sets. It is worth mentioning that Rüssmann in 1964 has shown that if the system of differential equations generated by (A) admits the analytic integral

$$G = \sum_{k=1}^{2} \beta_k (y_k^2 + x_k^2) + \cdots$$

with $\nu_1\beta_2 - \nu_2\beta_1 \neq 0$, then the Birkhoff Transformation converges. This justifies the fact that Integrable Hamiltonian systems in the vicinity of an equilibrium point can be defined to be those for which Birkhoff's Transformation converges. Being exceptional, confirms Poincaré's conjecture, but in a very precise sense.

The relation between Hamiltonian Systems and measure preserving mappings is easily established. Given such a mapping, say M, if M^o = identity, $M^1 = M$, $M^{t+s} = M^t M^s$, then M can be generated by the solution of a Hamiltonian system

$$\dot{y} = H_x(y;x), \quad \dot{x} = -H_y(y;x),$$

that is,

$$M^t \begin{cases} y = y(y_o;x_o;t) \\ x = x(y_o;x_o;t) \end{cases} .$$

In such a case M has an integral, $H(y;x)$, which is invariant with respect to M. Reciprocally, if M possesses an integral, then

$$M^o = \text{identity}, \quad M^1 = M, \quad M^{t+s} = M^t M^s$$

and, also, M can be reduced to a normal form by a convergent Birkhoff Transformation. The foundations for these results were laid down by Birkhoff in 1922. For more details and references, Moser's paper in 1961 on the integrability of mappings is suggested.

As far as the reduction to normal form of a system

$$\dot{x}_k = f_k(x), \qquad k = 1,2,\ldots,n$$

in the vicinity of an equilibrium point $x = 0$, the convergence of a transformation, produced by a series, can be easily established.

If

$$f_k(x) = \sum_{\ell=1}^{n} c_{k\ell} x_\ell + \text{higher order terms}$$

and the eigenvalues $\lambda_1, \lambda_2, \ldots, \lambda_n$ of the matrix $C = \{c_{k\ell}\}$ are all distinct, there exists a constant nonsingular matrix P such that

$$P^{-1} CP = \operatorname{diag}(\lambda_1, \lambda_2, \ldots, \lambda_n) = \lambda$$

and in the neighborhood of $x = 0$, we can assume the form

$$\dot{x} = \lambda x + \Phi(x).$$

If $\sum_k j_k \lambda_k - \lambda_\ell \neq 0$ with $\sum_k j_k \geq 2$, j_k integers, then there exists a transformation

$$x = y + \psi(y)$$

such that the original equation reduces to

$$\dot{y} = \lambda y.$$

Convergence can be established when Re λ_k have all the same sign, for in this case one shows that the small divisors of the problem, i.e., $\sum j_k\lambda_k - \lambda_\ell$, are bounded away from zero. Obviously, the above hypotheses cannot be met by a Hamiltonian system. But for nonconservative systems in general, Normalization is obviously a simpler affair then the application of Lie Series Methods and they both accomplish the same results.

A clear connection between reduction to a normal form and stability, is given by the following theorem (see Siegel-Moser, 1971). Let D be a bounded domain of $\mathbb{C}^n$ containing the origin and let $\dot{z} = f(z)$, $z = (z_1,\ldots,z_n)$, hold in D with $f(z)$ analytic in D, $f(0) = 0$.

Theorem: "If the solution of $\dot{z} = f(z)$, $z(0) = z_o \in D$, lies in D for all t and for all $z_o \in D$, there exists a coordinate transformation $z = u(\zeta) = \zeta$ + higher order terms, in a neighborhood D' of 0, $D' \subseteq D$, such that the transformed differential equation has the form $\dot{\zeta} = A\zeta$, where A is a constant matrix, diagonalizable and its eigenvalues have zero real part."

The Moser Theorem on invariant curves is of simpler understanding when stated for two dimensional systems, that is, in a plane. Consider an annulus A given by $A\{1 \leq R \leq 2\}$, where $R = x^2 + y^2$. If θ is the polar angle

$$dx\,dy = \frac{1}{2}\,dR\,d\theta.$$

The Twist Mapping is defined by

$$M_o \begin{cases} R^* = R \\ \theta^* = \theta + \gamma(R) \end{cases}.$$

M_o is obviously area preserving and circles R = constant are invariant under M_o.

It is important that the following twist condition be verified, $\gamma'(R) \neq 0$, $R \in A$. The basic properties of M_0 are that each circle for which $\gamma(R)$ is commensurable with 2π, $\gamma/2\pi = p/q$, consists of fixed points of M_0^q. Each circle for which γ is incommensurable with 2π is densely covered by the images of M_0^q of any point of the circle. Moser considers the perturbed twist mapping

$$M_\epsilon \begin{cases} R^* = R + \epsilon f(R,\theta,\epsilon) \\ \theta^* = \theta + \gamma(R) + \epsilon g(R,\theta,\epsilon) \end{cases}$$

where f,g are periodic of period 2π in θ. M_ϵ is defined in A. The presence of the small parameter ϵ is not necessary but here we choose this form for convenience. If M_ϵ is area preserving, given any rational number p/q between $\gamma(1)/2\pi$ and $\gamma(2)/2\pi$, there exist $2q$ fixed points of M_ϵ^q satisfying

$$R^q = R$$
$$\theta^q = \theta + 2\pi p$$

for ϵ sufficiently small. Moser's Theorem deals with the perturbation of those circles $R = \text{const.}$ for which $\gamma(R)/2\pi$ is irrational, and can be stated as follows (case of "small" perturbation):

Theorem: "Let $\gamma'(R) \neq 0$, $R \in A$, and let any curve Γ surrounding $R = 1$ and its image curve $\Gamma^* = M_\epsilon(\Gamma)$ intersect each other. The functions f,g are assumed to be sufficiently smooth. Then, given any number ω, between $\gamma(1)$ and $\gamma(2)$, incommensurable with 2π, and satisfying

$$|q\omega - 2\pi p| \geq c|q|^{-3/2}$$

for all integers p,q, $q \neq 0$, there exists a differentiable closed curve

$$\Gamma \begin{cases} R = F(\phi,\epsilon) \\ \theta = \phi + G(\phi,\epsilon) \end{cases}$$

with F,G of period 2π in ϕ, which is invariant under M_ϵ, provided ϵ is sufficiently small. Moreover, the image point of a point on Γ is obtained by replacing Φ by $\Phi + \omega$."

Clearly, if M_ϵ is area preserving, if Γ is a curve surrounding $R = 1$, its image $M_\epsilon(\Gamma)$ necessarily intercepts Γ, for in view of the area preserving condition, it cannot lie inside or outside Γ, for M_ϵ sufficiently near M_0.

The theorem is readily applicable for a straightforward proof of Arnol'd's Stability Theorem of elliptic equilibrium points of two-dimensional systems. An important application was also indicated by Kyner in 1968, in the problem of motion of a satellite of an oblate planet with revolution symmetry. In the same publication of the AMS, Moser applies his theorem to the problem of Adiabatic Invariants, in magnetic fields with closed slowing varying magnetic surfaces. For these applications we suggest the original works.

From the geometric point of view, Kolmogorov's Theorem takes an interesting aspect. In fact, consider a Hamiltonian $H(y;x;\epsilon)$, analytic at $\epsilon = 0$, for $x \in D \subset R^n$ and $|\text{Im } y| < \rho$. Also $H(y+2\pi;x;\epsilon) = H(y;x;\epsilon)$ and $H(y;x;0) = H_0(x)$. Let $x = x_0 \in D$ and define

$$\omega = \frac{\partial H}{\partial x} = (\omega_1, \omega_2, \ldots, \omega_n)$$

for $\epsilon = 0$, $x = x_0$. Suppose further that x_0 can be chosen so that at $\epsilon = 0$, $\det H_{xx} \neq 0$ and moreover the ω_k $(k = 1,2,\ldots,n)$ satisfy the usual irrationality condition.

Under the above hypotheses, Kolmogorov's Theorem is equivalent to state the existence of analytic (vector) functions u,v such that

$$y = \theta + u(\theta)$$
$$x = c + v(\theta)$$

represent an invariant torus. The motion on the torus is given by $\dot{\theta} = \omega$, where $\theta = (\theta_1, \theta_2, \ldots, \theta_n)$. The reduction of the above problem to a measure preserving mapping is easily accomplished. In fact, considering for simplicity $n = 2$, and

$$H = H_o(x_1,x_2) + \epsilon H_1(y_1,y_2,x_1,x_2,\epsilon)$$

let $x_1 = r$, $y_1 = \theta$, $y_2 = \tau$, $x_2 = x$. Moreover, suppose $\partial H_o/\partial x_2 \neq 0$. The elimination of t from the differential equations of motion and of x via the use of the energy integral, yields

$$\frac{dr}{d\tau} = -\frac{\partial H}{\partial \theta} \Big/ \frac{\partial H}{\partial x} = F(r,\theta,\tau,\epsilon) \; ,$$
$$\frac{d\theta}{d\tau} = -\frac{\partial H}{\partial r} \Big/ \frac{\partial H}{\partial x} = G(r,\theta,\tau,\epsilon) \; , \qquad \text{(C)}$$

where, obviously, F and G have period 2π in τ. Moreover, if (r_o,θ_o) are initial conditions corresponding to a given value of the energy and to $\tau = 0$, $H(\theta_o,0,r_o,x) = h$, the solution

$$\theta = \theta(\theta_o,r_o,\tau,\epsilon)$$
$$r = r(\theta_o,r_o,\tau,\epsilon)$$

for $\tau = 2\pi$, maps the point (θ_o,r_o) in the plane $\tau = 0$, into the point (θ,r) in the plane $\tau = 2\pi$. (These planes are normal to the τ axis.) Therefore, the mapping

$$M_\epsilon \begin{cases} \theta^* = \theta(\theta_o,r_o,2\pi,\epsilon) \\ r^* = r(\theta_o,r_o,2\pi,\epsilon) \end{cases}$$

is area preserving because of the Hamiltonian character of system (C) and, for $\epsilon = 0$, one easily sees that

$$M_o \begin{cases} \theta^* = \theta_o + 2\pi \dfrac{\omega_1}{\omega_2} \\ r^* = r_o \end{cases}$$

where

$$\frac{\omega_1}{\omega_2} = \frac{\partial H_o}{\partial x_1} \Big/ \frac{\partial H_o}{\partial x_2} = \frac{\partial H_o}{\partial r} \Big/ \frac{\partial H_o}{\partial x} = \frac{1}{2\pi}\,\alpha(r),$$

since x can be eliminated by using the energy integral. Then M_0 is given by

$$M_0 \begin{cases} \theta^* = \theta_0 + \alpha(r_0) \\ r^* = r_0 \end{cases}$$

and for $\epsilon \neq 0$,

$$M_\epsilon \begin{cases} \theta^* = \theta_0 + \alpha(r_0) + \epsilon\varphi(\theta_0, r_0, \epsilon) \\ r^* = r_0 + \epsilon\psi(\theta_0, r_0, \epsilon) \end{cases} .$$

Moser's Theorem on the "small" twist mapping can now be applied by properly restricting the value of $x_1 (= r)$, $0 < a \leq r_0 < b$.

A subject very closely related to the type of perturbations methods we have been discussing is the concept and utilization of Adiabatic Invariants. Such a field is much too extensive to be included in the present lectures, but we wish to mention some of the basic results available in the literature. The earliest reference to a properly defined situation is a paper by Andronov and others in 1928, mentioned by Arnol'd. Classical and extensive works are those of Kasuga in 1961 and Kruskal in 1961. Arnol'd studied part of the problem in 1962 and, quite extensively, again in 1963 (Chapter II, p. 111-124). Recent works, related one way or another to the subject, are by Kartsatos (1971), Wasow (1971) and Hallam (1972).

The term "Adiabatic" has the classical meaning, common in thermodynamics, of a very slow change of the parameters which specify the physical configuration of a system. Consider a dynamical system corresponding to a Hamiltonian

$$H = H(y;x;\tau) .$$

where $\tau = \epsilon t$ and ϵ is a small parameter. We have the following definition.

"A function $I(y;x;\tau)$ is called an Adiabatic Invariant of the above system if, for any $\delta > 0$, it is possible to find an $\epsilon_0 > 0$ such that when $0 < \epsilon < \epsilon_0$ then for all t in the interval $0 < t < 1/\epsilon$, one verifies

$$|I(t) - I(0)| < \delta. \text{"}$$

We have set $I(t) = I(y(t);x(t);\epsilon t)$ and $y(t),x(t)$ are integral curves of the system generated by H above.

Clearly, any integral is an adiabatic invariant, in fact, a perpetual invariant. The most typical example in systems with one degree of freedom is the following, and is clearly suggested by Quantum Mechanics. Consider $H = H(y,x,\tau)$, and for τ = const., the equation $H(y,x,\tau)$ = const. = $E(y_0,x_0,\tau)$ is an instantaneous configuration of the energy level lines in the (y,x) plane. We assume that for values of τ and E in some range, these lines form closed trajectories. In this case they enclose a certain domain, the area of which we write $2\pi\, I(y_0,x_0,\tau)$, where (y_0,x_0) is any point of the corresponding trajectory. Then one can show that I is an adiabatic invariant, in the above sense. For instance, consider a simple pendulum with slowly varying length, that is, the Hamiltonian is

$$H = \frac{1}{2}x^2 + \frac{1}{2}\omega^2y^2$$

where $\omega = \omega(\tau) > 0$ for all $\tau = \epsilon t$. Then,

$$I = \frac{H}{\omega} = \frac{1}{2\omega}x^2 + \frac{1}{2}\omega y^2 .$$

Although in a finite time $I(t)$ may vary little, it may have large (even secular) variation over a long period of time. Nevertheless, it has been shown (see Arnol'd, 1963, Chapter II, §2) that:

"For a slow periodic variation of the Hamiltonian $H(y,x,\tau)$ of a non-linear oscillating system with one degree of freedom, an adiabatic invariant is perpetually conserved. That is, for any $\delta > 0$ it is possible to find an $\epsilon_0(\delta) > 0$ such that for $|\epsilon| < \epsilon_0$ it follows that $|I(t) - I(t_0)| < \delta$ for all t finite."

This important fact is a result of the nonlinearity of the system, that is, the property of the frequency being amplitude dependent. The same result cannot be obtained in a linear system due to its classical instability at resonance.

In order to show the strong connection between Adiabatic Invariants and

Formal Integrals produced by asymptotic perturbation methods, we consider a system, with one degree of freedom, defined by

$$H(y,x,\tau) = H(y,x,\tau+2\pi), \quad \tau = \epsilon t.$$

For τ = const., the system is autonomous and directly integrable. In terms of the action-angle variables (I,ω) defined by the Hamilton-Jacobi generator $S = S(y,I,\tau)$, we have

$$x = \frac{\partial S}{\partial y}, \quad \omega = \frac{\partial S}{\partial I}$$

where

$$S(y,I,\tau) = \oint_h x(h,y,\tau)dy.$$

The above integral is computed along the closed trajectory defined by

$$h = H_o(I,\tau)$$

where the function $h = H_o$ is the inverse of

$$I(h,\tau) = \frac{1}{2\pi} \oint_{H=h} x(h,y,\tau)dy$$

and $x(h,y,\tau)$ is obtained from

$$H(y,x,\tau) = h.$$

For a modern and clear exposition of the action-angle treatment we suggest the book of Meirovitch (1970, Chapter 9).

The quantity I is obviously a constant and the angle ω, on the circle I = const., varies uniformly

$$\dot{\omega} = \dot{\omega}(I,\tau) = \frac{\partial H_o}{\partial I} .$$

Up to now we have considered τ to be constant. If now τ varies with time, $\tau = \epsilon t$, the above description is only an approximation (adiabatic) during $t \sim 1/\epsilon$. In general, however, we can write

$$H(I,\omega,\tau) = H_0(I,\tau) + \epsilon\, H_1(I,\omega,\tau)$$

where, since the transformation $(y,x) \to (\omega,I)$ is canonical but time dependent explicitly through τ,

$$H_1 = \frac{\partial S}{\partial \tau} = H_1(I,\omega,\tau).$$

The last is a single-valued function of ω and τ, with period 2π with respect to both. The classical picture of the situation in the phase-space (y,x,τ) is the following. One identifies points with coordinates τ, $\tau + 2\pi$ and ω, $\omega + 2\pi$. The equation $I = \text{const.}$ determines a two-dimensional torus, with angular coordinates τ and ω. If $\epsilon = 0$, the phase point moves along lines $\tau = \text{const.}$ with angular motion $\omega(I,\tau)$. For $\epsilon \neq 0$, a slow motion $(\tau = \epsilon t)$ takes place normally to the aforementioned, on the torus, and one has two frequencies. However, in the adiabatic approximation, the phase point remains on the invariant torus defined by $I = \text{constant}$. The true motion is close to the adiabatic approximation for $t \sim 1/\epsilon$. If the system is nonlinear, the adiabatic approximation is valid for all times. This is actually contained in Kolmogorov's theorem.

The mean frequency is defined by

$$\bar{\omega}(I) = \frac{1}{2\pi} \int_0^{2\pi} \omega(I,\tau)d\tau = \frac{d\bar{H}_0}{dI}$$

where

$$\bar{H}_0 = \frac{1}{2\pi} \int_0^{2\pi} H_0(I,\tau)d\tau.$$

The <u>nonlinear character</u> is given by the (non-degeneracy) condition

$$\frac{d^2\, \bar{H}_0}{dI^2} = \frac{d\bar{\omega}}{dI} \neq 0.$$

This condition implies the perpetual adiabatic invariance of I, that is, one can find invariant tori for the system corresponding to a variable τ, and they are close to the $I = \text{const.}$ tori for ϵ sufficiently small and all times. More precisely one can assert the following

1. "For any $\delta > 0$ it is possible to find $\epsilon_o > 0$ such that if $|\epsilon| < \epsilon_o$ then any point (y_o, x_o, τ_o) lies between two invariant tori T_1, T_2 where

$$|I(y_1, x_1, \tau_1) - I(y_2, x_2, \tau_2)| < \delta,$$

provided $(y_1, x_1, \tau_1) \in T_1$ and $(y_2, x_2, \tau_2) \in T_2$."

2. "Consider the Hamiltonian

$$H(I, \omega, \tau) = H_o(I, \tau) + \epsilon\, H_1(I, \omega, \tau)$$

such that $\partial H_o/\partial I \not\equiv 0$, $\partial^2 H_o/\partial I^2 \not\equiv 0$ and H is analytic for $|I - I_o| \leq \rho$. For every $\delta > 0$ it is possible to find $\epsilon_o > 0$ such that if $|\epsilon| < \epsilon_o$ and $|I(t_o) - I_o| \leq \rho^{-\delta}$, then for all t finite $|I(t) - I(t_o)| < \delta$."

By means of a Transformation Lemma, the Hamiltonian H is brought to the form

$$K(q, p, \theta) = \epsilon\, K_o(p) + \epsilon^2 K_1(q, p, \theta) + \cdots$$

where $\theta = \omega$ is the new independent variable and q is essentially τ (see Arnol'd, 1963, pp. 116-118). To the above Hamiltonian one can apply Kolmogorov's Theorem and the perpetual invariance of I follows.

Adiabatic invariance in systems with two degrees of freedom is also studied by Arnol'd (1963, Chapter 2, §3) and applied to the classical problem of Magnetic Traps. The result is that charged particles (say electrons) spiraling around magnetic field lines are trapped by the magnetic field, if it varies slowly or if the velocity of the particles is small. The basic Lemma put forth by Arnol'd to prove such a result is the following:

"Given the Hamiltonian $H(y_1', y_2, x_1, x_2)$ where $y_1' = \epsilon\, y_1$, analytic, suppose that for fixed values of (y_1', x_1) it defines an oscillatory system with action $P(y_1', x_1, h)$ and angle $\Omega(y_1', y_2, x_1, x_2)$. Then there exists an analytic transformation giving (y_1', y_2, x_1, x_2) in terms of the new variables $(y, \omega, p_y, p_\omega)$

such that:

a) The functions y_1', y_2, x_1, x_2 have period 2π with respect to ω and as $\epsilon \to 0$ one has $y \to y_1'$, $\omega \to \Omega$, $P_y \to x_1$, $P_\omega \to P$.

b) Along the solution of the system generated by H, the following canonical equations generated by $I(y, P_y, \omega, h)$ are satisfied

$$\frac{dP_y}{d\omega} = -\frac{\partial I}{\partial y}, \quad \frac{dy}{d\omega} = \frac{\partial I}{\partial P_y},$$

where h is a parameter (the energy integral $H = h$).

c) I has the form $I = -\epsilon P_\omega$, where

$$P_\omega = I_o(P_y, y, h) + \epsilon I_1(P_y, y, \omega, h) + \cdots$$

is an analytic function with period 2π with respect to ω, and

$$I_o(P_y, y, h) = P(y, P_y, h). \text{ "}$$

From the previous study of one-dimensional systems, it follows the perpetual adiabatic invariance of P and ϵx_1. The various stages of reduction to a Hamiltonian of the above type, which itself turns to be an adiabatic invariant, can be a perturbation technique in nonlinear systems.

In 1966, Contopoulos has compared perturbation techniques leading to a Third Integral (constructed by means of successive approximations to the Poisson's Parenthesis condition) and an Adiabatic Invariant constructed for the system generated by

$$H = \frac{1}{2}(x_1^2 + x_2^2 + \omega_1^2 y_1^2 + \omega_2^2 y_2^2) - \epsilon y_1 y_2^2 \sin \omega t = H_o + \epsilon H_1,$$

that is,

$$\dot{y}_k = H_{x_k}, \quad \dot{x}_k = -H_{y_k}.$$

The work, beside analytic developments of great interest, contains a numerical check of the results. The main conclusions are that, if no resonance exists among the frequencies ω_1, ω_2, then both methods give approximately equal good results. Nevertheless, if ω_1 and ω_2 are close to be commensurable, the adiabatic invariant degenerates quite rapidly (in time) and ceases to be valid at exact commensurability. In this case, on the other hand, a proper modification of the Third Integral can be found which holds good. As Contopoulos says, the non-validity of an adiabatic invariant at resonance was known in pioneer works, as referenced by Sommerfield, but it is seldom mentioned in modern literature.

We conclude these notes by mentioning that the present chapter falls short on the theory of existence of invariant manifolds of perturbed systems, along the lines defined by early works of Levinson (1950), Diliberto (1956), Kyner (1958) and Loud (1959). The description of such work, however, would be a repetition of the very extensive and master paper by Hale in 1960. The basic idea is to consider the autonomous system

$$\dot{x} = f(x), \quad x = (x_1, x_2, \ldots, x_n),$$

with a periodic solution

$$x = p_o(t) = p_o(t+T_o).$$

In the n-dimensional space (x,θ) the cylinder $x = p_o(\theta)$ is an invariant surface, that is, every solution which belongs to the cylinder at a certain time it will remain on it for all times.

Considering the perturbed system

$$\dot{x} = f(x) + \epsilon\, g(x,t,\epsilon),$$

where $g(x,t,\epsilon) = g(x,t+T,\epsilon)$, it can be shown that if $f(x)$ and $g(x,t,\epsilon)$ are at least C^3, there exists for sufficiently small ϵ a surface $x = p(t,\theta,\epsilon) \in C^1$ in the (x,t) space, lying near the cylinder $x = p_o(\theta)$, which is an invariant surface of the perturbed system. One also verifies that $p(t,\theta,\epsilon)$ has period T with respect to t and T_o with respect to θ, and $p(t,\theta,0) = p_o(\theta)$.

The extension of the above results to almost periodic (in t) functions

$g(x,t,\epsilon)$ is described by Hale in 1960. The result is analogous, with the remark that $p(t,\theta,\epsilon)$ is also almost periodic in t. In 1966, Pliss has given conditions for the existence of invariant manifolds independent of the concept of perturbation, that is, for systems of the form

$$\dot{x} = X(x,y,t), \quad \dot{y} = Y(x,y,t)$$

where x,y are n-dimensional vectors. The basic assumptions are $X,Y \in C^1$ in some domain $\{x \in R^n,\ t \in R,\ \|y\| \leq K\}$ and X,Y have period T in t. The proof makes use of a reduction of the problem to a contraction mapping and then to the existence of a fixed point. This same line of reasoning is extensively explored by Kartsatos in 1972. In this paper, many pertinent interesting references can be found, but their discussion will not be undertaken here.

REFERENCES

1. Andronov, A. A. et al., 1928, "A Contribution to the Theory of Adiabatic Invariants", Zh. Russ. Khim. Obsh. 60, 413-457.

2. Arnol'd, V. I., 1962, "On the Behavior of An Adiabatic Invariant Under a Slow Periodic Change of the Hamiltonian", Dokl. Akad. Nauk USSR 142, 758-761.

3. ____________, 1963, "Small Denominators and Problems of Stability of Motion in Classical and Celestial Mechanics", Uspeki Mat. Nauk USSR 18, 91-192 (specific references to pages of this article refer to translation in Russ. Math. Surveys 18, No. 6, 85-191(1963)).

4. Birkhoff, G. D., 1922, "Surface Tranformations and their Dynamical Applications", Acta Math. 43, 1-119.

5. ____________, 1927, "Dynamical Systems", Amer. Math. Soc. Colloquium Publ. IX, Providence, R.I.

6. Contopoulos, G., 1966, "Adiabatic Invariants and the Third Integral", J. Math. Phys. 7, 788-797.

7. Danby, J. M. A., 1970, "Wild Dynamical Systems" in "Periodic Orbits, Stability and Resonances" Ed. G. E. O. Giacaglia, D. Reidel Pub. Co., Dordrecht, Holland.

8. Diliberto, S. P., 1956, "An Application of Periodic Surfaces. Solution of a Small Divisor Problem", Contrib. to the Theory of Nonlinear Oscill., Vol. 3, Ann. of Math. Studies 36, 257-261.

9. ____________ and Hufford, G., 1956, "Perturbation Theorems for Nonlinear Ordinary Differential Equations", Ann. Math. Studies 36, 207-237.

10. Gustavson, F. A., 1966, "On Constructing Formal Integral of a Hamiltonian System near an Equilibrium Point", Astron. J. 71, 670-686.

11. Hale, J. K., 1960, "Integral Manifolds of Perturbed Differential Equations", Ann. Math. 73, 496-531.

12. Hallam, T. G., 1972, "Convergence of Solutions of Perturbed Nonlinear Differential Equations" (To appear in Ann. Mat. Pura Appl.).

13. Hènon, M. and Heiles, C., 1964, "The Applicability of the Third Integral of Motion; Some Numerical Experiments", Astron. J. 69, 73-79.

14. Kartsatos, A. G., 1971, "On the Maintenance of Oscillations of the n-th Order Equations under the Effect of a Small Forcing Term", J. Diff. Eq. 10, 355-363.

15. ____________, 1972, "On the Relationship between a Nonlinear System and its Nonlinear Perturbation" (Preprint from Dept. of Math., Univ. of South Florida, Tampa, Fla.).

16. Kasuga, T., 1961, "On the Adiabatic Theorem for the Hamiltonian System of Differential Equations in Classical Mechanics", Proc. Japan Acad. 37, 366-382.

17. Kolmogorov, A. N., 1954, "On the Conservation of Quasiperiodic Motions for a Small Change in the Hamiltonian Function", Dokl. Akad. Nauk USSR 98, 527-530.

18. Kruskal, M., 1961, "Adiabatic Invariants", Princeton University Press, Princeton.

19. Kyner, W. T., 1958, "Small Periodic Perturbations of an Autonomous System of Vector Equations", Ann. of Math. Studies 41, 111-125.

20. __________, 1968, "Rigorous and Formal Stability of Orbits about an Oblate Planet", Mem. Amer. Math. Soc. 81, 1-27 (second part).

21. Levinson, N., 1950, "Small Periodic Perturbations of an Autonomous System with a Stable Orbit", Ann. of Math. 52, 727-738.

22. Loud, W. S., 1959, "Periodic Solutions of a Perturbed Autonomous System", Ann. of Math. 70, 490-529.

23. Meirovitch, L., 1970, "Methods of Analytical Dynamics", McGraw-Hill, New York.

24. Moser, J., 1955, "Nonexistence of Integrals for Canonical Systems of Differential Equations", Comm. Pure Appl. Math. 8, 409-436.

25. ________, 1961, "On the Integrability of Area-preserving Cremona Mappings near an Elliptic Fixed Point", Bol. Soc. Mat. Mexicana, 176-180.

26. ________, 1962, "On Invariant Curves of Area-preserving Mappings of an Annulus", Nachr. Akad. Wiss. Gottingen Math.-Phys. Kl. II a, No. 1, 1-20.

27. ________, 1964, "Hamiltonian Systems", Lecture Notes, New York University.

28. ________, 1966, "On the Theory of Quasi-periodic Motions", SIAM Rev. 8, 145-172.

29. ________, 1968, "Lectures on Hamiltonian Systems", Mem. Amer. Math. Soc. 81, 1-60.

30. Pliss, V. A., 1966, "On the Theory of Invariant Surfaces", Diff. Eq. 2, 1139-1150.

31. Poincaré, H., 1912, "Sur une théorème de Géométrie", Rend. Circ. Mat. Palermo 33, 375-407.

32. Rüssmann, H., 1964, "Über das Verhalten analytischer Hamiltonscher Differentialgleichungen in der Nähe einer Gleichgewichtslösung", Math. Ann. 154, 285-300.

33. Siegel, C. L., 1941, "On the Integrals of Canonical Systems", Ann. Math. 42, 806-822.

34. Siegel, C. L., 1954, "Über die Existenzeiner Normalform analytischer Hamiltonscher Differentialgleichungen in der Nähe einer Gleichgewichtsloung", Math. Am. 128, 144-170.

35. ____________, 1956, "Vorlesungen über Himmelsmechanik", Springer-Verlag, Berlin.

36. ____________ and Moser, J. K., 1971, "Lectures on Celestial Mechanics", Springer-Verlag, New York.

37. Sommerfeld, A., 1951, "Atombau und Spektrallinien", Vol. I, 7th Ed., Vieweg, Braunschweig (p. 370 and p. 698).

38. Wasow, W., 1971, "Adiabatic Invariance of a Simple Oscillator", Not. Am. Math. Soc. 18, No. 7 (Abstract).

39. Whittaker, E. T., 1937, "A Treatise on the Analytic Dynamics of Particles and Rigid Bodies", Cambridge Univ. Press, London (1960, Reprint of 7th Edition).

CHAPTER V

RESONANCE

1. Introduction.

The earliest verion of the problem along the lines to be described here is found in Bohlin's works (Poincaré, 1898, Vol. II), in von Zeipel's (1911) monograph on the theory of motion of asteroids and in Whittaker's problem of series solutions and adelphic integrals (1927), all of them directly or indirectly dealing with problems of celestial mechanics. In the theory of nonlinear and linear oscillations, the problem has been dealt with initially and studied by Lyapunov (1966), Bogoliubov (1945), Mitropolsky (1962) and Krylov (1934). Modern literature, that is, after the middle of the century is plentiful of works on resonance, generalizations and many different definitions and approaches.

Although it is a subject of general knowledge, its proper definition depends today very strongly on the particular taste of the author and on his field of research. We do not escape from such a bias, though we try to pose the problem in terms general enough to make their application as wide as possible.

The physical assumption is that we are given a differential system describing the behavior of a mechanism, either mechanic or electric. The mechanism is an oscillator in the sense that it can be described by a well defined set of action and angle variables In the general case, the angle variables have time frequencies rationally independent, while rational dependence is an exceptional but possible case. We assume, therefore, that an oscillator has a discrete spectrum of frequencies and that it is band limited. These are basic assumptions.

Typically, the problem we are interested in, is related to the behavior of an oscillator when small changes in its structure are introduced and/or external factors (or perturbations) are acting on the system. Such changes and perturbations produce effects which are drastically connected with the frequencies of the oscillator. If this moves in resonance (periodic solution), that is, there exists at least an integer linear combination of the frequencies which is zero, what is the effect of a small disturbance (internal or external) in the system. Or, if an external action is an·oscillator itself with frequencies which are rational with the frequencies of the system, what is the resulting motion. Classically, a harmonic oscillator (linear) subjected to a forcing action in resonance with the oscillator, will increase its amplitude with no bound. But in natural problems this is basically impossible since there exists no purely linear system, nor dissipative forces can be completely eliminated, or else, the system breaks down when the amplitude of oscillation reaches a value sufficient to destroy the system.

For a linear oscillator the frequency does not depend on the corresponding amplitude while the amplitude factor goes to infinity at exact resonance. For a nonlinear oscillator the frequency depends on the corresponding amplitude (and vice versa) so that, as the amplitude changes, the oscillator is driven away from resonant conditions. Nevertheless, when two nonlinear oscillators are in resonance, it is a common phenomenon that the resulting system is stationary with bounded amplitudes and fixed resonances. The oscillators are "locked in resonance". Such a situation is generally stable, in the sense that small changes in the configuration will produce oscillations about the stationary situation. Asymptotic stability can only occur in the presence of dissipative forces, but never in conservative systems, of course.

A typical problem taken as an example to illustrate resonance, much to the taste of authors working in celestial mechanics, is the simple pendulum. It was used intensively by Brown (1932) and, in a more sophisticated way, recently by Kyner (1967).

We shall not, here, follow this example again, but rather will approach the problem from a more general point of view.

In any event, it is quite important to realize that the behavior of a system under perturbations and resonance conditions can only be fully understood if the singular points of the associated differential equations are known and their character quite well specified. For systems with one degree of freedom this is generally an easy task and the characterization of the singular points (Fuchsian Theory) is well-known. For systems with two degrees of freedom such characterization has been generalized (Nemytskii, 1967) but the geometric visualization becomes practically impossible. Moreover, much of the geometric theory available for systems with one degree of freedom (e.g , the Birkhoff's Fixed Point Theorem, the Theorem of Bendixon-Poincaré, the theory of Limit Cycles, and similar geometric problems) is not generalizable to systems with two or more degrees of freedom.

Most of the description we shall give to problems of resonance, in non-linear system, is simply heuristic, except for few exceptional cases. The final outcome will, in most cases, be in the form of a generalized Bohlin's equation, to which we arrive by applying a principle generally called of minimum energy or conservation of stable stationary solutions. Historically, we wish to mention an interesting paper by Baker (1915) who encountered the problem of resonance (small divisors) in problems of celestial mechanics and made quite an interesting discussion of the overall situation. Much physical and mathematical hint can be drawn from that work. Modernly, the question has been tackled by many and one of the best reseaches still remains the work by Arnold (1963). The problem cannot, moreover, be disconnected from the question of structural stability, but for systems with more than two degrees of freedom, even the results on this field are very scarce (Peixoto, 1967). The known results shed some light on the theory of perturbations: Hamiltonian systems (time independent) can be approximated by a structurally stable system. In other words, given a conservative system, one can find another system which is "very close" to the given one and which is structurally stable ("very close" in the C^1 sense). In this respect we mention again the potential consequences of the fact that any system of ordinary differential equations, reducible to normal form, can be written in Hamiltonian form (Chapter 2, §4). In our opinion such a possibility should be explored in all aspects due to the special characteristics of a Hamiltonian system, and especially in regard to the properties of the pertinent invariant manifolds.

2. Motion in the Neighborhood of an Equilibrium Point.

In this section we give a short resume of the problem, its solution in terms of formal series and the approach necessary when the normal modes of oscillations are linearly independent over the set of integers (Chapter 3, §4).

Consider the Hamiltonian $H(y;x)$ to be analytic in a certain domain D of the phase space and the pertinent equations of motion

$$\dot{y} = H_x^T \qquad \dot{x} = -H_y^T \tag{5.2.1}$$

where x,y are n-vectors. We also admit the existence of an isolated stationary solution in D, corresponding to $(x^o;y^o)$, that is

$$H_{x^o} = H_{y^o} = 0 \tag{5.2.2.}$$

while we suppose the matrix $\frac{\partial^2 H}{\partial y \partial x}$ not to be singular at $x = x^o$, $y = y^o$. It follows, of course, that such a point is a maximum or a minimum for H which implies that the quadratic part of the Taylor series of H about that point can be reduced to a normal form, or, that is the same, the corresponding equations are those of n uncoupled oscillators. We shall assume (x^o,y^o) to be a point of minimum so that the above mentioned reduction can be achieved by means of real transformations. Let $\delta > 0$ be given. In correspondence one can find an $\epsilon > 0$ such that for

$$|x_o - x^o|, \quad |y_o - y^o| < \epsilon$$

with $(x_o;y_o) \in D$, one has, for all times

$$|x - x^o|, \quad |y - y^o| < \delta$$

where $(x;y)$ is the solution of (5.2.1) corresponding to the initial conditions $(x_o;y_o)$. Let us, therefore, define $q = y - y^o$ and $p = x - x^o$ so that one can assume p,q to be bounded for all times. Taking (q,p) as new variables and expanding $H(y^o + q; x^o + p)$ in Taylor series, in view of the hypotheses, and dropping constants, one gets

$$H = H_2 + H_3 + \cdots \tag{5.2.3}$$

where $H_k(q;p)$ is a homogeneous polynomial of degree k in the components of q,p. Series (5.2.3) is absolutely convergent. In particular, we have

$$H_2 = \frac{1}{2} q^T A q + q^T B p + \frac{1}{2} p^T C p \tag{5.2.4}$$

where $A^T = A$, $C^T = C$, and, evidently, H_2 is by hypothesis, a positive definite quadratic form, reducible by a linear symplectic real transformation to the normal form

$$H_2 = \frac{1}{2} (\eta^T \eta + \xi^T D^2 \xi) \tag{5.2.5}$$

where $D^2 = \mathrm{diag}(\omega_1^2, \omega_2^2, \ldots, \omega_n^2)$, the ω_j^2 $(j = 1,2,\ldots,n)$ are the eigenvalues of the problem, and the Jacobian J of the linear transformation

$$(q,p) \to (\eta,\xi)$$

that is

$$\frac{\partial(q,p)}{\partial(\eta,\xi)} = J$$

is constant, real and symplectic (Siegel, 1956). The relation (5.2.5) is equivalently written

$$H_2 = \frac{1}{2} \sum_{k=1}^{n} (\eta_k^2 + \omega_k^2 \xi_k^2)$$

so that H_2 generates n uncoupled harmonic oscillators, representing the limit motion around the equilibrium point when the amplitude of oscillation tends to zero. Applying the transformation $(q,p) \to (\eta,\xi)$ to the full Hamiltonian, one finds

$$H = \frac{1}{2} (\eta^T \eta + \xi^T D^2 \xi) + H_3 + H_4 \cdots$$

where H_k are homogeneous polynomials of degree k in the components of (η,ξ) and the equations can be written

$$\dot{\eta} = H_\xi = D^2\xi + \Phi(\xi,\eta)$$
$$\dot{\xi} = -H_\eta = -\eta + \Psi(\xi,\eta) \tag{5.2.6}$$

where Φ and Ψ are series of homogeneous polynomials in the components of η, ξ of minimum degree equal to 2. It is also true that, for all t, writing $H = H_2 + H_3 + H_4 + \cdots$, each H_k will be bounded by a quantity $O(\delta^k)$. The same observation applies to the series Φ and Ψ, and, evidently, (5.2.6) have unique solution in D.

We now introduce the homogeneous completely canonical transformation

$$x_k = \frac{1}{2\omega_k}(\eta_k^2 + \omega_k^2\xi_k^2)$$
$$\tan y_k = \eta_k/2\omega_k, \quad \omega_k > 0 \tag{5.2.7}$$

or

$$\xi_k = (2x_k/\omega_k)^{\frac{1}{2}} \cos y_k$$
$$\eta_k = (2x_k\omega_k)^{\frac{1}{2}} \sin y_k . \tag{5.2.8}$$

It follows that, since $|\eta_k|$, $|\xi_k|$ are bounded by some quantity $O(\delta)$ and all the ω_k $(k = 1,2,\ldots,n)$ are positive and finite, then $|x_k|$ is bounded by some quantity $O(\delta^2)$ and $|y_k| < \pi/2$ necessarily. Therefore, y_k is completely defined by (5.2.7) through the value of its tangent.

With the above transformation the Hamiltonian takes the form (Whittaker, 1937)

$$H = \omega_1 x_1 + \omega_2 x_2 + \cdots + \omega_n x_n + H_3 + H_4 + \cdots \tag{5.2.9}$$

where H_k has a finite number of terms corresponding to a trigonometric polynomial of maximum degree k in the y variables, that is

$$H_k = \sum_m \sum_\nu A_k^{m,\nu} x_1^{m_1} x_2^{m_2} \cdots x_n^{m_n} \exp[i(\nu_1 y_1 + \nu_2 y_2 + \cdots + \nu_n y_n)] \tag{5.2.10}$$

where

$$m_1 + m_2 + \cdots + m_n = \frac{1}{2} k$$
$$|\nu_k| \leq 2m_k \tag{5.2.11}$$
$$|\nu_1| + |\nu_2| + \cdots + |\nu_n| \leq k.$$

The relations (5.2.11) are called the D'Alembert Characteristics of H_k (or H). The m_k are positive half integers and the ν_k are integers. Since $x_k = O(\delta^2)$, it follows that $H_k = O(\delta^k)$, $k = 3,4,\ldots$. The integration of the system defined by the Hamiltonian (5.2.10) is now reduced to the perturbation of the solution

$$x_k = x_k^o = \text{const.}$$
$$y_n = \omega_k t + y_k^o \tag{5.2.12}$$

corresponding to $H = H_2 = \omega_1 x_1 + \omega_2 x_2 + \cdots + \omega_n x_n$. In fact, (5.2.12) represent infinitesimal oscillations in the neighborhood of the stable equilibrium point (x^o, y^o), which, in the new system of variables, corresponds to $x = y = 0$.

It is initially evident that H is degenerate in the sense that the matrix $\left\{ \frac{\partial^2 H_2}{\partial x_i \partial x_j} \right\}$ is singular. On the other hand, H_3 cannot contain secular terms (terms independent of the y_j variables) since it is an odd function of the vector y and periodic or period 2π in each component of this vector. It is also very important to remember that every H_k is a collection of a finite number of terms (k finite). The number of such terms increases with k according to the conditions (5.2.11). The form of H_2 is such that none of the theorem's discussed in Chapter 3 are applicable to the case. Nonetheless we can show, to begin with, that there exist, in any event, a formal series which normalizes H, that is, reduces it to the form

$$H = K(X;0)$$

where all angle variables are ignorable. In fact, we will show that a transformation exists, defined by a finite number of trigonometric polynomials of the same form that H, such that the above normalization can be achieved to any degree of approximation, although the limiting situation might lead to a divergent series. In a different problem, such reduction has been indicated by Deprit (1970) making use of Lie's series. We shall, at present, use the von Zeipel type approach, which, in fact, has been shown to be equivalent, in the sense discussed in Chapter 2.

3. Solution by Formal Series.

Initially, let us consider the case where $\omega_1, \omega_2, \ldots, \omega_n$ are linearly independent over the integers, i.e., the condition

$$j_1\omega_1 + j_2\omega_2 + \cdots + j_n\omega_n = 0, \qquad (5.3.1)$$

where $j_1, j_2, \ldots, j_n$ are integers, is satisfied if and only if all j's are zero. It implies also that none of the ω's can be zero, or, in other words, none of the variables are ignorable. But this is a direct consequence of the assumption that the equilibrium point is stable.

The von Zeipel's procedure cannot be used directly, without some previous considerations and extreme care on the definition of the order of a term. In fact, the Hamiltonian is

$$H = \sum_{k=1}^{n} \omega_k x_k + H_3 + H_4 + \cdots \qquad (5.3.2)$$

where $H_p = O(\delta^p)$ as a consequence that the degree of H_p in terms of $x_1, x_2, \ldots, x_n$ is $p/2$ and $x_j = O(\delta^2)$. It follows that differentiation with respect to an x variable will lower the order of a term by a factor 2, so that, for example

$$\frac{\partial^p H_k(x,y)}{\partial x_1^{p_1}\, \partial x_2^{p_2} \cdots \partial x_1^{p_n}}, \quad p_1 + p_2 + \cdots + p_n = p,$$

is $O(\delta^{k-2p})$. In the equations resulting from the generalized Hamilton-Jacobi equation, we shall, nevertheless, never have a term of negative order in δ. The generating function of the canonical transformation $(x,y) \to (X,Y)$ is chosen to be

$$S = X^T y + S_3(X;y) + S_4(X;y) + \cdots \tag{5.3.3}$$

so that

$$x_k = S_{y_k} = X_k + S_{3y_k} + S_{4y_k} + \cdots \tag{5.3.4}$$

and the normalized Hamiltonian is

$$K(X) = \sum \omega_k X_k + K_3(X) + K_4(X) + \cdots . \tag{5.3.5}$$

All series involved are, at this stage, purely formal, with the exception of (5.3.2) which is uniformly convergent. Inserting (5.3.4) into (5.3.2) and expanding in Taylor's series, the first few approximations are given by

$$\sum_k \omega_k \frac{\partial S_3}{\partial y_k} + H_3(X;y) = K_3(X) \tag{5.3.6}$$

$$\sum_k \omega_k \frac{\partial S_4}{\partial y_k} + \sum_k \frac{\partial H_3(X;y)}{\partial X_k} \frac{\partial S_3}{\partial y_k} + H_4(X;y) = K_4(X) \tag{5.3.7}$$

$$\begin{aligned} \sum_k \omega_k \frac{\partial S_5}{\partial y_k} &+ \sum_k \frac{\partial H_3(X;y)}{\partial X_k} \frac{\partial S_4}{\partial y_k} + \frac{1}{2} \sum_k \sum_j \frac{\partial^2 H_3(X;y)}{\partial X_k \partial X_j} \frac{\partial S_3}{\partial y_k} \frac{\partial S_3}{\partial y_j} \\ &+ \sum \frac{\partial H_4(X;y)}{\partial X_k} \frac{\partial S_3}{\partial y_k} + H_5(X;y) = K_5(X). \end{aligned} \tag{5.3.8}$$

In the event of non-resonance, the formal solution presents no problems in general. In fact, since

$$H_p(X;y) = \sum_{\nu^p} A_p^{\nu^p}(X) \exp i[\nu_1^p y_1 + \nu_2^p y_2 + \cdots + \nu_n^p y_n] \tag{5.3.9}$$

where $\nu^p = (\nu_1^p, \ldots, \nu_n^p)$ and $A_p^{\nu^p}(X)$ is a homogeneous function in $X_1, X_2, \ldots, X_n$ of degree $p/2$, it follows that, for example

$$K_3(X) = A_3^0(X)$$

$$S_3 = \sum_{\nu^p} A_p^{\nu^p}(X)\{i(\nu_1^3\omega_1 + \nu_2^3\omega_2 + \cdots + \nu_n^3\omega_n)\}^{-1} \qquad (5.3.10)$$

$$\cdot \exp i[\nu_1^3 y_1 + \cdots + \nu_n^3 y_n] + F_3(X;0)$$

where $F_3(X;0)$ is an arbitrary function of X. It can be set equal to zero without affecting higher order approximations, since the functions $S_3, S_4, \ldots$ appear only as derivatives with respect to the y variables. The equation of any order of approximation is of the same kind and solved in a similar way.

Now, let us consider the case in which there is one (and only one) set of integers (irreducible) $j_1, j_2, \ldots, j_n$, not all zero, such that (5.3.1) is satisfied. In such case, it is possible that some of the divisors present in (5.3.10) are zero, that is, some $(\nu_1^3, \nu_2^3, \ldots, \nu_n^3)$ is a multiple of the set $(j_1, j_2, \ldots, j_n)$. This does not necessarily happen in a finite number of approximations since H_p has a finite number of terms, for p finite. Nevertheless, it is clear that as p increases the denominators in question may become arbitrarily small. For rationally independent ω's, convergence may possibly be achieved by considering lower bounds already discussed in Chapters 3 and 4. In fact, Whittaker (1960) mentions the case that the series

$$\sum_{m=1}^{\infty} \sum_{n=1}^{\infty} \frac{q_1^m q_2^n}{m - n\alpha}$$

where $|q_1|, |q_2| < 1$ and α is an irrational number of order ≥ 2, do actually converge. For rationally dependent ω's, sooner or later a zero divisor will appear and therefore the method, as is, cannot be used.

In the case where H_3 contains no terms $\nu_1^3 y_1 + \cdots + \nu_n^3 y_n$ such that $\nu_1^3\omega_1 + \cdots + \nu_n^3\omega_n = 0$, all angular variables can still be eliminated by observing that in this case one can write S_3 as in (5.3.10) with the exception that the arbitrary function can now include the y-variables in the critical combination $j_1 y_1 + j_2 y_2 + \cdots + j_n y_n$, that is

$$F_3 = F_3(X;j_1y_1 + j_2y_2 + \cdots + j_ny_n). \tag{5.3.11}$$

An arbitrary function of such critical argument will not have any effect on (5.3.6) as it is easily verified. The arbitrary function is now used to cancel any term containing the critical argument (or a multiple) in the next approximations. Here, the important role played by the next to the quadratic form in the transformation involved in Birkhoff's fixed point theorem, is also felt with respect to the Hamiltonian.

In fact, one assumes

$$F_3(X;j^Ty) = \sum_\alpha B_3^\alpha(X)\exp\,[i\alpha(j^Ty)] \tag{5.3.12}$$

where α is an integer taking values to be determined in the next stage of approximation, and $B_3^\alpha(X)$ are functions of $X_1,X_2,\ldots,X_n$, homogeneous of degree $3/2$.

By defining

$$H_3(X;y) = H_{3s}(X) + H_{3p}(X;y) \tag{5.3.13}$$

it follows that

$$B_3^\alpha(X) = -\frac{i}{\alpha}A_4^\alpha(X)\left\{\sum_k j_k\frac{\partial H_{3s}(X)}{\partial X_k}\right\}^{-1} \tag{5.3.14}$$

where $A_4^\alpha(X)$ are the coefficients of the critical arguments (αj^Ty) in H_4. It is seen that extra trouble is possible in the vicinity and at points $X = \overline{X}$ such that

$$\sum_k j_k\frac{\partial H_{3s}(X)}{\partial X_k}\bigg|_{X=\overline{X}} = 0 \tag{5.3.15}$$

but this, in general, will not be the case. The solution proceeds the same way to every stage by introducing an arbitrary function $F_p(X;j^Ty)$ into $S_p(X;y)$, to be determined at the next stage. <u>It is also possible that the "secular" part of</u> H_3

or some other approximation be zero. In this case one is forced to introduce the secular part of a higher approximation, thus introducing an amplification of the size of the corresponding terms of the series, which nevertheless can be in many instances acceptable. If $H_{3s} = 0$, displacing H_{4s} one would get, instead of (5.3.14),

$$B_3^{\alpha}(X) = -\frac{i}{\alpha} A_4^{\alpha}(X) \left\{ \sum_k j_k \frac{\partial H_{4s}(X)}{\partial X_k} \right\}^{-1}$$

so that actually $B_3^{\alpha}(X)$ is a second order quantity. Such cases and the case where H_3 contains critical arguments are nevertheless best dealt with by the following procedure which reduces the system to one having a single degree of freedom, and where the only angular variable appearing is the critical argument. Therefore, formally, the system is reduced to a quadrature.

In the above mentioned situation, consider initially a canonical transformation to new variables $(x';y')$,

$$\begin{aligned} y'_k &= y_k \quad (k = 1,2,\ldots,n-1) \\ y'_n &= j_1 y_1 + j_2 y_2 + \cdots + j_n y_n \\ x_k &= x'_k + j_k x'_n \quad (k = 1,2,\ldots,n-1) \\ x_n &= j_n x'_n \quad (j_n \neq 0) \end{aligned} \tag{5.3.16}$$

In terms of these new variables, the Hamiltonian becomes

$$H = \omega_1 x'_1 + \omega_2 x'_2 + \cdots + \omega_{n-1} x'_{n-1} + \sum_{\alpha} A_3^{\alpha}(x') \exp i\alpha y'_n + H_3^* + H_4 + \cdots$$

where H_3^* is free from terms containing the critical argument y'_n alone. Under this new form the system is highly degenerate and, unless all $A_3^{\alpha}(x')$ are identically zero and H_3^* has a non-zero secular part, one can only eliminate the angles $y'_1, y'_2, \ldots, y'_{n-1}$, that is, those corresponding to the momenta present in the linear part of H. We now eliminate the primes and write H in the new form

$$H = \omega_1 x_1 + \omega_2 x_2 + \cdots + \omega_{n-1} x_{n-1} + \sum_\alpha A_3^\alpha(x) \exp i\alpha y_n$$
$$+ H_3^*(x;y) + H_4(x;y) + \cdots \tag{5.3.17}$$

where H_3^* can be written as

$$H_3^* = \sum_\nu A_3^\nu(x) \exp [i(\nu_1 y_1 + \nu_2 y_2 + \cdots + \nu_n y_n)]$$

where $\nu_1, \nu_2, \ldots, \nu_{n-1}$ are not simultaneously zero if $\nu_n \neq 0$. Therefore, in view of the hypotheses of a unique set of integers not all zero, satisfying (5.3.1), the condition

$$\nu_1 \omega_1 + \nu_2 \omega_2 + \cdots + \nu_n \omega_n = 0$$

cannot be verified by any set $\nu = (\nu_1, \nu_2, \ldots, \nu_n)$ in H_3^*.

Now, one only requires that a new Hamiltonian K be found in terms of new coordinates $(X;Y)$, such that

$$K = K(X;Y_n) = K_2 + K_3 + K_4 + \cdots ,$$

a formal series in general, so that $X_1, X_2, \ldots, X_{n-1}$ are constant and the system has a single degree of freedom. The generating function is again defined by a formal series

$$S = X^T y + S_3(X;y) + S_4(X;y) + \cdots$$

and the transformations relevant to the solution are

$$x_k = X_k + \frac{\partial S_3}{\partial y_k} + \frac{\partial S_4}{\partial y_k} + \cdots$$

$$Y_n = y_n + \frac{\partial S_3}{\partial X_n} \quad \frac{\partial S_4}{\partial X_n} + \cdots$$

which give, as always, $K_2 = \omega_1 X_1 + \omega_2 X_2 + \cdots + \omega_{n-1} X_{n-1}$. One considers the Taylor's expansion of the generalized Hamilton-Jacobi equation

$$H(X + \partial S_3/\partial y^T + \cdots ; y) = K(X; y_n + \partial S_3/\partial X_n + \cdots) =$$

$$= \sum_k \omega_k X_k + K_3(X; Y_n) + \cdots$$

Taking into account a decrease in order by a factor 2 with every derivative with respect to X, the first few approximations are given by

$$\sum_{k=1}^{n-1} \omega_k \frac{\partial S_3}{\partial y_k} + H_3(X;y) = K_3(X;y_n),$$

$$\sum_{k=1}^{n-1} \omega_k \frac{\partial S_4}{\partial y_k} + \sum_k \frac{\partial H_3(X;y)}{\partial X_k} \frac{\partial S_3}{\partial y_k} + H_4(X;y) = K_4(X;y_n) + \frac{\partial K_3(X;y_n)}{\partial y_n} \frac{\partial S_3}{\partial X_n},$$

$$\sum_{k=1}^{n-1} \omega_k \frac{\partial S_5}{\partial y_k} + \sum_k \frac{\partial H_3(X;y)}{\partial X_k} \frac{\partial S_4}{\partial y_k} + \frac{1}{2} \sum_k \sum_j \frac{\partial^2 H_3(X;y)}{\partial X_k \partial X_j} \frac{\partial S_3}{\partial y_k} \frac{\partial S_3}{\partial y_j} +$$

$$+ \sum_k \frac{\partial H_4(X,y)}{\partial X_k} \frac{\partial S_3}{\partial y_k} + H_5(X;y) = K_5(X;y_n) + \frac{\partial K_3(X;y_n)}{\partial y_n} \frac{\partial S_4}{\partial X_n} +$$

$$+ \frac{1}{2} \frac{\partial^2 K_3(X;y_n)}{\partial y_n^2} \left(\frac{\partial S_3}{\partial X_n} \right)^2 + \frac{\partial K_4(X;y_n)}{\partial y_n} \frac{\partial S_3}{\partial X_n},$$

and so on. At every stage, the equation can evidently be written as

$$\sum_{k=1}^{n-1} \omega_k \frac{\partial S_\beta}{\partial y_k} + H^*_\beta(X;y) + \sum_{\alpha_\beta} A_\beta^{\alpha_\beta}(X) \exp(i\alpha_\beta y_n) = K_\beta(X;y_n) \tag{5.3.18}$$

where H^*_β has the properties described previously for H^*_3 in (5.3.17). We shall define

$$K_\beta(X;y_n) = H^*_{\beta s}(X) + \sum_{\alpha_\beta} A_\beta^{\alpha_\beta}(X) \exp(i\alpha_\beta y_n) \tag{5.3.19}$$

where

$$H^*_\beta(X;y) = H^*_{\beta s}(X) + H^*_{\beta p}(X;y)$$

and, being S_β defined by

$$\sum_{k=1}^{n-1} \omega_k \frac{\partial S_\beta}{\partial y_k} + H^*_{\beta p}(X;y) = 0,$$

no zero divisors will be present, because of the choice of H^*_β . This partial normalization can be achieved up to any order of approximation, so that, by eventually setting a bound in time, we can write the new Hamiltonian as

$$K = \sum_{k=1}^{n-1} \omega_k X_k + K_3(X;Y_n) + K_4(X;Y_n) + \cdots + K_\beta(X;Y_n) + O(|X|^{\beta+1})$$

and, "within $O(|X|^{\beta+1})$ ", the system is integrable by quadrature.

4. Equivalence with the Problem of Perturbation of a Linear System.

The relation between the previous sections and the classical problem of perturbations of a linear autonomous system studied by Cesari (1940), Hale (1954) and several others is easily established. In fact, consider (5.2.6) written in the form

$$\begin{pmatrix} \dot{\eta} \\ \dot{\xi} \end{pmatrix} = \begin{pmatrix} 0 & D^2 \\ -I & 0 \end{pmatrix} \begin{pmatrix} \eta \\ \xi \end{pmatrix} + \begin{pmatrix} 0 & I \\ -I & 0 \end{pmatrix} \left(\frac{\partial \Delta H}{\partial(\eta,\xi)} \right)^T \qquad (5.4.1)$$

where $\Delta H = H_3 + H_4 + \cdots$. Let us introduce the linear transformations

$$z = \begin{pmatrix} -iD^{-1} & -I \\ iD^{-1} & -I \end{pmatrix} \begin{pmatrix} \eta \\ \xi \end{pmatrix} \qquad (5.4.2)$$

or, its inverse,

$$\begin{pmatrix} \eta \\ \xi \end{pmatrix} = \frac{1}{2} \begin{pmatrix} iD & -iD \\ -I & -I \end{pmatrix} z \qquad (5.4.3)$$

where z is a 2n-vector. It follows that z satisfies the equation

$$z = \begin{pmatrix} iD & 0 \\ 0 & -iD \end{pmatrix} z + 2i \begin{pmatrix} 0 & I \\ -I & 0 \end{pmatrix} \begin{pmatrix} D^{-1} & 0 \\ 0 & D^{-1} \end{pmatrix} \left(\frac{\partial \Delta H}{\partial z}\right)^T ,$$

or

$$\begin{aligned} \dot{z}_k &= i\omega_k z_k + f_k(z) \\ \dot{z}_{n+k} &= -i\omega_k z_{n+k} + f_{n+k}(z) \\ &\text{for } k = 1,2,\ldots,n, \end{aligned} \tag{5.4.4}$$

where

$$\begin{aligned} f_k &= \frac{2i}{\omega_k} \frac{\partial \Delta H}{\partial z_{n+k}} \\ f_{n+k} &= \frac{-2i}{\omega_k} \frac{\partial \Delta H}{\partial z_k} . \end{aligned} \tag{5.4.5}$$

It might be worth noting that (5.4.4) can also be written

$$\begin{aligned} \dot{z}_k &= \frac{2i}{\omega_k} \frac{\partial \tilde{H}}{\partial z_{n+k}} \\ \dot{z}_{n+k} &= -\frac{2i}{\omega_k} \frac{\partial \tilde{H}}{\partial z_k} \end{aligned} \tag{5.4.6}$$

for $k = 1,2,\ldots,n$, and where

$$\tilde{H} = \frac{1}{2} \sum \omega_k^2 z_k z_{n+k} + \Delta H.$$

In any event, the equation of the fundamental system of solutions of (5.4.4) can be written

$$\dot{Z} = AZ + \Phi(Z) \tag{5.4.7}$$

where $\Phi(Z)$ is a matrix whose elements are series of homogeneous functions beginning with degree $3/2$. In our current notation

$$\Phi = \Phi_3 + \Phi_4 + \cdots .$$

We can now express the averaging method given in the previous section in terms of the following process of successive approximations. Let B be a constant diagonal matrix with unknown elements λ_k $(k = 1,2,\ldots,2n)$. We can actually assume

$$B = \text{diag}\,(i\tau_1, i\tau_2, \ldots, i\tau_n, -i\tau_1, -i\tau_2, \ldots, -i\tau_n)$$

where the constants $\tau_1, \tau_2, \ldots, \tau_n$ are unknown and to be determined as the result of an averaging method. Let us define the auxiliary equation

$$\dot{Z} = BZ + G(Z) \tag{5.4.8}$$

where

$$G(Z) = \Phi(Z) + (A-B)Z. \tag{5.4.9}$$

We observe that for $G(Z) = 0$, the solution of (5.4.8) is

$$Z^{(0)} = e^{Bt}C \tag{5.4.10}$$

where C is a constant matrix which we can take equal to unity.

Let us introduce the transformation

$$Z = e^{Bt}Y$$

so that (5.4.8) changes to

$$\dot{Y} = e^{-Bt}G(e^{Bt}Y) \tag{5.4.11}$$

and $Y^{(0)} = \text{const.}$ is the zero-th approximation. The integral of

$$e^{-Bt}G(e^{Bt}Y^{(0)})$$

will not, in general, be free from secular terms but will contain such terms and a quasi-periodic function of t $(\tau_1, \tau_2, \ldots, \tau_n$ supposed to be linearly independent over the integers). Thus, we define the averaging operation

$$P[M(t)] = \lim_{T \to \infty} \frac{1}{T} \int_0^T M(t)dt$$

so that the matrix

$$M(t) - P[M(t)] = (J-P)M(t)$$

is quasi-periodic or, exceptionally, periodic. We thus define

$$\dot{Y}^{(1)} = (J-P)e^{-Bt}G(e^{Bt}Y^{(0)})$$

which, upon integration, yields a quasi-periodic matrix. In general, we define the process of successive approximation and average by

$$Y^{(m)} = Y^{(0)} + \int^t (J-P)e^{-B\theta}G(e^{B\theta}Y^{(m-1)})d\theta.$$

Reverting to the Z matrix we obtain

$$Z^{(m)} = e^{Bt}\, Y^{(0)} + e^{Bt} \int^t (J-P)e^{-B\theta} [\Phi(Z^{(m-1)}) + (A-B)Z^{(m-1)}]d\theta.$$

If the process converges, the sequence $Z^{(m)}$ will have a limit Z satisfying the integral equation

$$Z = e^{Bt}Y^{(0)} + e^{Bt} \int^t (J-P)e^{-B\theta} [\Phi(Z) + (A-B)Z]d\theta.$$

By differentiation

$$\dot{Z} = Be^{Bt}Y^{(0)} + B(Z-e^{Bt}Y^{(0)}] + e^{Bt}(J-P)e^{-Bt} [\Phi(Z) + (A-B)Z]$$

or

$$\dot{Z} = A(Z) + \Phi(Z) - e^{Bt}P\{e^{-Bt}[\Phi(Z) + (A-B)Z]\}$$

which is the solution of (5.4.7) if and only if

$$P\{e^{-Bt}[\Phi(Z) + (A-B)Z]\} = 0$$

or

$$\lim_{T \to \infty} \frac{1}{T} \int_0^T e^{-Bt}[\Phi(Z(t)) + (A-B)Z(t)]dt = 0. \qquad (5.4.13)$$

The method of successive approximations defined by (5.4.12) has not been shown to converge or diverge. The convergence theorem proved by Cesari (1940) and Hale (1954) in more general terms, for the definition of a periodic solution is not easily generalizable to this case since the contraction principle which would allow the application of Banach's fixed point theorem is clearly not valid. In fact, the completeness of the space of all quasi-periodic functions is not verified. A direct proof like the one given by Cesari (1940) seems more likely to apply in this case. It is our conjecture that the method converges for at least a set of frequencies $\omega_1, \omega_2, \ldots, \omega_n$ satisfying a convenient irrationality condition and, probably, excluding some finite number of relations among the ω_k which would lead to the classical problem of parametric instability (moser, 1958; Gelfand and Lidskii, 1958; Giacaglia, 1971).

In components form, (5.4.12) can be written

$$Z_{kj}^{(m)} = e^{i\tau_k t} Z_{kj}^{(0)} + e^{i\tau_k t} \int^t (J-P)e^{-i\tau_k \theta} [\Phi_{kj}(Z^{(m-1)}) + i(\omega_k - \tau_k)Z_{kj}^{(m-1)}]d\theta$$

while condition (5.4.13) defines the constants $Z_{kj}^{(0)}$ and must be reducible to a system of equations for the unknown $\tau_1, \tau_2, \ldots, \tau_n$ in terms of $\omega_1, \omega_2, \ldots, \omega_n$. The equivalent relations for the original system (5.4.4) are

$$z_k^{(m)} = e^{i\tau_k t} z_k^{(0)} + e^{i\tau_k t} \int^t (J-P)e^{-i\tau_k \theta} [f_k(z^{(m-1)}) + i(\omega_k - \tau_k)z_k^{(m-1)}]d\theta$$

$$z_{n+k}^{(m)} = e^{\,i\tau_k t} z_{n+k}^{(0)} + e^{-i\tau_k t} \int^t (J-P)e^{i\tau_k \theta} [f_{n+k}(z^{(n-1)}) - i(\omega_k - \tau_k)z_{n+k}^{(m-1)}]d\theta$$

of, for the z vector,

$$z^{(m)} = e^{Bt} z^{(0)} + e^{Bt} \int^{t} (J-P) e^{-iB\theta} [f(z^{m-1}) - (A-B) z^{(m-1)}] d\theta \qquad (5.4.14)$$

The fact that close to linear integer relations among the frequencies $\omega_1, \omega_2, \ldots, \omega_n$ can and will produce, if not zero, small divisors it is seen immediately by observing the result of the application of the operator $P[f(t)]$ when $f(t) = \exp i[k_1 y_1 + k_2 y_2 + \cdots + k_n y_n]$ where $y_k^{(0)} = \omega_k t + y_{k0}$, $k = 1,2,\ldots,n$.

Hale (1963) discusses the existence of almost periodic solutions for the equation (system)

$$\dot{y} = Ay + q(t,y,\epsilon) \qquad (5.4.15)$$

requiring that q be almost periodic in t and the system $\dot{y} = Ay$ be non-critical, that is, all eigenvalues of A have non-zero real part. The case we are discussing here, corresponds evidently to a critical system, although it is possible to define the transformation so as to make all elements of diagonal matrix A real. For non-critical cases one can show the existence of almost periodic solution of (5.4.15) with the same frequencies of q, in t. This result, in dynamical systems, is quite important when studying perturbations of almost periodic solutions. If the resulting variational system is normalized up to quadratic terms and results to be non-critical, almost periodic solutions will exist in a conveniently restricted neighborhood of the starting solution. The problem is altogether related to stability in the Lyapunov's sense, to structural stability with respect to perturbations and with properties of invariant manifolds or integral manifolds. The main results on the subject were obtained by Diliberto (1960) and Bogoliubov and Mitropolski (1958). Main results on quasi-periodic solutions were obtained by Malkin (1952, 1956) and Roseau (1966).

5. Nonlinear Resonance.

Up to now we have dealt with the problem of constructing a solution in the neighborhood of an equilibrium point, starting from the limiting harmonic oscillations corresponding to the normal modes. We now deal with a more general problem, that is, we assume the "reference" oscillator to be nonlinear, in the sense that the frequencies depend on the amplitudes. This problem, as mentioned before, was dealt in some details by Poincaré (1893) following suggestions put forth by Bohlin (1887).

The study and effects of perturbations of a nonlinear oscillator require a precise knowledge of the singular points in phase space, the definition of separatrix, of saddles and centers, of the regions of circulation and of libration. All these concepts are of course well known and good references to the point are, for example, Bhatia-Szegö (1970), Nemytskii-Stepanov (1960), Urabe (1967), Lefschetz (1967), Nemytskii (1962), Elsgolts (1962), Andronov et al. (1966). These references are basic also to most of the subjects we are dealing with in the present chapter, although Moser's (1966, 1967, 1968) are the best available. A specific set of notes on the subject was prepared by Kyner (1968), but, unfortunately, not made generally available. We begin considering an autonomous Hamiltonian system defined by $H(q;p)$ in some domain D of the phase space. In D, H has a finite number of singular points (centers or saddle points). A separatrix is simply defined here as a trajectory connecting, in the limiting sense, two saddle points, which can, eventually, coincide. In the region interior to a separatrix there is always a center. For systems with more than one degree of freedom some of these concepts are not immediately generalizable. In the previous sections we have described series solutions, eventually only formal, of motion in the vicinity of a stable equilibrium point. Here we enlarge somehow the problem by obtaining formal series solutions in the vicinity of a center, or in a libration region, and in the vicinity of a circulatory motion, or in a circulation region. In the first case one or more of the angle variables are restricted to lower and upper limits, in a total interval less than 2π, while in the second case, all angle variables are unrestricted.

We assume that the Hamiltonian can be partitioned into a finite or countably infinite number of parts, that is,

$$H = H_0 + H_1 + H_2 + \cdots$$

where, for simplicity, $H_k = O(\epsilon^k)$ will be used as a convention, although the presence of a "small parameter" ϵ is not essential. Its use, nevertheless, simplifies the issue in many instances. The following hypotheses are also considered:

(a) H_0 is integrable and therefore, if one wishes, can be written in terms of momenta (or coordinates) only.

(b) H_k, $k > 0$, is a collection of a finite number of terms of the form

$$H_k = \sum_{\nu^k} A_k^{\nu^k}(p) \exp i\,(\nu_1^k q_1 + \nu_2^k q_2 + \cdots + \nu_n^k q_n).$$

(c) The set of invariant manifolds of the system generated by H is not essentially different from that of the system generated by H_0. In other words, we assume that there is a continuous transformation bringing one set into the other and such that, for $\epsilon \to 0$, it reduces to the identity.

The last hypothesis is, in essence, the result of Kolmogorov's Theorem where, of course, linear dependence among the frequencies corresponding to H_0 has to be excluded. Since these frequencies are assumed to depend continuously on the amplitude, it is possible to exclude a set of initial conditions so as to avoid the trouble. The question is how can one describe the motion in the vicinity of a resonance region (a critical point).

We begin our study with a system having a single degree of freedom, that is, reducible in principle to quadrature. The discussion, therefore, serves the only purpose of future generalizations.

Thus, let us assume the Hamiltonian

$$H = H_0(p) + H_1(p,q) + H_2(p,q) + \cdots$$

with p,q scalar quantities, $H_k = O(\epsilon^k)$ and $H_k = \sum_\nu A_k^\nu(p) \exp(i\nu q)$. We assume that H is analytic in some domain D of the phase plane (p,q) and that for $\epsilon = 0$, $\partial^k H_0/\partial p^k = 0$ for $p = p_0 \in D$ and $k = 1,2,\ldots,m$ (finite).

We also assume that for $\epsilon \neq 0$ the Hamiltonian H has a point of minimum, that is, one can solve the problem

$$\frac{\partial H(p,q)}{\partial q} = 0$$

$$\frac{\partial H(p,q)}{\partial p} = 0$$

and the Hessian is non-negative in the vicinity of a solution of (5.5.1). Because of the hypotheses, (5.5.1) has certainly at least two solutions, a maximum and a minimum in D. Let the point of minimum be defined by $(\bar{p},\bar{q})$ where

$$\begin{aligned} \bar{p} &= p_0 + \epsilon p_1 + \epsilon^2 p_2 + \cdots \\ \epsilon\bar{q} &= \epsilon q_1 + \epsilon^2 q_2 + \epsilon^3 q_3 + \cdots . \end{aligned} \tag{5.5.2}$$

The analytic character of H in D implies that there exists a $\delta = \delta(H,\epsilon) > 0$ such that for $|p - p_0| \leq \delta$, $p \in D$ and $0 < q < 2\pi$,

$$\left|\frac{\partial^k H_0(p)}{\partial p^k}\right| \leq \Omega_0 \epsilon^{(m+1-k)s} \tag{5.5.3}$$

with $s > 0$, Ω_0 finite and independent of ϵ,δ and $k = 1,2,3,\ldots,m$.

The goal is the elimination of q from the Hamiltonian, that is, reduce it to a normal form in the vicinity of the singular point p_0. This goal will be enlarged in the following sections.

Let the canonical normalizing transformation be generated by

$$S(P,q) = Pq + \Delta S(P,q)$$

where ΔS is, in some domain Ω of P and for $0 \leq q < 2\pi$, of some order ϵ^r $(r > 0)$. The value of r depends on s, on the order α of the lowest term of H which contains the angle q and on m. Let us also assume that the new Hamiltonian K(P) can be written as

$$K(P) = K_0(P) + \Delta K(P)$$

where $\Delta K(P)$ is, in Ω, of some order ϵ^β, $\beta > 0$. All the real numbers s,r,β are "a priori" unknown and should be determined in terms of α,m.

The energy equation

$$H\left(P + \frac{\partial\Delta S}{\partial q}, q\right) = K(P) \tag{5.5.4}$$

gives, using Taylor's expansion,

$$H_0(P) + \sum_{k=1}^{m+1} \frac{1}{k!} \frac{\partial^k H_0(P)}{\partial P^k} \left(\frac{\partial \Delta S}{\partial q}\right)^k + \cdots$$

$$+ H_1(P) + \sum_{k=1} \frac{1}{k!} \frac{\partial^k H_1(P,q)}{\partial P^k} \left(\frac{\partial \Delta S}{\partial q}\right)^k + \cdots$$

$$+ H_2(P) + \sum_{k=1} \frac{1}{k!} \frac{\partial^k H_2(P,q)}{\partial P^k} \left(\frac{\partial \Delta S}{\partial q}\right)^k + \cdots \tag{5.5.5}$$

$$= K_0(P) + \Delta K(P)$$

so that, as usual $K_0(P) = H_0(P)$. The next approximation of S, ΔS, must be used to eliminate the angle q in $H_\alpha(P,q)$, assuming $H_1, H_2, \ldots, H_{\alpha-1}$ independent of q. In view of the hypotheses it follows that for ΔS to be defined however small ϵ is, one has to choose

$$\sum_{k=1} \frac{1}{k!} \frac{\partial^k H_0(P)}{\partial P^k} \left(\frac{\partial \Delta S}{\partial q}\right)^k + H_\alpha(P,q) = K_\alpha(P) \tag{5.5.6}$$

where $\alpha \geq 1$, $m \geq 1$ are given integers. Also

$$\Delta K = K_1(P) + K_2(P) + \cdots + K_{\alpha-1}(P) + K_\alpha(P) + \cdots \tag{5.5.7}$$

where

$$K_j(P) = H_j(P), \quad j = 1,2,\ldots,\alpha-1.$$

Equation (5.5.6) implies that

$$(m+1-k)s + kr = \alpha, \quad k = 1,2,\ldots,m \tag{5.5.8}$$

so that, necessarily,

$$r = s = \frac{\alpha}{m+1} \tag{5.5.9}$$

which satisfies (5.5.8) for all k.

<u>These conditions define what we call the region of libration, that is, the region containing a center and limited by a closed separatrix.</u>

In the case where $\alpha = 1$ and $m = 1$, we obtain the classical result $r = s = 1/2$, that is, the expansions of both $S(P,q)$ and $K(P)$ are in powers of the square root of the small parameter ϵ. The idea of the $1/2$ power is quite old and, as mentioned before, follows from Weierstrass' Theorem on the factorization of power series. It comes out naturally in Bohlin's works (1887) on oscillatory motions. In our derivation we have basically assumed that near p_o, H behaves like

$$H \sim (p-p_o)^{m+1} f(p,q) + \epsilon g(p,q)$$

with $f(p_o,q) \neq 0$. The classical assumption implies $m = 1$.

Although not strictly necessary we shall describe the simple and, in fact, most common situation where $\alpha = 1$, $m = 1$, so that we can write,

$$S = Pq + S_{1/2}(P,q) + S_1(P,q) + \cdots$$

$$K = H_o(P) + K_1(P) + K_{3/2}(P) + K_2(P) + \cdots ,$$

since it is easy to verify that $K_o(P) = H_o(P)$, $K_{1/2}(P) = 0$. It follows from (5.5.5) that the first order equation, defining $S_{1/2}$ and K_1, is

$$H_o' \frac{\partial S_{1/2}}{\partial q} + \frac{1}{2} H_o'' \left(\frac{\partial S_{1/2}}{\partial q}\right)^2 + H_1(P,q) = K_1(P) \qquad (5.5.10)$$

where primes indicate derivatives with respect to P. <u>In order to define</u> $K_1(P)$, <u>we require that the center</u> $(\overline{p},\overline{q})$ <u>be a fixed point of the transformation defined by</u> S. Since $H_1(P,q)$ is continuous and periodic in q, we may define

$$K_1(P) = \min_{\{q\}} H_1(P,q) = H_1(P,\overline{q}_1(P)) \qquad (5.5.11)$$

where $\overline{q}_1(P)$ is such that

$$\left.\frac{\partial H_1}{\partial q}\right|_{q=\bar{q}_1} = 0$$

$$\left.\frac{\partial^2 H_1}{\partial q^2}\right|_{q=\bar{q}_1} > 0$$

for all $P \in \Omega$. Evidently, since H is analytic, $\bar{q} - \bar{q}_1(\bar{p}) = O(\epsilon)$ or $\bar{q} - \bar{q}_1(p_o) = O(\epsilon)$.

Let

$$F_1(P,q) = H_1(P,q) - K_1(P)$$

so that, obviously, $F_1(P,q)$ is positive for $P \in \Omega$ and for $0 \leq q < 2\pi$ except at $q = \bar{q}_1(P)$ where it is zero. The function $S_{1/2}$ is now defined by Bohlin's equation

$$H_o' \frac{\partial S_{1/2}}{\partial q} + \frac{1}{2} H_o'' \left(\frac{\partial S_{1/2}}{\partial q}\right)^2 + F_1(P,q) = 0. \tag{5.5.12}$$

At $q = \bar{q}_1(P)$ and $P = p_o$ it follows that

$$\frac{\partial S_{1/2}}{\partial q} = 0$$

so that, since to this order

$$p = P + \frac{\partial S_{1/2}}{\partial q}, \qquad Q = q + \frac{\partial S_{1/2}}{\partial P},$$

the center is a fixed point if $S_{1/2}$ is chosen so that $\partial S_{1/2}/\partial P = 0$ for $P = p_o$.

For the sake of simplicity let us write Equation (5.5.12) as

$$A \frac{\partial W}{\partial q} + \frac{1}{2} B\left(\frac{\partial W}{\partial q}\right)^2 + F(P,q) = 0 \tag{5.5.13}$$

where $F(P,q) > 0$, $F(P,\bar{q}_1(P)) = 0$ and $A = O(\epsilon^{1/2})$ while B is bounded from below, however small ϵ is taken. Under these conditions, since $F(P,q)$ is periodic in q, if it is also an even function of this variable,

$$F(P,q) = F(P,-q)$$

and, therefore, it can be written as

$$F = \sum_{j=1}^{\infty} \alpha_j(P) \sin^{2j} q.$$

Let, in any case,

$$\frac{A^2}{2B} \frac{1}{\sigma^2} = \max_{\{q\}} F(P,q) \tag{5.5.14}$$

with $\sigma > 0$ and a function of P. The function

$$\Psi(P,q) = \frac{2B\sigma^2}{A^2} F(P,q)$$

is such that

$$\max_{\{q\}} \psi(P,q) = 1 \qquad (q = \pm\pi/2)$$

$$\min_{\{q\}} \psi(P,q) = 0 \qquad (q = 0)$$

$$\psi(P,q) = \psi(P,-q)$$

and can, therefore, be written as

$$\psi(P,q) = \sum_{j=1}^{\infty} \beta_j(P) \sin^{2j} q$$

where

$$\beta_j = \frac{2B\sigma^2}{A^2} \alpha_j(P)$$

and, we shall assume that

$$B_1 \simeq 1, \quad |\beta_j| < 1, \qquad j \geq 2.$$

Excluding the case $A = 0$, we obtain

$$\frac{\partial W}{\partial q} = R(P)\{-1 \pm [1 - \frac{1}{\sigma^2} \psi(P,q)]^{1/2}\} \tag{5.5.15}$$

where $R = A/B$. We have also assumed $A > 0$. The case $A < 0$ corresponds to an analogous discussion by replacing $|A|$ for A with proper changes of signs. The possible cases are, obviously

$\sigma < 1$: Libration

$\sigma > 1$: Circulation

$\sigma = 1$: Separatrix or saddle points

The last case however, since it implies an equality, is meaningful only in the limit of the approximation to S if such series converges at all.

When $\sigma < 1$, $\psi(P,q)$ cannot reach its maximum value (unity), that is, there are values $q = q_1$, $q = q_2$ such that

$$\psi(P,q_1) = \psi(P,q_2) = \sigma^2$$

and, in the above considered case, $q_1 = -q_2$. The values (q_1,q_2) are the end points of the libration. If $\sigma > 1$, $\psi(P,q)$ takes all possible values, the angle q is unrestricted and $\dot{q}$ has, in the mean, a constant sign.

Let us now define a modulus k by $k = \min(\sigma,\sigma^{-1})$ and introduce the elliptic integral u defined by

$$\psi(P,q) = k^2 \operatorname{sn}^2 u = \sigma^2 \operatorname{sn}^2 u \quad \text{(Libration)}$$

or

$$\psi(P,q) = \operatorname{sn}^2 u \text{ (Circulation)}$$

where $\operatorname{sn} u = \operatorname{sn}(u,k)$ is the Jacobian elliptic sn function modulus k and amplitude ϕ given by

$$\text{am } u = \phi_L = \text{arc } \sin(\tfrac{1}{\sigma}\sqrt{\psi}) \quad \text{(Libration)}$$

or

$$\text{am } u = \phi_C = \text{arc } \sin(\sqrt{\psi}) \quad \text{(Circulation)}$$

In both cases the maximum amplitude is $\pi/2$ which coincides with $q = q_1$ or q_2 in libration and $q = \pi/2$ in circulation. The variable u covers a full cycle of period $4K$, where K is the complete elliptic integral of the first kind $u(\frac{\pi}{2}, k)$. The modulus k is a function of P.

The equation for W can now be written

$$\left(\frac{\partial W}{\partial q}\right)_L = R(-1 + \text{cn } u)$$

and

$$\left(\frac{\partial W}{\partial q}\right)_C = R(-1 + \text{dn } u)$$

in libration and circulation respectively. The $\pm$ sign is not necessary, of course, since cnu changes sign every half cycle $(2K)$ and, along the real axis u, dnu is always positive.

Let us assume the case where ψ is an even function, so that $q_2 = -q_1$ and the libration is a symmetric process about the libration center. It follows that

$$\psi_L(P,q) = \sigma^2 \text{ sn}^2 u = \sum_{j=1}^{\infty} \beta_j(P) \sin^{2j} q$$

and

$$(\sin^2 q)_C = \sum_{j=0}^{\infty} A_j \text{ sn}^{2(2j+1)} u$$

and in both cases we can write

$$\sin^2 q = \sum_{j=0}^{\infty} B_j \text{ sn}^{2(2j+1)} u$$

where $|B_j| < B_o$, $j \geq 1$, $B_o \simeq \sigma^2$ in Libration and $B_o \simeq 1$ in Circulation. We also find

$$\sin q = \sum_{j=0}^{\infty} C_j \ \mathrm{sn}^{4j+1} u$$

where $C_o(L) = \sigma$ and $C_o(c) = 1$. Also $|C_j| < C_o$, $j \geq 1$. The expression for cosq depends on the particular motion. We have, in general

$$\cos^2 q = 1 - \sin^2 q = 1 - (B_o \ \mathrm{sn}^2 u + B_1 \ \mathrm{sn}^6 u + \cdots).$$

In case of Libration, $B_o \simeq \sigma^2 = k^2$, so that, adding and subtracting $\sigma^2 \mathrm{sn}^2 u$

$$\cos^2 q = \mathrm{dn}^2 u - [(B_o - \sigma^2)\mathrm{sn}^2 u + B_1 \ \mathrm{sn}^6 u + \cdots]$$

and all coefficients $|B_o - \sigma^2|$, $|B_1|$, $|B_2|, \ldots$ are small compared with unity. Because $k = \sigma < 1$, from the above series we obtain the convergent expansion

$$(\cos q)_L = \mathrm{dn}\ u \left[1 + \sum_{j=1}^{\infty} D_j \ \mathrm{sn}^{2j} u\right]$$

with $|D_j| < 1$.

In case of Circulation, $B_o \simeq 1$, so that adding and subtracting $\mathrm{sn}^2 u$ to $\cos^2 q$,

$$\cos^2 q = \mathrm{cn}^2 u - [(B_o - 1)\mathrm{sn}^2 u + B_1 \ \mathrm{sn}^6 u + \cdots]$$

where $|B_o - 1|$, $|B_1|$, $|B_2|, \ldots$ are small compared with unity. Therefore

$$(\cos q)_C = \mathrm{cn}\ u \left[\sum_{i=1}^{\infty} \sum_{j=0}^{\infty} E_{ij} \ \mathrm{tn}^{2i} u \ \mathrm{sn}^{4j} u + 1\right]$$

where $|E_{ij}| < 1$. Finally, by considering the Fourier Series of snu, cnu, tnu, dnu it is possible to write cosq, sinq and $\psi(P,q)$ as Fourier Series in the argument u. The function $q = q(u)$ is easily obtained, with the aid of the foregoing relations, and the identity

$$\frac{d}{du}(\sin q) = \cos q \frac{dq}{du} .$$

In case of Libration one finds

$$\frac{dq}{du} = \text{cn } u \sum_{j=0}^{\infty} F_j \, \text{sn}^{2j} u$$

where $F_0 \simeq \sigma$. It follows that

$$W_L = -Rq + R \int \text{cn } u \left(\frac{dq}{du}\right)_L du = -Rq + R \sum_{j=0}^{\infty} F_j \int \text{sn}^{2j} u \, \text{cn}^2 u \, du$$

which is, in general, an elliptic integral representable in a convergent Fourier Series in $\sin(j\pi u/2K)$ plus a linear term in u. Of the same character is the relation between q_L and u, that is,

$$q_L = \sum_{j=0}^{\infty} F_j \int \text{cn } u \, \text{sn}^{2j} u \, du$$

with $F_0 \simeq \sigma$. The essential features in the p variable are found from

$$p = P + \frac{\partial S_{1/2}}{\partial q} + O(\epsilon)$$

and in case of Libration

$$p = P + R(-1+\text{cn } u) = P - \frac{H_0'}{H_0''} + \frac{H_0'}{H_0''} \text{cn } u$$

so that p oscillates about the mean value

$$\bar{p} = P - \frac{H_0'(P)}{H_0''(P)}$$

reaching its maximum value at $u = 0, 4K$ ($q = 0$ increasing) and its minimum value at $u = 2K$ ($q = 0$ and decreasing).

In case of circulation, one finds

$$\left(\frac{dq}{du}\right)_C = \mathrm{dn}\, u \sum_{i=0}^{\infty} \sum_{j=0}^{\infty} G_{ij}\, \mathrm{tn}^{2i} u\, \mathrm{sn}^{4j} u$$

where $G_{00} \simeq 1$ and $|G_{ij}| < 1$ for all other values. Therefore,

$$W_C = -Rq + R \sum_{i=0}^{\infty} \sum_{j=0}^{\infty} G_{ij} \int \mathrm{tn}^{2i} u\, \mathrm{sn}^{4j}\, \mathrm{dn}^2 u\, du$$

which is also an elliptic integral expressible in a Fourier Series in u. In this case one finds

$$p = P - R(1-\mathrm{dn}\, u) = P - \frac{H_o'}{H_o''} + \frac{H_o'}{H_o''}\, \mathrm{dn}\, u$$

and the mean value of p is given by

$$\bar{p} = P - \frac{H_o'}{H_o''} + \frac{H_o'}{H_o''} \frac{1 + \sqrt{1-k^2}}{2}$$

where

$$k^2 = \frac{1}{\sigma^2} = [\psi(P,q_1)]^{-1} = [\psi(P,q_2)]^{-1}.$$

The maximum value of p corresponds to $u = 0,\ 2K,\ 4K\ (q = 0)$ and the minimum value to $u = K,\ 3K\ (q = q_1$ or $q_2)$.

6. Asymptotic Expansion to Any Order.

In order to validate the formal approach of first order described in the previous section, one must show that higher-order solutions can be obtained. Whatever is the approximation reached we require ΔS to be stationary at the libration center.

The equation of order $\epsilon^{3/2}$ is found to be

$$\left(H_o' + H_o'' \frac{\partial S_{1/2}}{\partial q}\right) \frac{\partial S_1}{\partial q} + \frac{1}{6} H_o''' \left(\frac{\partial S_{1/2}}{\partial q}\right)^3 + H_1' \frac{\partial S_{1/2}}{\partial q} = K_{3/2}(P).$$

Nevertheless, we note that at the libration center, the coefficient of $\partial S_1/\partial q$ vanishes so that S_1 becomes undetermined at that point. We must therefore add to the equation the term of least order $(\geq 3/2)$ which contains $(\partial S_1/\partial q)^2$. We thus rewrite the above equation as

$$\left(H_0' + H_0'' \frac{\partial S_{1/2}}{\partial q}\right) \frac{\partial S_1}{\partial q} + \frac{1}{2} H_0'' \left(\frac{\partial S_1}{\partial q}\right)^2 + U_{3/2}(P,q) = K_{3/2}(P) \tag{5.6.1}$$

where $U_{3/2}$ is defined from the previous equation. Since $H_{3/2}$ is absent, the error in assuming $\bar{q}_1(P)$ a point of minimum of $U_{3/2}(P,q)$ is one order $(\epsilon^{1/2})$ of magnitude smaller than the equation we are dealing with and can therefore be tolerated. Thus, we define

$$K_{3/2}(P) = U_{3/2}(P,\bar{q}_1(P))$$

and let

$$U_{3/2} - K_{3/2} = F_{3/2}(P,q).$$

Defining now

$$\Delta S^{(1)} = S_{1/2} + S_1 \tag{5.6.2}$$

and

$$F^{(3/2)} = F_1 + F_{3/2},$$

it follows from (5.5.12) and (5.6.1) that

$$H_0' \frac{\partial \Delta S^{(1)}}{\partial q} + \frac{1}{2} H_0'' \left(\frac{\partial \Delta S^{(1)}}{\partial q}\right)^2 + F^{(3/2)}(P,q) = 0. \tag{5.6.3}$$

which defines, as in the first approximation, $\Delta S^{(1)}$ or S_1. Far from the libration center, the quadratic term $(O(\epsilon^2))$ in (5.6.1) introduces no error since it is an order of magnitude higher. As we approach the center, every term in (5.6.1) approaches zero as $\epsilon^{1/2}$. Since in the limit $F_{3/2} = 0$, the location of the center is not changed. Nevertheless, the amplitude of libration may change by a quantity which can be as large as $O(\epsilon)$. This will not be the case in the next order equation defining $S_{3/2}$, since $H_2 \neq 0$ in general. Such equations will have the form

$$H_o' \frac{\partial \Delta S^{(3/2)}}{\partial q} + \frac{1}{2} H_o'' \left(\frac{\partial \Delta S^{(3/2)}}{\partial q} \right)^2 + U^{(2)}(P,q) = K^{(2)}(P) \qquad (5.6.4)$$

and the location of the liberation center is now displaced to $\bar{q}_2(P)$, the solution of

$$\frac{\partial U^{(2)}(P,q)}{\partial q} = 0$$

$$\frac{\partial^2 U^{(2)}(P,q)}{\partial q^2} > 0$$

and, in view of the initial hypotheses of analyticity

$$\bar{q} - \bar{q}_2 = O(\epsilon^2).$$

The previous approximations are corrected accordingly and (5.6.4) yields the improved solution. One can easily show that the process can be repeated to any order of approximation so that, in general

$$H_o' \frac{\partial \Delta S^{(n)}}{\partial q} + \frac{1}{2} H_o'' \left(\frac{\partial \Delta S^{(n)}}{\partial q} \right)^2 + F^{(n+1/2)}(P,q) = 0.$$

If the process converges, we evidently obtain

$$\frac{\partial \Delta S}{\partial q} = \frac{H_o'}{H_o''} (-1 + \text{cn } u)$$

or

$$\frac{\partial \Delta S}{\partial q} = \frac{H_o'}{H_o''} (-1 + \text{dn } u)$$

according to when

$$2 \frac{H'_o}{H''^2_o} \max_q F^{\infty}(P,q) = \frac{1}{\sigma^2} = \frac{1}{k^2} > 1$$

or

$$2 \frac{H'_o}{H''^2_o} \max_q F^{\infty}(P,q) = \frac{1}{\sigma^2} = k^2 < 1$$

respectively.

It is only in this limit approximation, if it exists, that asymptotic motion can be defined, the separatrix being defined by the condition

$$\frac{2H'_o}{H''^2_o} \max_q F^{\infty}(P,q) = 1.$$

The proper transformation, in this case, is obviously

$$\frac{2H'_o}{H''^2_o} F^{\infty}(P,q) = \tanh u$$

which gives

$$\frac{\partial \Delta S}{\partial q} = \frac{H'_o}{H''_o} (-1+\operatorname{sech} u)$$

and $q = 0$ corresponds to $u = \infty$. It follows that

$$p = P - \frac{H'_o}{H''_o} (1-\operatorname{sech} u)$$

and the limit value of p is given by

$$p_{lim} = P - \frac{H'_o}{H''_o} ,$$

exactly the mean value of the libration motion. In general, as mentioned before, the convergence of the series can only be obtained in a finite time $T = O(\epsilon^{-1})$ whatever is the order of approximation reached. (Kyner, 1967).

7. Extended Theory and the Ideal Resonance Problem.

We have considered a Hamiltonian

$$H = H_0(x) + H_1(x,y) + H_2(x,y) + \cdots = \text{const.}$$

where $H_1, H_2, \ldots$ are periodic of period say 2π in y. Consider the topology of the phase plane (y,x) with the trajectories $H(x,y) = \text{const.}$ We suppose that in this plane we have a center and two saddle points, which, with a proper transformation we suppose to be located as follows:

$$\begin{aligned} &\text{center: } x = \bar{x}, \quad y = \pi \\ &\text{saddles: } x = \bar{x}, \; y = 0, (2\pi). \end{aligned} \tag{5.7.1}$$

This is not without reason. If one considers a point x_0 and a Taylor expansion of H around this point, one finds

$$\begin{aligned} H = {} & H_0(x_0) + H_0'(x_0)(x-x_0) + \frac{1}{2} H_0''(x_0)(x-x_0)^2 + \cdots \\ & + H_1(x_0,y) + H_1'(x_0,y)(x-x_0) + \cdots = \text{const.} \end{aligned} \tag{5.7.2}$$

where primes indicate derivative with respect to x. Here we follow the approach basically introduced by Hori (1960) and structured in rigorous terms by Jupp (1969, 1970). We also assume, as usual, that $H_p(x,y)$, $p \geq 1$, are developable in Fourier Series (which we suppose are rapidly convergent) and that H is an even function of y, although this restriction could be removed. It follows that $H_p(x,y)$, $p \geq 1$, can be written as

$$H_p(x,y) = \sum_{n=0} A_p^n(x) \cos ny. \tag{5.7.3}$$

We suppose the classical case $|x-x_0| = O(\epsilon^{1/2})$ although this is related to the behavior of $H_o(x)$ in the neighborhood of x_o, as mentioned in earlier sections. It follows that the "dominant" part of H is, eliminating constants,

$$\begin{aligned} \tilde{H}_1 &= B\delta + \frac{1}{2} B'\delta^2 + A_1^1(x_o)\cos y \\ &= B\delta + \frac{1}{2} B'\delta^2 + A \cos y \end{aligned} \tag{5.7.4}$$

where A,B,B' are constants and $\delta = x - x_o$. We shall assume $A > 0$. For $A < 0$ the formalism is basically the same except by evident modifications. The remaining part of the Hamiltonian can be written as a Fourier Series with coefficients being polynomials in δ:

$$H = \tilde{H}_1 + \tilde{H}_2 + \tilde{H}_3 + \cdots . \tag{5.7.5}$$

Each $\tilde{H}_p$ is made up of a finite number of terms as far as δ is concerned but the number of trigonometric terms can be infinite. It should also be observed that (5.7.5) may be brought into form (5.7.4) which is termed the <u>ideal resonance problem</u>. Such reduction will be sketched at the end of this section.

Let us now consider the problem generated by (5.7.4) and write

$$\tilde{H}_1 = F = B\delta + \frac{1}{2} B'\delta^2 + A \cos y = C(\text{constant}) \tag{5.7.6}$$

neglecting terms higher than $O(\epsilon)$. Note that we are assuming the maximum value of B to be $O(\epsilon^{1/2})$, B' to be finite and not small, i.e., O(1) and $A = O(\epsilon)$. The equations of motion generated by (5.7.6) are

$$\begin{aligned} \dot{\delta} &= \partial F/\partial y \\ \dot{y} &= -\partial F/\partial \delta . \end{aligned} \tag{5.7.7}$$

Evidently (5.7.7) have the equilibrium solutions

$$y = 0,\ \pi,\ (2\pi)$$
$$\delta = -\frac{B}{B'}$$

and evidently $y = \pi$, $\delta = -B/B'$ is a center while $y = 0$, 2π and $\delta = -B/B'$ are unstable (saddle) points. Let at $t = 0$, $\delta = 0$ and $y = \pi$, and define $\eta' = y - \pi$. Thus (5.7.6) becomes, with $C = -A + \text{const.}$,

$$F = B\delta + \frac{1}{2} B'\delta^2 + 2A \sin^2 \frac{1}{2} \eta' \tag{5.7.8}$$

which is the new "energy integral". Finally, let

$$2A = \omega^2 = O(\epsilon)$$

$$\delta = \frac{1}{2} \xi - B/B'$$

$$\eta' = 2\eta$$

which is a canonical transformation $(\delta, \eta') \rightarrow (\xi, \eta)$. It follows that

$$F = \xi^2 + \omega^2 \sin^2 \eta \tag{5.7.9}$$

where, corresponding to the initial conditions previously set, that is, at $t = 0$, $\eta = 0$ and $\xi = 2B/B'$, the value of F is completely determined. Evidently, it is now clear the importance of the analogy with the pendulum. The equilibrium points of (5.7.9) are

$$\text{stable (center): } \eta = 0,\ \xi = 0$$

$$\text{unstable (saddles): } \eta = \pm\pi/2,\ \xi = 0$$

and we consider $|\xi| = O(\epsilon^{1/2})$.

If the same transformations defined above are introduced in the full Hamiltonian, we shall find

$$H = F + \sum_k F_k(\xi, 2\eta) \tag{5.7.10}$$

where $F_k(\xi, 2\eta)$ is a Fourier Series in 2η and the coefficients are polynomials in ξ.

The solution of the main (ideal) problem generated by F is well known in terms of the von Zeipel procedure (e.g., Jupp, 1970) while the pendulum analogy is explored in details by Kyner (1967). The first paper presents the solution to any order of approximation, although in general, it leads to hyperelliptic integrals.

The formalism of the first order $(\epsilon^{1/2})$ is basically as it was presented in Section 5 of this chapter.

The basic difference, from the previous approach, is that instead of using the libration center of the original system as a reference, one can use any point x_0 to perform the expansion indicated in (5.7.2), such point x_0 being eventually anywhere in the libration or circulation region. It cannot, however, be taken on the separatrix or at the unstable points for, in this case, the expansion cannot be convergent and it is actually meaningless. Such motions can only be obtained as limiting cases of circulation or libration. Such limiting cases were originally discussed by Poincaré (1893) but for a more recent review we refer to the article by Garfinkel and others (1971). It is interesting to see how the problem is handled with Lie's Series. Consider for instance the Hori-Lie approach. The dominant term in H is F, and this is the generator of the auxiliary system

$$\frac{d\xi}{d\tau} = \frac{\partial F}{\partial \eta} = \omega^2 \sin 2\eta$$
$$\frac{d\eta}{d\tau} = - \frac{\partial F}{\partial \xi} = -2\xi \ , \qquad (5.7.11)$$

so that we obtain the pendulum equation

$$\frac{d^2\eta}{d\tau^2} + 2\omega^2 \sin 2\eta = 0.$$

The distinction between libration, circulation and asymptotic motion depends evidently on the initial conditions or, let us say, the energy integral (5.7.9) and the value of ξ or η at some instant τ. It is well known from the study of the simple pendulum that the value of the energy defines the type of motion. The uniform representation of all types of motion as a reference for the true motion as defined by the perturbations (5.7.10) to F is something of doubtfull significance. Such perturbations will be properly handled in the region of libration or circulation, far enough from the separatrix. Perturbations of the asymptotic motion or saddle points is something else and should be treated differently. Such kind of motions are unlikely to be conserved while a libration or circulation will most likely be conserved under small enough perturbations. Arnol'd's Theorem gives us a positive support on the reasoning. The reference orbit is, in this case, defined by elliptic functions or integrals and perturbations technique might be quite involved, but because the "best" reference is used, as opposed to the use of an equilibrium point (necessarily a center), the successive approximation method is expected to have a better chance of convergence. The Hori-Lie equations are, in the present formulation, exactly the same as for nonresonant systems, the lowering of order in the derivatives with respect to ξ being exactly balanced by multiplication with respect to η, which, upon substitution of the auxiliary system solution, brings out a small factor $O(\epsilon^{1/2})$. One can indeed treat (5.7.10) as a normal series, with $F = F_0$ and the rest made up of terms of increasing order, eventually the order being a fraction $p/2$ with p integer. The character of the expansion is pretty much the same as in Jupp (1970, 1972) and not worthwhile to be repeated here. To second order is discussed by Jupp (1972).

The reduction of the general case, mentioned at the beginning of the section, to the ideal form (5.7.9) can be accomplished in the following way (see Giacaglia, 1970).

Let

$$H = A_0(x) + \sum_{j=1}^{\infty} A_j(x) \cos jy = H(x,y) \tag{5.7.12}$$

where we assume $A_j(x)$, $j = 1,2,\ldots$ to be bounded by a small quantity order ϵ. As usual, we write

$$A_j(x) = O(\epsilon), \qquad j = 1,2,\ldots$$

for x in some interval D of R. We also assume $A_0(x) = O(1)$, $A_0'(x) = O(\epsilon^{1/2})$ in that same interval, while y is defined in $[0,2\pi]$. The ideal resonance problem is defined by

$$H_I = A(x) + B(x) \cos y \tag{5.7.13}$$

and under certain conditions there exists a canonical transformation, defined by a formal series, which reduces (5.7.12) to the form (5.7.13), that is, assuming the completely canonical transformation $(x,y) \to (\xi,\eta)$,

$$H(x(\xi,\eta),\ y(\xi,\eta)) = K(\xi,\eta) = P(\xi) + Q(\xi)\cos\eta. \tag{5.7.14}$$

We assume initially that:

A) For any x in D and $0 \leq y \leq 2\pi$ there are only two solutions $y = 0$ and $y = \pi$ of the equation $\dot{x} = -H_y = 0$.

B) $A_0(x),\ A_1(x) > 0$ for $x \in D$.

C) The maximum value of H is attained, for $x \in D$, at $y = 0$.

D) $M(x) = \sum_{j=1}^{\infty} A_j(x) > 0$

E) The minimum value of H is attained, for $x \in D$, at $y = \pi$.

F) $m(x) = \sum_{j=1}^{\infty} A_j(x)(-1)^j = 0.$

G) The generating function of the transformation $(x,y) \to (\xi,\eta)$ has the classical asymptotic form

$$S(\xi,y) = \xi y + S_{1/2}(\xi,y) + S_1(\xi,y) + \cdots .$$

H) The coefficients $P(\xi)$ and $Q(\xi)$ have the similar asymptotic form

$$P(\xi) = P_0(\xi) + P_{1/2}(\xi) + P_1(\xi) + \cdots ,$$

$$Q(\xi) = Q_0(\xi) + Q_{1/2}(\xi) + Q_1(\xi) + \cdots .$$

Equating terms of same order in the Taylor's expansion of

$$H(\xi + \partial S_{1/2}/\partial y + \cdots, y) = K(\xi, y + \partial S_{1/2}/\partial\xi + \cdots)$$

we find

$$O(1): \quad P_o(\xi) = A_o(\xi)$$

$$Q_o(\xi) = 0$$

$$O(\epsilon^{1/2}): \quad P_{1/2}(\xi) = Q_{1/2}(\xi) = 0$$

$$O(\epsilon): \quad A_o'(\xi) \frac{\partial S_{1/2}}{\partial y} + \frac{1}{2} A_o''(\xi)\left(\frac{\partial S_{1/2}}{\partial y}\right)^2 + \sum_{j=1}^{\infty} A_j(\xi) \cos jy$$

$$= P_1(\xi) + Q_1(\xi) \cos y$$

$$O(\epsilon^{3/2}): \quad \left[A_o'(\xi) + A_o''(\xi) \frac{\partial S_{1/2}}{\partial y}\right] \frac{\partial S_1}{\partial y} + \frac{1}{2} A_o''(\xi)\left(\frac{\partial S_1}{\partial y}\right)^2 +$$

$$+ \frac{1}{6} A_o'''(\xi)\left(\frac{\partial S_{1/2}}{\partial y}\right)^3 + \frac{\partial S_{1/2}}{\partial y} \sum_{j=1}^{\infty} A_j'(\xi) \cos jy =$$

$$= P_{3/2}(\xi) + Q_{3/2}(\xi) \cos y - Q_1(\xi) \frac{\partial S_{1/2}}{\partial \xi} \sin y$$

and so on. In general, the equation to be considered is

$$\left[A_o'(\xi) + A_o''(\xi) \frac{\partial S_{1/2}}{\partial y}\right] \frac{\partial S_n}{\partial y} + \frac{1}{2} A_o''(\xi)\left(\frac{\partial S_n}{\partial y}\right)^2 = P_{n+1/2}(\xi) + Q_{n+1/2}(\xi)\cos y + R_{n+1/2}(\xi,y) \tag{5.7.15}$$

for $n = 1,2,3,\ldots$ and where $R_{n+1/2}(\xi,y)$ depends on the knowledge of the previous approximations. We may also write (5.7.15), including higher order terms, in the form

$$\left[A_o'(\xi) + A''(\xi) \frac{\partial S^{(n-1/2)}}{\partial y}\right] \frac{\partial S_n}{\partial y} + \frac{1}{2} A_o''(\xi)\left(\frac{\partial S_n}{\partial y}\right)^2 = P_{n+1/2}(\xi) + Q_{n+1/2}(\xi) \cos y + R_{n+1/2}(\xi,y) \tag{5.7.16}$$

where

$$S^{(p)} = S_{1/2} + S_1 + S_{3/2} + \cdots + S_p .$$

Summing Equation (5.7.16) from $n = 1$ to $n = p$ and adding the $O(\epsilon)$ equation, one finds

$$A_o'(\xi)\,\frac{\partial S^{(p)}}{\partial y} + \frac{1}{2}A_o''(\xi)\left(\frac{\partial S^{(p)}}{\partial y}\right)^2 = P^{(p+1/2)}(\xi) + Q^{(p+1/2)}(\xi)\cos y + \\ + R^{(p+1/2)}(\xi,y) \tag{5.7.17}$$

for $p = \frac{1}{2}, 1, \frac{3}{2}, \ldots$ and where

$$P^{(k)} = P_1 + P_{3/2} + \cdots + P_k$$

$$Q^{(k)} = Q_1 + Q_{3/2} + \cdots + Q_k$$

$$R^{(k)} = R_{1/2} + R_1 + \cdots + R_k .$$

Consider the case $p = \frac{1}{2}$. Solving for $S_{1/2,y}$ we find

$$S_{1/2,y} = -\frac{A_o'}{A_o''} \pm \left\{\left(\frac{A_o'}{A_o''}\right) + \frac{2}{A_o''}\left[P_1 - (A_1 - Q_1)\cos y - \sum_{j=2}^{\infty} A_j \cos jy\right]\right\}^{1/2}$$

and we choose

$$P_1(\xi) = Q_1(\xi) = A_1(\xi)$$

so that

$$S_{1/2,y} = -\frac{A_o'}{A_o''} \pm \left\{\left(\frac{A_o'}{A_o''}\right)^2 + \frac{2}{A_o''}\left[2A_1 \cos^2\frac{y}{2} - \sum_{j=1}^{\infty} A_j \cos jy\right]\right\}^{1/2}$$

If $y = 0$, the quantity under square root becomes

$$\left(\frac{A_o'}{A_o''}\right)^2 + \frac{2}{A_o''}\,[2A_1 \quad M(\xi)].$$

If such quantity is positive $S_{1/2,y}$ is always real and y undergoes circulation.

If it is negative, y cannot reach the value $y = 0$ and undergoes a libration about $y = \pi$. If $y = \pi$, that same quantity becomes $(A_o'/A_o'')^2$ and we choose the sign in such a way that $S_{1/2,y} = 0$ and we can choose $S_{1/2}$ such that $S_{1/2,\xi} = 0$, at that point. These conditions are readily verified if $S_{1/2,y}$ is a sine series of some integer multiple of y, but this will not be the general case. Consider now the equation

$$\dot{y} = H_x = 0$$

that is

$$A_o'(x) + \sum_{j=1}^{\infty} A_j'(x) \cos jy = 0.$$

For $y = \pi$,

$$A_o'(\bar{x}) + \sum_{j=1}^{\infty} (-1)^j A_j'(\bar{x}) = 0$$

and because of the orders of magnitude involved we suppose that an approximate (good to order $\epsilon^{1/2}$) solution is given by

I) $\bar{x} \simeq \bar{x}_{1/2}$, <u>where</u> $A_o'(\bar{x}_{1/2}) = 0$. <u>The point</u> $(\bar{x}, y = \pi)$ <u>is a libration center and the point</u> $(\bar{x}_{1/2}, y = \pi)$ <u>is a first approximation of its location.</u>

For $y = 0$,

$$A_o'(x^*) + \sum_{j=1}^{\infty} A_j'(x^*) = 0$$

and again

II) $x^* = x^*_{1/2}$, <u>where</u> $A_o'(x^*_{1/2}) = 0$. <u>The point</u> $(x^*, y = 0)$ <u>is a saddle point and the point</u> $(x^*_{1/2}, y = 0)$ <u>is a first approximation of its location.</u>

For $y = \pi$, $x = \bar{x}$, the first approximation $S = \xi y + S_{1/2}$ to the transformation is the identity, since $S_{1/2,y} = 0$ and $S_{1/2,\xi} = 0$, by construction.

In the next approximation, in order to have $S_{1,y} = 0$, we shall choose again $P_{3/2} = Q_{3/2}$ and, now, $Q_{3/2}$ is defined by the coefficients of the known terms in $\cos y$, that is,

$$Q_{3/2}(\xi) = \frac{1}{\pi}\int_0^{2\pi}\left\{\frac{1}{6}A_0'''\left(\frac{\partial S_{1/2}}{\partial y}\right)^3 + \frac{\partial S_{1/2}}{\partial y}\sum_{j=1}^{\infty} A_j'(\xi)\cos jy + \right.$$

$$\left. + Q_1(\xi)\frac{\partial S_{1/2}}{\partial \xi}\sin y\right\}\cos y\, dy.$$

The choice will, in general, be

$$P^{(k)}(\xi) = Q^{(k)}(\xi) = -\frac{1}{\pi}\int^{2\pi} R^{(k)}(\xi,y)dy$$

and by recurrence, one easily establishes that $R^{(k)}(\xi,\pi) = 0$, and

$$\frac{\partial S^{(p)}}{\partial y} = -\frac{A_0'}{A_0''} \pm \left\{\left(\frac{A_0'}{A_0''}\right)^2 + \frac{2}{A_0''}\left[2P^{(p+1/2)}\cos^2\frac{y}{2} + R^{(p+1/2)}(\xi,y)\right]\right\}^{1/2}$$

and, for $y = \pi$, $S_y^{(p)} = 0$. Libration, Circulation or Asymptotic motion is decided at every stage according to whether

$$\left(\frac{A_0'}{A_0''}\right)^2 + \frac{2}{A_0''}\left[2P(\xi)^{(p+1/2)} + R^{(p+1/2)}(\xi,0)\right] \gtreqless 0,$$

although the asymptotic case can only be established in the limit, if it exists, $p \to \infty$. <u>We have now defined the formal series</u> $S(\xi,y)$ <u>which reduces the Hamiltonian to the form</u>

$$K(\xi,\eta) = (P_0 + P_1 + P_{3/2} + \cdots) + (Q_1 + Q_{3/2} + \cdots)\cos\eta$$
$$= A_0(\xi) + 2P(\xi)\cos^2\frac{\eta}{2} \qquad (5.7.18)$$

where $A_0(\xi) = O(1)$, $A_0'(\xi) = O(\epsilon^{1/2})$, $P(\xi) = O(\epsilon)$. For $\eta = \pi$, $K(\xi,\eta)$ has a minimum, assumed of course, $A_0(\xi) > 0$, $P(\xi) > 0$. This is precisely the form which defines the ideal resonance problem.

We have presented here the simplest situation that one can encounter. More complex topology of the phase plane (x,y) is evidently possible with a number of centers and at most twice as many minus one points. The discussion is evidently impossible for the most general case. The simplest case after the one we have considered here corresponds to the more general ideal resonance problem where two centers and three saddle points are present in the interval $0 \leq y \leq 2\pi$. Here, the dominant Hamiltonian takes the form

$$F = A_o(x) + A_1(x)\cos y + B_1(x)\cos 2y \tag{5.7.19}$$

where $A_o'(x) = O(\epsilon^{1/2})$ for $x \in D$ and $A_1(x)$, $B_1(x)$ are $O(\epsilon)$ in D. The discussion of this case has been given in detail by Giacaglia (1970).

8. Several Degrees of Freedom.

Up to the present moment we have dealt with problems of one degree of freedom. In fact, in general, system with m resonant and rationally independent frequency conditions can be reduced to systems having m degrees of freedom. If $m > 1$ the situation is of little hope. As generally accepted, very little is known about systems with two degrees of freedom and, as mentioned before, the visualization of critical points is quite cumbersome. The problem can, however, be formulated, in some generality, as follows.

Consider again the system generated by the Hamiltonian

$$H(x;y) = H_o(x) + H_1(x,y) + \cdots \tag{5.8.1}$$

where, for $k \geq 1$,

$$H_k = \sum_p A_k^p(x) \exp(i\, p^T y) \tag{5.8.2}$$

and the number of terms in each H_k is supposed to be finite. We assume, as usual, H to be analytic in $x \in D$ where D is an n-dimensional differentiable manifold. Series (5.8.1) is supposed to be uniformly convergent as a power series of a "small parameter" ϵ, which, for all purposes serves only to simplify matters, although, in some instances we can show the validity (convergence) of the formal series to be constructed for $0 < \epsilon < \epsilon_o$, ϵ_o sufficiently small. We consider (5.8.1) irreducible, in the sense that all angle variables are slow moving. All fast variables of the system are supposed to have been eliminated one way or another (see Chapter 2). The non-linear resonance hypothesis corresponds now to consider the existence of singular points for the equations

$$\dot{x} = -H_y^T$$
$$\dot{y} = +H_x^T$$

that is $H_x = 0$, $H_y = 0$ for some $x = x^o$, $y = y^o$. Solutions of this type are "centers" (the characteristic polynomial of the first variational system has purely imaginary roots only) or "saddles" (the characteristic polynomial has at least two roots with non-zero real part, one positive and, necessarily, one negative). The "centers" are not necessarily stable points as is well known. The "saddles" are certainly unstable. For conservative systems, however, it has been recently proved that the converse of Lagrange-Dirichlet Theorem holds under quite general conditions (Hagedorn, 1971) so that if H is time independent, analyticity is more than enough to ensure stability at a point of minimum for the potential and instability at a point of maximum.

An approximation to the nonlinear resonance condition is evidently given by

$$x = x^o = \text{const.}$$
$$y = H_{ox^o}^T t + y^o ,$$

where, given a $\mu > 0$, we assume there exists an $\epsilon > 0$ such that $\|\partial H_o/\partial x_k\|_{x=x^o} = O(\epsilon^{1/2})$ for $\|x-x^o\| \leq \mu$. The determinant of the second derivatives is supposed not to be singular and bounded away from zero by a quantity $O(1)$, that is, non vanishing with and independent of ϵ.

We perform an n-dimensional Taylor expansion about some point x_o arriving at the dominant part of H defined by

$$F(\delta;y) = a^T\delta + \delta^T A\delta + H_1(x_0;y) \tag{5.8.3}$$

where

$$a = \left.\frac{\partial H_0}{\partial x}\right|_{x=x_0},$$

$$\delta = x - x_0 ,$$

$$A = \left.\frac{\partial^2 H_0}{\partial x\,\partial x}\right|_{x=x_0},$$

$$H_1(x_0;y) = \sum_p A_1^p(x_0)\exp i(p^T y).$$

One can show that it is possible to obtain a formal canonical transformation that brings the general problem to the main problem defined by (5.8.3). The procedure is much alike the one shown for the one-dimensional case. We are again assuming $\|\delta\| = O(\epsilon^{1/2})$, $\|a\| = O(\epsilon^{1/2})$ and $|H_1| = O(\epsilon)$. Now, evidently, the matrix A is symmetric and therefore diagonalizable by a similarity transformation. But such a transformation would introduce noninteger coefficients in H_1 when expressed in terms of the new angle variables and, therefore, is not very convenient. As in most cases, let us assume that the "dominant" term in H_1 corresponds to a single combination of the angle variables y_k $(k = 1,2,\ldots,n)$ and let

$$z = \bar{p}_1 y_1 + \bar{p}_2 y_2 + \cdots + \bar{p}_n y_n$$

be that combination. The equations generated by (5.8.3) are

$$\dot{\delta}_k = -\frac{\partial F}{\partial y_k} = -iA_1^{\bar{p}}\,\bar{p}_k \exp(i\,\bar{p}^T y) + i(A_1^{\bar{p}})^*\,\bar{p}_k \exp(-i\,\bar{p}^T y) \tag{5.8.4}$$

$$\dot{y}_k = \frac{\partial F}{\partial \delta_k} = a_k + 2\sum_{j=1}^{n} A_{kj}\,\delta_j \tag{5.8.5}$$

where $*$ means complex conjugate. It follows that

$$\ddot{z} = \left[-2i\, A_1^{\bar{p}} \sum_{j=1}^{n} \sum_{k=1}^{n} A_{kj}\, \bar{p}_j \bar{p}_k\right] \exp\,(iz) +$$

$$+ \left[2i (A_1^{\bar{p}})^* \sum_{j=1}^{n} \sum_{k=1}^{n} A_{kj}\, \bar{p}_j \bar{p}_k\right] \exp\,(-iz)$$

or

$$\ddot{z} = \omega e^{iz} + \omega^* e^{-iz} \tag{5.8.6}$$

where z is a real quantity and ω is necessarily complex and given by

$$\omega = -2i\, A_1^{\bar{p}} \sum_{j=1}^{n} \sum_{k=1}^{n} A_{kj}\, \bar{p}_j \bar{p}_k\ .$$

In fact, we can write (5.8.6) in the real form

$$\ddot{z} = \omega_1 \cos z - \omega_2 \sin z \tag{5.8.7}$$

where

$$\omega_1 = 2\,\mathrm{Re}[\omega],$$

$$\omega_2 = 2\,\mathrm{Im}[\omega].$$

The solution of (5.8.7) is an elliptic integral of the first kind, easily reduced to the normal form by the introduction of

$$\zeta = Z + \alpha$$

$$\sin \alpha = \omega_1 / \sqrt{\omega_1^2 + \omega_2^2}$$

$$-\cos \alpha = \omega_2 / \sqrt{\omega_1^2 + \omega_2^2}$$

so that, with $m = \sqrt{\omega_1^2 + \omega_2^2}$,

$$\ddot{\zeta} = m \sin \zeta \tag{5.8.8}$$

which is again the simple pendulum equation. The variable ζ is, therefore, subjected to the usual discussion and the "main argument" z can undergo libration, circulation or an asymptotic motion.

The complete integration of this equation (5.8.4) and (5.8.5) is now easily done by first obtaining the δ's from (5.8.4), by simple quadrature, since

$$\dot{\delta}_k = ip_k[-A_1^{\bar{p}} e^{iz} + (A_1^{\bar{p}})^* e^{-iz}]$$

and afterwards, (5.8.5) gives each angle y_k, again by simple quadratures.

Evidently, the case we have discussed above is actually equivalent to a one-dimensional case. The problem can be handled in similar way if the dominant part of H_1 is function of an angle $z = \bar{p}_1 y_1 + \cdots + \bar{p}_n y_n$ and a finite number of multiples of z, although the equation for z, in this case, might lead to an hyper-elliptic integral, $\ddot{\zeta} = m_1 \sin \zeta + m_2 \sin 2\zeta + \cdots + m_p \sin p\zeta$.

When several combinations, linearly independent, $z_1, z_2, \ldots, z_p$ are present, the known methods of handling the problem can only be applied if one can define non-overlapping domains where each z_k corresponds to a dominant term. The solution can, in this case, be obtained by a matching of solutions valid locally in each of the above domains. One of the most efficient methods is a multiple variable expansion procedure, asymptotically valid in the domains above mentioned. Such method, in case of $p = 2$ has been developed by Kevorkian (1965) and later by Cole (1968). Here, we shall not endeavor in such procedures. In the case of p combinations, the system to be solved will have the form

$$\ddot{z}_k = \sum_{j=1}^{p} (A_{kj} \cos z_j + B_{kj} \sin z_j)$$

or

$$\ddot{\zeta}_k = \sum_{j=1}^{p} m_{kj} \sin \zeta_j \qquad (5.8.9)$$

for $k = 1, 2, \ldots, p$. For "small" oscillations in the vicinity of $\zeta_j = 0$, the above system is linear and the solution is immediate. Otherwise, system (5.8.9) is far from being trivial. This is equivalent to say that if all angles y_k librate around some equilibrium point, for small librations, the solution can be approximated as well as one wishes. But if at least one angle y_k circulate, the solution is not easily obtained. The same reasoning is transferred to the ζ variables.

As an example consider the case $p = 2$, so that one can write (assuming F even in y_1, y_2),

$$F = a_1\delta_1 + a_2\delta_2 + a_{11}\delta_1^2 + 2a_{12}\delta_1\delta_2 + a_{22}\delta_2^2 + \\ + A_1^{\alpha\beta} \cos(\alpha y_1 + \beta y_2) + A_1^{pq} \cos(py_1 + qy_2) \qquad (5.8.10)$$

In this case we find

$$\ddot{z}_1 = -k_{11} \sin z_1 - k_{12} \sin z_2 \\ \ddot{z}_2 = -k_{21} \sin z_1 - k_{22} \sin z_2 \qquad (5.8.11)$$

where

$$z_1 = \alpha y_1 + \beta y_2 , \quad z_2 = py_1 + qy_2 ,$$
$$k_{11} = \tilde{k}_{11}A_1^{\alpha\beta} , \quad k_{12} = \tilde{k}_{12}A_1^{pq} ,$$
$$k_{21} = \tilde{k}_{12}A_1^{\alpha\beta} , \quad k_{22} = \tilde{k}_{22}A_1^{pq} ,$$

and

$$\tilde{k}_{11} = 2(\alpha^2 a_{11} + 2\alpha\beta a_{12} + \beta^2 a_{22})$$
$$\tilde{k}_{12} = 2(\alpha p a_{11} + \beta q a_{12} + \alpha q a_{12} + \beta q a_{22}),$$
$$\tilde{k}_{22} = 2(p^2 a_{11} + 2pqa_{12} + q^2 a_{22}).$$

The transformations

$$\sin \frac{z_1}{2} = k_1 \operatorname{sn}(u_1, k_1)$$

and

$$\sin \frac{z_2}{2} = k_2 \operatorname{sn}(u_2, k_2),$$

lead to the equations

$$\frac{d}{dt} [(\dot{u}_1^2 - k_{11})\operatorname{cn}^2 u_1] = \alpha_1 \operatorname{cn}u_1 \operatorname{sn}u_2 \operatorname{dn}u_2 \dot{u}_1 \\ \frac{d}{dt} [(\dot{u}_2^2 - k_{22})\operatorname{cn}^2 u_2] = \alpha_2 \operatorname{cn}u_2 \operatorname{sn}u_1 \operatorname{dn}u_1 \dot{u}_2 , \qquad (5.8.12)$$

where

$$\alpha_1 = -2k_{12}\, k_2/k_1 \ ,$$

and

$$\alpha_2 = -2\, k_{21}\, k_1/k_2 \ .$$

The system uncouples if $\alpha_1 = \alpha_2 = 0$, that is, it is necessary and sufficient that

$$\alpha p a_{11} + (\beta q + \alpha q) a_{12} + \beta q a_{22} = 0 \tag{5.8.13}$$

in which case the solution of (5.8.12) is immediate. One of the possible cases is to have

$$a_{11} = a_{12} = a_{22} = 0 \tag{5.8.14}$$

but, then, we are not in a case of resonance. Another integrable case is, of course, when

$$b_{11}b_{22} - b_{12}b_{21} = 0 \tag{5.8.15}$$

for, in this case, z_1 and z_2 are such that one is equal to an integer multiple of the other and we fall back into a one-dimensional case.

Other than these particular solutions, it is also possible to have $k_1 = k_2$, that is, z_1 and z_2 are periodic (along the real t-axis) with same period. In fact, it is easily verified that if

$$k_{11} + k_{12} = k_{22} + k_{21}$$

we have as a particular solution $\dot{u}_1 = \dot{u}_2$ so that z_1 and z_2 are simply dephased periodic functions with same period.

In general, however, z_1 and z_2 will be quasiperiodic functions of t (real branch) and, eventually, their exponential Fourier nonperiodic series can be obtained from the original system (5.8.11), by setting

$$\sin \frac{1}{2} z_j = \sum_k \sum_\ell a_j^{k\ell} \exp[i(k\omega_1 t + \ell\omega_2 t)]$$

where, of course, ω_1 and ω_2 are unknown frequencies. Evidently, since for $k_{12} = 0$ one has

$$\sin \frac{1}{2} z_1 = k_1 \operatorname{sn}(u_1, k_1) = \sum_{k=0}^{\infty} b_1^k(k_1) \sin \left[(2k+1) \frac{\pi u_1}{2K(k_1)}\right],$$

with $u_1 = \sqrt{k_{11}}\, t + u_{10}$, and for $k_{21} = 0$

$$\sin \frac{1}{2} z_2 = k_2 \operatorname{sn}(u_2, k_2) = \sum_{\ell=0}^{\infty} b_2^\ell(k_2) \sin \left[(2\ell+1) \frac{\pi u_2}{2K(k_2)}\right],$$

with $u_2 = \sqrt{k_{22}}\, t + u_{10}$, where the moduli k_1, k_2 depend on the initial conditions, the zero-th order terms of the expansions of ω_1 and ω_2 in terms of k_{11}, k_{22}, k_1, k_2 are easily obtained, and the higher order terms eventually found by recurrence. This procedure is typical of a case where the coupling of the system is weak, that is, $|k_{12}|, |k_{21}| \ll |k_{11}|, |k_{22}|$. Evidently, one recognizes the difficulty of the problem, for the foregoing discussion relates simply to the main problem, the solution of which would serve to a basis for the extension of the solution to higher orders or, in case of Lie's series approach, to the definition of the auxiliary system. In both cases, the main problem being excessively complicated, little hope there is for any farther approximation.

If one takes the point, about which the expansion is made, to coincide with the center, for $p = 2$, the von Zeipel's equivalent of the previous problem leads to the equation

$$\begin{aligned} &a_1 S_{1/2,y_1} + a_2 S_{1/2,y_2} + a_{11}(S_{1/2,y_1})^2 + 2a_{12} S_{1/2,y_1} S_{1/2,y_2} + \\ &\quad + a_{22}(S_{1/2,y_2})^2 + A_1^{\alpha\beta} \cos(\alpha y_1 + \beta y_2) + A_1^{pq} \cos(p y_1 + q y_2) = \\ &\quad = K_1(\delta_1', \delta_2') \end{aligned} \tag{5.8.16}$$

where $S_{1/2}$ is the first order $(\epsilon^{1/2})$ approximation of the generating function of a canonical transformation $(\delta_1,\delta_2,y_1,y_2) \to (\delta_1',\delta_2',y_1',y_2')$. Of course nothing is gained, since the solution of the partial differential equation (5.8.16) is little understood in general. Also, we lack a principle for the definition of $K_1(\delta_1',\delta_2')$, the $O(\epsilon)$ part of the new Hamiltonian. The solution is trivial for the normal (non resonance!) case $a_{11} = a_{12} = a_{22} = 0$, or, better say, when S = identity + ΔS, and ΔS can be taken $O(\epsilon)$. With the same definitions of z_1 and z_2, Equation (5.8.16) changes into

$$(\alpha a_1 + \beta a_2)S_{1/2,z_1} + (pa_1 + qa_2)S_{1/2,z_2} +$$
$$+ \frac{1}{2}\tilde{k}_{11}(S_{1/2,z_1})^2 + \tilde{k}_{12}\, S_{1/2,z_1}\, S_{1/2,z_2} + \frac{1}{2}\tilde{k}_{22}(S_{1/2,z_2})^2 +$$
$$+ A_1^{\alpha\beta} \cos z_1 + A_1^{pq} \cos z_2 = K_1 \tag{5.8.17}$$

which again has a simple uncoupled Jacobian solution for $k_{12} = 0$, while, in general, is not simpler than the original one.

It might be useful to observe that, in certain situations, it will be possible to choose the reference point so that

$$\alpha p a_{11} + (\alpha q+\beta p)a_{12} + \beta q a_{22} = 0 \tag{5.8.18}$$

without a_{11},a_{12},a_{22} being zero. In fact, the integers α,β,p,q are given, but, since

$$a_{ij} = \left.\frac{\partial^2 H_o}{\partial x_i\, \partial x_j}\right|_{\substack{x_1=x_{10}\\ x_2=x_{20}}}$$

it may happen that a specific choice of x_{10},x_{20} uncouples the system. For this to be true, H_o has to belong to the class of functions f defined by $(p = 2)$

$$k\,\frac{\partial^2 f}{\partial x^2} + 2\ell\,\frac{\partial^2 f}{\partial x\,\partial y} + m\,\frac{\partial^2 f}{\partial y^2} = 0 \tag{5.8.19}$$

with k, ℓ, m given integers. We consider the following important cases

I) $\alpha = \beta$, so that $\alpha q + \beta q = \beta q + \alpha p$ and, therefore, in (5.8.19), $2\ell = m + k$.

In this case one can write (5.8.19) as

$$\left(\frac{\partial}{\partial x} + \frac{\partial}{\partial y}\right)\left(k\frac{\partial f}{\partial x} + m\frac{\partial f}{\partial y}\right) = 0$$

which is readily solvable.

II) $p = q$. Same as above.

III) We know that if $km \gtreqless \ell^2$ the equation is elliptic, parabolic or hyperbolic. In each case the properties of the solutions for f are well known and thoroughly discussed in any text on partial differential equations.

The problem is actually less complicated, since f, i.e., $H_0(\delta_1, \delta_2)$, is a known given function and the question of whether (5.8.18) may be verified or not will reduce to the solution of an equation (in general non algebraic) in two variables. All possible solutions of this equation will give the regions where the resonance effects can be separated.

9. Coupling of Two Harmonic Oscillators.

We conclude this chapter by giving a brief discussion of the problem of nonlinear coupling of oscillators. This description is based on results obtained by Hori (1967) and serves as a concluding example on perturbation techniques based on Lie Series, for conservative systems. The system to be considered is specified by

$$\begin{aligned} \ddot{x}_1 + \omega_1^2 x_1 &= \epsilon x_2^2 \\ \ddot{x}_2 + \omega_2^2 x_2 &= 2\epsilon x_1 x_2 \end{aligned} \tag{5.9.1}$$

where $\omega_1, \omega_2, \epsilon$ are positive real constants. The existence of a third integral for the above system has been discussed in several papers by Contopoulos and associates (1960, 1962, 1963, 1965).

In Hamiltonian form, (5.9.1) can be written as

$$\dot{x}_k = F_{y_k} , \ \dot{y}_k = -F_{x_k} \qquad (k = 1,2) \tag{5.9.2}$$

where

$$F = F_0(x;y) + F_1(x), \tag{5.9.3}$$

and

$$F_0 = \frac{1}{2} (y_1^2 + y_2^2 + \omega_1^2 x_1^2 + \omega_2^2 x_2^2), \tag{5.9.4}$$

$$F_1 = -\epsilon x_1 x_2^2 . \tag{5.9.5}$$

We shall consider a second order theory and a Hori-Lie Generator $S(\xi;\eta;\epsilon)$, developable in power series of ϵ, for a canonical transformation $(x;y) \rightarrow (\xi;\eta)$, defined by

$$\begin{aligned} x_j &= \xi_j + \frac{\partial S_1}{\partial \eta_j} + \frac{\partial S_2}{\partial \eta_j} + \frac{1}{2} \left(\frac{\partial S_1}{\partial \eta_j}, S_1 \right) \\ y_j &= \eta_j - \frac{\partial S_1}{\partial \xi_j} - \frac{\partial S_2}{\partial \xi_j} - \frac{1}{2} \left(\frac{\partial S_1}{\partial \xi_j}, S_1 \right) , \end{aligned} \tag{5.9.6}$$

where we have neglected terms $O(\epsilon^3)$, and

$$S = S_1 + S_2 + \cdots .$$

To the same order the mapping via S of a function $f(\xi;\eta)$ is

$$f(x;y) = f(\xi;\eta) + (f,S_1) + (f,S_2) + \frac{1}{2} ((f,S_1),S_1). \tag{5.9.7}$$

This relation can be used to obtain the perturbation formulae as derived in Chapter 2, that is,

$$\Phi(\xi;\eta) = F(x;y) \tag{5.9.8}$$

where Φ is the new (transformed) Hamiltonian. Note that in this example x and ξ stand for coordinates and y and η for momenta.

Since

$$\Phi_o = F_o = \frac{1}{2}(\eta_1^2 + \eta_2^2 + \omega_1^2\xi_1^2 + \omega_2^2\xi_2^2)$$

the solution of the auxiliary system is

$$\begin{aligned} \xi_j &= c_j \cos(\omega_j\tau + c_j') \\ \eta_j &= -c_j\omega_j \sin(\omega_j\tau + c_j') \end{aligned} \tag{5.9.9}$$

for $j = 1,2$.

Non-resonance Case.

Suppose ω_1 and ω_2 are linearly independent over the integers. The solution of the equations

$$\Phi_1 = \lim_{T\to\infty} \frac{1}{T}\int_0^T F_1(\xi(\tau);\eta(\tau))d\tau \triangleq F_{1s} \tag{5.9.10}$$

$$S_1 = \int [F_1(\xi(\tau);\eta(\tau)) - \Phi_1]d\tau \tag{5.9.11}$$

$$\Phi_2 = \frac{1}{2}(F_1 + \Phi_1, S_1)_s \tag{5.9.12}$$

$$S_2 = \int \{\frac{1}{2}(F_1 + \Phi_1, S_1) - \Phi_2\}d\tau \tag{5.9.13}$$

$$\begin{aligned} \Phi_3 = {} & \frac{1}{2}(F_1 + \Phi_1, S_2)_s + \frac{1}{2}(F_2 + \Phi_2, S_1)_s + \\ & + \frac{1}{12}((F_1 - \Phi_1, S_1), S_1)_s \end{aligned} \tag{5.9.14}$$

gives the generator S to second order included and the new Hamiltonian to third order included.

One actually finds:

$$S_1 = \frac{\epsilon}{\omega_1^2(\omega_1^2 - 4\omega_2^2)} \{(\omega_1^2 - 2\omega_2^2)\eta_1\xi_2^2 - 2\omega_1^2\xi_1\xi_2\eta_2 - 2\eta_1\eta_2^2\}$$

$$\Phi_1 = 0,$$

$$S_2 = \frac{\epsilon^2}{2\omega_1^2(\omega_1^2 - 4\omega_2^2)} \left\{ \frac{\xi_1\eta_1}{\omega_2^2 - \omega_1^2} [(2\omega_1^2 + \omega_2^2)\xi_2^2 - 3\eta_2^2] + \right.$$

$$+ \xi_2\eta_2 \left[\frac{\omega_1^2(4\omega_2^2 - \omega_1^2)}{\omega_2^2(\omega_1^2 - \omega_2^2)} \xi_1^2 - \frac{\omega_1^2 + 2\omega_2^2}{\omega_2^2(\omega_1^2 - \omega_2^2)} \eta_1^2 + \right.$$

$$\left.\left. + \frac{5\omega_1^2 - 8\omega_2^2}{8\omega_2^2} \xi_2^2 + \frac{3\omega_1^2 - 8\omega_2^2}{8\omega_2^4} \eta_2^2 \right] \right\}, \tag{5.9.15}$$

$$\Phi_2 = \frac{\epsilon^2}{2\omega_1^2(\omega_1^2 - 4\omega_2^2)} \left[(\omega_2^2 - \frac{3}{8}\omega_1^2)\left(\xi_2^2 + \frac{\eta_2^2}{\omega_2^2}\right)^2 + \omega_1^2\left(\xi_1^2 + \frac{\eta_1^2}{\omega^2}\right)\left(\xi_2^2 + \frac{\eta_2^2}{\omega_2^2}\right) \right],$$

$$\Phi_3 = 0.$$

According to the general results stated in Chapter 2, the following quantities are integrals of motion for the differential equations generated by $\Phi(\xi;\eta)$:

$$\begin{aligned} \Phi(\xi,\eta) &= \text{const.} \\ \Phi_0(\xi,\eta) &= \text{const.} \end{aligned} \tag{5.9.16}$$

It is easily seen that (5.9.16) imply

$$\xi_j^2 + \frac{1}{\omega_j^2}\eta_j^2 = \text{const.} = c_j^2 , \tag{5.9.17}$$

for $j = 1,2$. The solution of the system generated by Φ is therefore

$$\begin{aligned} \xi_j &= c_j \cos(\omega_j^* t + c_j'), \\ \eta_j &= -c_j\omega_j \sin(\omega_j^* t + c_j'), \end{aligned} \tag{5.9.18}$$

where the <u>adjusted frequencies</u> ω_j^* are given, <u>neglecting terms of order</u> ϵ^4, by

$$\omega_1^{*2} = \omega_1^2 \left\{ 1 + \frac{\epsilon^2 c_2^2}{\omega_1^2(\omega_1^2 - 4\omega_2^2)} \right\},$$

$$\omega_2^{*2} = \omega_2^2 \left\{ 1 + \frac{\epsilon^2}{\omega_1^2\omega_2^2(\omega_1^2 - 4\omega_2^2)} [\omega_1^2 c_1^2 + (2\omega_2^2 - \frac{3}{4}\omega_1^2)c_2^2] \right\}. \qquad (5.9.19)$$

The relations among $(x;y)$ and $(\xi;\eta)$ are easily obtained from (5.9.6). <u>The inverse transformation</u> of (5.9.7) is given by

$$f(\xi;\eta) = f(x;y) - (f,S_1) - (f,S_2) + \frac{1}{2}((f,S_1),S_1) + \cdots \qquad (5.9.20)$$

where terms $O(\epsilon^3)$ and higher have been neglected. The application of (5.9.20) to the function

$$f = \xi_1^2 + \frac{1}{\omega_1^2}\eta_1^2 = c_1^2 \text{ (const.)}$$

gives, to that same precision, the <u>third integral</u> obtained by Contopoulos (1965),

$$x_1^2 + \frac{1}{\omega_1^2} y_1^2 + \cdots = c_1^2 = \text{const.}\ .$$

The choice of

$$f = \xi_2^2 + \frac{1}{\omega_2^2}\eta_2^2$$

leads to no new integral, independent of c_1^2 and the energy integral (Φ or F).

<u>Resonance Case.</u>

To the order we have considered, it is obvious that the cases

$$\omega_1^2 = 4\omega_2^2$$

$$\omega_1^2 = \omega_2^2$$

must be excluded, as seen from (5.9.15) and (5.9.19). In fact, going to higher orders of approximations will introduce divisors of the type $n\omega_1^2 - m\omega_2^2$, with n,m integers and increasing with the order of the theory. In fact, as was pointed out in earlier sections, such divisors do not even need to be zero for the theory to be applicable, but simply small enough. For instance, the first of (5.9.15) shows that if $\omega_1^2 - 4\omega_2^2 = O(\epsilon)$, S_1 is not anymore a first order quantity, as it is assumed from the beginning. In such cases the argument (critical) corresponding to the small divisors should be retained in Φ when expressed in terms of the solution of the auxiliary system (5.9.9).

When $\omega_1^2 \simeq 4\omega_2^2$, such approach leads to

$$\Phi_1 = -\frac{\epsilon}{4\omega_1\omega_2^2}[\omega_1\xi_1(\omega_2^2\xi_2^2 - \eta_2^2) + 2\omega_2\eta_1\xi_2\eta_2]$$

$$S_1 = \frac{\epsilon}{4\omega_1^2\omega_2^2(\omega_1 + 2\omega_2)}[\omega_2^2(3\omega_1 + 4\omega_2)\eta_1\xi_2^2 + $$

$$+ (\omega_1 + 4\omega_2)\eta_1\eta_2^2 + 2\omega_1^2\omega_2\xi_1\xi_2\eta_2] \tag{5.9.21}$$

$$\Phi_2 = -\frac{\epsilon^2}{32\omega_1^2\omega_2^2(\omega_1 + 2\omega_2)}\left[4\omega_1^2\omega_2\left(\xi_1^2 + \frac{\eta_1^2}{\omega_1^2}\right)\left(\xi_2^2 + \frac{\eta_2^2}{\omega_2^2}\right) + \right.$$

$$\left. + \omega_2^2(5\omega_1 + 8\omega_2)\left(\xi_2^2 + \frac{\eta_2^2}{\omega_2^2}\right)^2\right].$$

These relations are sufficient to justify the following construction of a third integral and establish the results obtained by Contopoulos (1963).

As suggested by Lindstedt's method, it is now possible to introduce a canonical transformation such that one of the coordinates becomes ignorable, precisely the one corresponding to a non-critical argument. Making use of the auxiliary system, the new coordinates should, therefore, be defined by

$$\begin{aligned} q_1 &= 2(\omega_2\tau + c_2') - (\omega_1\tau + c_1') \\ q_2 &= -(\omega_2\tau + c_2'). \end{aligned} \tag{5.9.22}$$

The second of these is chosen so as to generate an ignorable q_2 corresponding exactly to a p_2 (associate momentum) coincident with Contopoulos' third integral for this resonance case. Otherwise, one could choose q_2 as any linear combination (not multiple of the critical argument q_1) of $\omega_1\tau + c_1'$ and $\omega_2\tau + c_2'$. The momenta conjugate to q_1, q_2 defined by (5.9.22), come out to be

$$p_1 = \frac{\omega_1}{2} c_1^2$$

$$p_2 = \omega_1 c_1^2 + \frac{\omega_2}{2} c_2^2 \tag{5.9.23}$$

where q_1, q_2, p_1, p_2 are easily expressed in terms of $(\xi;\eta)$ via (5.9.9).

It is easily seen that the Hamiltonian, when expressed in terms of $(q;p)$ does not contain q_2, so that p_2 is constant, that is

$$\omega_1\left(\xi_1^2 + \frac{\eta_1^2}{\omega_1^2}\right) + \frac{\omega_2}{2}\left(\xi_2^2 + \frac{\eta_2^2}{\omega_2^2}\right) = p_2 = \text{const.} \tag{5.9.24}$$

Transformation of this by means of (5.9.20) gives Contopoulos' Third Integral in this case of resonance:

$$\omega_1\left(x_1^2 + \frac{y_1^2}{\omega_1^2}\right) + \frac{\omega_2}{2}\left(x_2^2 + \frac{y_2^2}{\omega_2^2}\right) \frac{2\epsilon}{\omega_1\omega_2(\omega_1 + 2\omega_2)} [\omega_2(\omega_1 + \\ + \omega_2)x_1x_2^2 + x_1y_2^2 - y_1x_2y_2] + \cdots = p_2 = \text{const.} \tag{5.9.25}$$

Complete equivalence is shown by Hori in the above referred paper, by assuming $\omega_1^2 = 4\omega_2^2$ exactly. The other resonance considered by Contopoulos (1965) and also present here is $\omega_1^2 - \omega_2^2 \simeq 0$. In this case the <u>critical argument</u> leading to a small (or zero) divisor in the general theory is $(\omega_1\tau_1 + c_1') - (\omega_2\tau + c_2')$. Keeping this argument and multiples in Φ_2 (Φ_1 is unchanged) leads to

$$\Phi_2 = \frac{\epsilon^2}{2\omega_1^2(\omega_1^2 - 4\omega_2^2)} \quad (\omega_2^2 - \frac{3}{8}\omega_1^2)\left(\xi_2^2 + \frac{1}{\omega_2^2}\eta_2^2\right)^2 + \\ + \omega_1^2\left(\xi_1^2 + \frac{1}{\omega_1^2}\eta_1^2\right)\left(\xi_2^2 + \frac{1}{\omega_2^2}\eta_2^2\right) + \\ + \frac{1}{2}\omega_1(\omega_1 + 2\omega_2)\left[\left(\xi_1^2 - \frac{1}{\omega_1^2}\eta_1^2\right)\left(\xi_2^2 - \frac{1}{\omega_2^2}\eta_2^2\right) + \\ + \frac{4}{\omega_1\omega_2}\xi_1\eta_1\xi_2\eta_2\right].$$

As in the previous case, by introduction of the appropriate set of canonical conjugate variables

$$q_1 = (\omega_2\tau + c_2') - (\omega_1\tau + c_1')$$
$$q_2 = -(\omega\tau + c_2')$$
$$p_1 = \frac{1}{2}\,\omega_1 c_1^2$$
$$p_2 = \frac{1}{2}\,\omega_1 c_1^2 + \frac{1}{2}\,\omega_2 c_2^2$$

the new Hamiltonian does not contain the coordinate q_2, so that p_2 is constant. It follows

$$\frac{1}{2}\,\omega_1\left(\xi_1^2 + \frac{1}{\omega_1^2}\,\eta_1^2\right) + \frac{1}{2}\,\omega_2\left(\xi_2^2 + \frac{1}{\omega_2^2}\,\eta_2^2\right) = p_2 = \text{const.}\,,$$

or making use of (5.9.20),

$$\frac{\omega_1}{2}\left(x_1^2 + \frac{1}{\omega_1^2}\,y_1^2\right) + \frac{\omega_2}{2}\left(x_2^2 + \frac{1}{\omega_2^2}\,y_2^2\right) -$$
$$- \frac{\epsilon}{\omega_1\omega_2\sqrt{\omega_1^2 - 4\omega_2^2}}\,[\omega_2(\omega_1^2 - 2\omega_1\omega_2 - 2\omega_2^2)x_1x_2^2$$
$$+ 2(\omega_1 - \omega_2)(x_1y_2^2 - y_1x_2y_2)] + O(\epsilon^2) = p_2 = \text{const.}\;.$$

This again can be shown to coincide exactly with Contopoulos' result in the case $\omega^2 = \omega_2^2$. <u>In both resonant cases considered here it can be easily seen that the new Hamiltonian, in terms of the</u> $(q;p)$ <u>variables, has a dominant part identical to that defining the ideal resonance problem.</u> More specifically, for the resonance $\omega_1^2 \simeq 4\omega_2^2$, one has

$$H = (\omega_1 - 2\omega_2)p_1 + \omega_2 p_2 - \frac{\epsilon}{\sqrt{2}\sqrt{\omega_1}\,\omega_2}\,(p_2 - 2p_1)\sqrt{p_1}\,\cos q_1$$

and for the resonance $\omega_1^2 \simeq \omega_2^2$,

$$H = (\omega_1 - \omega_2)p_1 + \omega_2 p_2 + \frac{\epsilon^2}{4\omega_1^2\omega_2^2(\omega_1^2 - 4\omega_2^2)} [(8\omega_2^2 - 3\omega_1^2)p_2^2 +$$

$$+ 2(3\omega_1^2 + 4\omega_1\omega_2 - 8\omega_2^2)p_1 p_2 - (3\omega_1^2 + 8\omega_1\omega_2 - 8\omega_2^2)p_1^2 +$$

$$+ 4\omega_2(\omega_1 + 2\omega_2)(p_2 - p_1)p_1 \cos 2q_1].$$

In both cases one sees that the derivative

$$\frac{\partial H_0}{\partial p_1} = 0$$

at exact resonance. In the second case, however, one may allow that derivative to be as small as $O(\epsilon)$ and still be able to develop a formal series solution in powers of ϵ. In the first case, the smallest order of magnitude allowed is $O(\epsilon^{1/2})$ and in this case the solution is a power series in $\epsilon^{1/2}$, as described in earlier sections.

Systems of differential equations of the type of (5.9.1) are widely studied in the literature and a comparison of classical methods of approach with Lie's Series formal approach might prove quite fruitful. Convergence of this last is perhaps the most intriguing of the problems.

NOTES

The concepts of linear and nonlinear resonance, from the physical point of view, can only reflect certain aspects of particular situations. For all practical purposes, linear resonance does not exist. Nonlinear resonance is the essential principle behind tuning, recovery of signals from a noisy background and, in natural systems, the cause of stable oscillatory configurations: the common denomination is the "lock-in" phenomenon. From the mathematical point of view, such phenomenon is always explained in particular systems of differential equations where nonlinearities have a "treatable" character. In more complex systems, resonance is only understood via numerical explorations. Good examples are given by the works of Hènon and Heiles in 1964 and of Gustavson in 1966. When possible, the method of surfaces of section (or Poincaré's Consequents), is very effective, as in the problem studied by Danby in 1970. As far as Hamiltonian systems are concerned, we face the immediate question of whether such systems are oscillators. The clear picture given by the action-angle variables is only applicable when Hamilton-Jacobi equation can be separated, so that it becomes too restrictive. Thus, we fall into the usual description of oscillations in the vicinity of a state of equilibrium, which we must assume to be stable. The nonlinear character is that, being such oscillations periodic in each variable which describes deviation from equilibrium, their period is a function of that deviation. A mathematical definition of nonlinear resonance based on the method used to develop asymptotic solutions is certainly not satisfactory, although often used (e.g., Kyner, 1968, p. 18). Such a definition, when based on the appearance of zero divisors in approximation methods, as described in this chapter, is nothing more than an equivalent statement of partial degeneracy. Before going into more sophisticated definitions, one should realize that the question under focus is the behavior of a nonlinear dynamical system under the influence of external (or internal) perturbations. Asymptotically stable solutions can eventually be obtained by asymptotic or purely qualitative methods, as indicated by Bogoliubov and Mitropolskii, Nemytskii and Stepanov, Cesari and Mitropolskii respectively in 1951,

1960, 1959 and 1964 (translated in 1966). The search for these solutions is quite frequently reducible to the determination of the location and nature of the singular points in phase space. The nature of these points is defined by the behavior of integral curves in their neighborhoods, and therefore, with this knowledge, we can establish the topological picture in these neighborhoods in phase space. However, in the large, the behavior of integral curves in phase space remain vague and numerical techniques are often times the only method available. Or else, one tries to understand the nonlinear effects by simplifying the system. The characteristic example, considered to contain typical nonlinear difficulties and resonance characters, is condensed in the equation

$$\ddot{x} + \omega^2 x + \alpha\dot{x} = \beta f(\tau) \qquad \text{(A)}$$

where ω is a function of $x,\dot{x}$ and so is α. The amplitude of the forcing term, β, is considered fixed and $\tau = \epsilon t$, where ϵ is small. The typical assumption is that ω^2 and α are functions of the "energy" E of the system, that is,

$$2E = \dot{x}^2 + \omega^2(E)x^2 \qquad \text{(B)}$$

and one looks for an oscillatory solution

$$x = a\cos(\nu t+\varphi)$$
$$\dot{x} = -a\nu\sin(\nu t+\varphi).$$

If both amplitude (a) and phase (φ) are slowly varying with time, over the period $T = 2\pi/\nu$, one can average the corresponding equations over such period and obtain a solution with a specific error bound [e.g., Kyner, 1968, pp. 17-18, Eqs. (2.19) and (2.20)].

Such averaged equations result to be

$$\dot{\alpha} = -\frac{\alpha}{2}a - \frac{\beta}{2\nu}\sin\varphi$$
$$\dot{\varphi} = \frac{1}{2\nu}(\omega^2 - \nu^2) - \frac{\beta}{2\nu a}\cos\varphi \qquad \text{(C)}$$
$$E = \frac{1}{4}a^2(\omega^2 + \nu^2)$$

and they show that the assumption of the slow varying amplitude is met when

$$|\alpha a|,\ |\beta/\nu| \ll \nu a.$$

Slow varying phase is verified in regions of the plane (a,φ) where

$$|\nu - (\omega^2 + \frac{\beta}{a}\cos\varphi)^{1/2}| \ll \nu.$$

In case of no dissipation $(\alpha = 0)$, the system is Hamiltonian and defined by

$$H(p,\theta) = \frac{1}{2\nu}\int^{p}[\omega^2(x) - \nu^2]dx - 2\epsilon\sqrt{p}\cos\theta$$

where $p = a^2$, $\epsilon = \beta/2\nu$.

It follows that the amplitude a is bounded if

$$\lim_{a\to\infty}\int_{a_0}^{a}|\omega^2(x^2) - \nu^2|x\,dx \qquad \text{(D)}$$

is finite and not zero. Such condition really implies that the system does not behave linearly for large amplitudes. Except for those integral curves having singular points as limit points, the trajectories in the phase plane (a,θ) are closed and do not intercept, if the points $\theta = 0$ and $\theta = 2\pi$ are identified. In absence of dissipation, the singular points of the system

$$\begin{aligned}\dot{a} &= -\epsilon\sin\varphi\\ \dot{\varphi} &= \frac{1}{2\nu}(\omega^2 - \nu^2) - \frac{\epsilon}{a}\cos\varphi,\end{aligned} \qquad \text{(E)}$$

which is of the type discussed in Chapter III, are given by

$$a = 0,\ \varphi = \frac{\pi}{2},\ \varphi = 3\frac{\pi}{2},$$

and $\dot{a} = \dot{\varphi} = 0$, i.e.,

$$\varphi = 0,\pi$$

$$\frac{a}{2\nu}(\omega^2 - \nu^2) = \pm\epsilon\ . \qquad \text{(F)}$$

The Equation (F) is the resonance curve of the system, in the plane (a,ν). That is, the coordinate a of a singular point for a given value of ν, is a root of

$$\nu_{\pm}(a) = (\omega^2 \pm \beta/a)^{1/2}$$

while the phase $\varphi = \pi$ for ν_+ and $\varphi = 0$ for ν_-. It is seen that $\nu_+(a)$ is a curve of centers when $\nu_+'(a) < 0$, saddles when $\nu_+'(a) > 0$ and of mixed type at $\nu_+'(a) = 0$. The situations are reversed for $\nu_-(a)$. Some important facts, generally true for nonlinear systems, are that

a) For any frequency ν there exists at least a center.

b) The number of centers is larger by one than the number of saddles.

c) Under a small variation every mixed point either disappears or splits into a pair of points: a center and a saddle.

d) A separatrix passes through every saddle and the exceptional singular points $a = 0$, $\varphi = \pi/2, 3\pi/2$.

Finally, the general characteristics of the motion without dissipation are:

e) The period, defined by

$$T(C) = \oint_{H=C} \frac{d\varphi}{\dot{\varphi}}$$

can be estimated by the following approximation if the integral curves are sufficiently close to a center. In fact, linearization of Equations (E) in the vicinity of a center C_k gives

$$\dot{a} = -\epsilon\varphi$$
$$\dot{\varphi} = -\alpha_k[a_k - a_k]$$

where

$$\varphi(C_k) = 0$$

$$a_k = a(C_k)$$

$$\alpha_k = \frac{\omega_k'\omega}{\nu} + \frac{\epsilon}{a^2}$$

$$\omega_k = \omega(a(C_k)), \quad \omega_k' = \omega'(a(C_k)).$$

Such linearized equations can obviously be written

$$\ddot{x} + \epsilon\alpha_k x = 0$$

where

$$x = a - a(C_k).$$

The frequency in the amplitude oscillation (x) is thus, in the limit $C = C(C_k)$, given by

$$\Omega_k = (\epsilon\alpha_k)^{1/2}$$

and the period is

$$T_k = 2\pi(\epsilon\alpha_k)^{1/2} .$$

The $1/2$ power comes out naturally in this system and the usual reference to boundary layer studies seems hardly necessary. Actually, as we have seen in Section 5 of this chapter, under different assumptions, different powers of ϵ are obtained (Equation 5.5.9). In regions not close to centers the integral for $T(C)$ should be computed exactly and, most of the times, this can only be done numerically. However, if $a >> \epsilon$, neglecting terms order $(\epsilon/a)^2$ and higher, one finds

$$\dot{x} = -\epsilon \sin \varphi$$

$$\dot{\varphi} = \alpha_k x$$

and the integral $T(C)$ gives, for $1 \geq C_k \geq -1$,

$$T(C) = 2(\epsilon\alpha_k)^{-1/2} K\left(\sqrt{\frac{1+C_k}{2}}\right),$$

and for $C_k > 1$,

$$T(C) = 2[\epsilon\alpha_k(1+C_k)/2]^{-1/2} K\left(\sqrt{\frac{2}{1+C_k}}\right),$$

where $K(k)$ is the complete elliptic integral of the first kind modulus k.

We can recover the result for a center by setting $C_k = -1$. If $C_k \to +1$, the period $T \to \infty$ and, in fact, $C_k = +1$ corresponds to a separatrix.

The amplitude variation Δa (amplitude of libration in the vicinity of a center) is defined by

$$\Delta a(C) = \max a(C,\varphi) - \min a(C,\varphi)$$

and one obtains

$$\Delta a = 2[2\Delta(1+C_k)]\alpha_k]^{1/2}$$

and

$$\Delta a = (2\epsilon/\alpha_k)^{1/2} [(1+C_k)^{1/2} - (C_k - 1)^{1/2}] \quad (C_k > 1).$$

These results are essentially described in our papers (Giacaglia, 1968, 1969, 1970) but here we have followed the direct description by Bakai in 1966.

The conclusion is that Δa increases with C_k, reaches a maximum value as $C_k \to 1-$ and then decreases abruptly as the type of motion changes from libration (finite phase variation) to circulation (unbounded phase variation).

When dissipation is introduced the picture of the phase trajectories is also not difficult to find, despite the lack of an energy integral (Hamiltonian). In fact, although the trajectories are not closed anymore they approach either a closed trajectory or a fixed point. Contrary to the previous case we have that

a) The phase φ of a singular point is not constant. For the singular points defined by the branch $\nu_+(a)$, $\pi/2 \le \varphi \le 3\pi/2$, while for those defined by the branch $\nu_-(a)$, $-\pi/2 \le \varphi \le \pi/2$.

(b) At the value $\underline{a}$ satisfying $a\alpha = 2\epsilon$, the branches $\nu_+(a)$ and $\nu_-(a)$ merge with the curve defined by $\nu = \omega(a)$. In the absence of friction $\nu_+(a)[\nu_-(a)]$ lies always above [below] that curve.

(c) The centers are changed into foci, which are stable for $d(\alpha a)/da > 0$ and unstable for $d(\alpha a)/da < 0$. Saddles and mixed points are conserved.

(d) Every phase trajectory tends to a stable focus or stable limit cycle surrounding it. The transition is through a phase trajectory having a saddle as its limit singular point. In this case, one branch approaches the focus and the other approaches the limit cycle.

To conclude, we wish to mention the important results obtained by <u>Moser</u> in 1967, for the development of quasi-periodic motions, in particular, in the vicinity of resonant systems.

The question lies on the possibility of applying, to cases of resonance, Moser's Theorem on quasi-periodic motions given at p. 200. This extension, for Hamiltonian Systems, is obtained by <u>Moser</u> in 1967 (pp. 170-175). Specifically, considering a system with N degrees of freedom, if n is the number of rationally independent frequencies of the unperturbed system, when $n = N$ Kolmogorov's or Moser's Theorems apply. When $n = 0$ we have equilibrium and $n = 1$ correspond to periodic motions, a case originally delt with by <u>Arnol'd</u> in 1961. All other cases, $1 < n < N$, are situations of resonance and the <u>order of resonance</u> is $N - n$. From the point of view seen in Chapter III (pp. 170-173), resonance has been delt with by <u>Arnol'd</u> in his long paper of 1963, as an extension of Kolmogorov's Theorem. Actually, the basic reasoning in Moser's work is present in Arnol'd's Theorem, a fact visible by comparing, for instance, $\overline{H}_{1s}$ (p. 172) with Moser's $\tilde{H}$ (1967, p. 172). In case of simple resonance, i.e., $n = N - 1$, the conclusions of Moser can be very clearly stated.

In fact, consider the Hamiltonian $H = H(y_1, y_2, \ldots, y_N \,;\, x_1, x_2, \ldots, x_N \,;\, \epsilon)$, with $H(y;x;0) = H_0(y_N;x)$, H 2π-periodic with respect to every angle $y_1, y_2, \ldots, y_n$, $n = N - 1$. Moreover, for fixed values of $x_1, x_2, \ldots, x_{N-1}$ restricted to some bounded domain, we assume that $y_N = x_N = 0$ is an equilibrium solution of H_0, that is,

$$\frac{\partial H_o}{\partial y_N} = \frac{\partial H_o}{\partial x_N} = 0.$$

For $\epsilon = 0$, we have on the manifold

$$\begin{aligned} x_k &= C_k = \text{const.} \quad (k = 1,2,\ldots,n) \\ y_N &= x_N = 0 \end{aligned} \tag{G}$$

an n-parameter family of quasi-periodic motions. Such motions, provided the equilibrium point is elliptic, can be continued under sufficiently small perturbations, that is, small ϵ. More precisely Moser (1967, p. 171, Theorem 6) shows that

"If the Hamiltonian $H(y;x;\epsilon)$ for $\epsilon = 0$ has the equilibrium point $x_N = y_N = 0$ and

1)

$$\det \begin{pmatrix} \frac{\partial^2 H}{\partial x_i \partial x_j} & \frac{\partial^2 H}{\partial x_i \partial y_N} \\ \frac{\partial^2 H}{\partial y_N \partial x_j} & \frac{\partial^2 H}{\partial y_N^2} \end{pmatrix} \neq 0$$

at $\epsilon = 0$, $x_N = y_N = 0$, $x_k = C_k$.

2) Being

$$\Delta = \begin{pmatrix} \frac{\partial^2 H}{\partial x_N^2} & \frac{\partial^2 H}{\partial x_N \partial y_N} \\ \frac{\partial^2 H}{\partial x_N \partial y_N} & \frac{\partial^2 H}{\partial y_N^2} \end{pmatrix} = \alpha^2 > 0$$

at $\epsilon = 0$, $x_N = y_N = 0$, $x_k = C_k$, then

$$\det \begin{pmatrix} \frac{\partial^2 H}{\partial x_i \partial x_j} & \frac{\partial \Delta}{\partial x_i} \\ \frac{\partial H}{\partial x_j} & \Delta \end{pmatrix} \neq 0$$

at $\epsilon = 0$, $x_N = y_N = 0$, $x_k = C_k$, $i,j,k = 1,2,\ldots,n$.

Then, for ϵ sufficiently small, there exists a quasi-periodic solution, with n frequencies, near the solution defined by (G)".

A Taylor expansion of H is made in the vicinity of (G) and the equilibrium point $x_N = y_N = 0$ is supposed to be elliptic, that is, the matrix

$$\begin{Bmatrix} \dfrac{\partial^2 H_o}{\partial x_N^2} & \dfrac{\partial^2 H_o}{\partial x_N \partial y_N} \\ \dfrac{\partial^2 H_o}{\partial x_N \partial y_N} & \dfrac{\partial^2 H_o}{\partial y_N^2} \end{Bmatrix}$$

is reducible to the diagonal form

$$\Delta = \begin{pmatrix} \alpha & 0 \\ 0 & \alpha \end{pmatrix}$$

with $\alpha \neq 0$. Thus, a final form is achieved

$$\tilde{H} = \sum_{j=1}^{n} a_j X_j + \frac{1}{2}(X_N^2 + Y_N^2) + O(\epsilon)$$

where

$$|X_j - x_j| \sim O(\epsilon^2) \qquad (j = 1,2,\ldots,n)$$

and

$$|X_N - x_N| \sim |Y_N - y_N| \sim O(\epsilon^4).$$

The differential equations take the form

$$\begin{aligned} \dot{y} &= a + O(\epsilon) \\ \dot{\xi} &= \Omega\xi + O(\epsilon) \end{aligned} \qquad \text{(L)}$$

where $y = (y_1, y_2, \ldots, y_n)$, $\xi = (x_1, x_2, \ldots, x_n, x_N, y_N)$. The eigenvalues of Ω are

$\Omega_1 = \Omega_2 = \cdots \Omega_n = 0$, $\Omega_N = -\Omega_{N+1} = \sqrt{-1}$.

In order to apply Moser's Theorem (p. 200), the modified system

$$\dot{y} = a + \lambda + O(\epsilon)$$

$$\dot{\xi} = \Omega\xi + M\xi + O(\epsilon)$$

must be considered. But as shown by Moser, it is possible to choose $M = \sigma\Omega$, $\sigma \neq 0$ real, so that by a time "stretching"

$$d\tau = (1+\sigma)dt$$

one finds

$$y' = (1+\sigma)^{-1}(a+\lambda) + O(\epsilon)$$

$$\xi' = \Omega\xi + O(\epsilon)$$

and the constants

$$a_j = (1+\sigma)^{-1}(a_j + \lambda_j)$$

can be obtained corresponding to values of c_k $(k = 1,2,\ldots,n)$ which satisfy the hypotheses of the Theorem.

Then, for system (L), quasi-period motions exist.

As far as periodic solutions in the vicinity of an equilibrium, when the normal modes are rationally dependent, the question has been treated originally by Siegel. He has given a counterexample to the possible extension of Lyapunov's Theorem. More recently, however, an important theorem has been proved by Berger (1969) for Hamiltonian systems in the absence of giroscopic terms. On the other hand Henrard (1969) has given a formal normalization in a specific problem where such terms are present and shown by numerical techniques that periodic orbits do seem to exist for a certain resonance relation among the normal modes at an equilibrium point. Finally, Roels in 1969 gave an application of a normalization technique showing periodic orbits corresponding to rational modes in the Restricted Problem of Three Bodies, and in 1971 produced a proof of a theorem which generalizes Lyapunov's results. More precisely, there exist periodic solutions in the neighborhood of an equilibrium, when one eigenvalue $\lambda_i = k\lambda_1$ ($k \geq 4$, integer; λ_1 pure imaginary), under certain conditions depending on k and on higher order terms of the expansion of the Hamiltonian about the equilibrium. Here we limit to mention the most important results, by Berger and Roels.

Berger's generalization of Lyapunov's Theorem is as follows:

Theorem: "Given the Hamiltonian

$$H(y;x) = \frac{1}{2} x^2 + \frac{1}{2} y^T A y + F(y)$$

where y and x are n-vectors, $F(y)$ is even and C^1 for $0 \leq y_k < \geq 2\pi$, $\lim_{\|y\| \to 0} \|\frac{\partial F}{\partial y}\|/\|y\| = 0$ and A is a self-adjoint constant $n \times n$ matrix, with eigenvalues $\lambda_1^2, \lambda_2^2, \ldots, \lambda_n^2$. If there exist p eigenvalues not necessarily distinct but equal to $k^2\lambda_j^2$ (k integer) for some $1 \leq j \leq n$, then there exist at least p distinct one-parameter families $u_i(\epsilon)$, $i = 1,2,\ldots,p$,, of periodic orbits with periods $T_i(\epsilon)$. As $\epsilon \to 0$, the families tend to equilibrium and the periods to $2\pi/|\lambda_j|$. Moreover, if $F(y)$ is real analytic, $u_i(\epsilon)$ and $T_i(\epsilon)$ are continuous in a neighborhood of the equilibrium".

Obviously the theorem does not apply when the Hamiltonian contains linear terms in x, the "velocity" vector. In this case Roel's result applies and under the following hypotheses.

Theorem: "Given the Hamiltonian $H(y;x)$ real analytic in the vicinity of the equilibrium point $(0;0)$, if all the eigenvalues of the linear variational system are distinct, if one of them, say λ_1, is pure imaginary and if there exists one other eigenvalue, say λ_2, such that $\lambda_2 = k\lambda_1$ ($k \geq 4$, integer), then there exists a family of real periodic solutions depending analytically on a real parameter ϵ in the neighborhood of the equilibrium point, provided the number $R(k) \neq 0$. This number is a function of k and of coefficients of the development of H in the neighborhood of $(0;0)$, up to the fourth order. If $\epsilon \to 0$, those solutions tend to the equilibrium and their periods, analytic functions of ϵ, tend to $2\pi/|\lambda_1|$. The first approximation of the solution is the same as given by Lyapunov's method in the nonresonant case". Of course, the list of results in many particular situations could be extended to an endless limit. We wish to mention the following outstanding works: by Hale (1969, mostly sections III.5 through III.11 and Chapters IV, V), by Sansone and Conti (1964, especially sections VII.1 and VIII.7), by Cesari (1959, §8 and 9), by Andronov and others (1966, sections I.2 through I.5, Chapters II and VI), by Andronov and others (1967, from the important point of view of structural stability, Chapters II, III, IV), by Nemytskii and Stepanov (1960, Chapters IV,V), by LaSalle and Lefschetz (1961, Chapter 4, from the point of view of an extended Lyapunov Stability Method), by several authors in the Proceedings of an International Symposium on Differential Equations and Dynamical Systems (Ed. by Hale and LaSalle, 1967), by Urabe (1967), Chapters 4 and 5), by Roseau (1966, Chapters 9,10,12,13,18 and several examples along the entire book), by several authors in an AMS Translation (1962), in several examples of vibrations in nonlinear mechanics by Chen in 1966.

One of the best works dealing with Oscillations in Electric Circuit Analysis, where the problem of resonance is analysed under several circumstances and with several examples, by Blaquière in 1966.

Finally, many general situations and special cases in Celestial Mechanics are collected in the Proceedings of an International Symposium on Periodic Orbits, Stability and Resonances (Ed. by Giacaglia, 1970).

It is often stated that, as in fact is "mathematically generally true", asymptotic methods cannot predict accurately orbits for a period of time which does not exceed $\sim 1/\epsilon$, where ϵ is the small parameter of the problem. Nevertheless, both in nonresonant and resonant situations, such methods, mostly based on Lindstedt's Device (Poincaré or von Zeipel Methods) have proved to be able to predict very accurately the observations over a very extended period of time. For instance, long period perturbations computed for earth's artificial satellites should hold good for a time not to exceed about 2×10^3 hours (about 84 days) and even then degenerate very rapidly at the end of such period. However, a precision of better than one part in a million can be maintained for periods of well over 100 days, by proper use of averaging techniques. The same is true also for the verification of validity of normal forms obtained by "generally divergent" series. In this respect, an observation by Moser (1956, p. 291) is quite interesting, as he says "... As in the case of two degrees of freedom such a transformation (Birkhoff's Normalization) can only be given by divergent series. But in the discussion of stability for practical purposes these series give a sufficient picture ...". The same kind of observation is made by Kyner (1968, p. 41) when he states "... the (pendulum) model is valid over a time proportional to $1/\sqrt{J_{22}}$ (J_{22} is the small parameter in his problem), e.g. over a libration period. Statements of this type must be accompanied by the phrase-if J_{22} is sufficiently small. However, numerical tests show that the theory can be used for synchronous satellites of the earth...". We wish to add for a time well in excess of the mathematical interval of validity obtained by estimation.

With respect to the representation of motion in the vicinity of singular points, by means of asymptotic averaging methods, several works, beginning with Hori's theory in 1960, have been published by Garfinkel (see 1972 for a list of references), by Giacaglia (1968, 1969, 1969 bis, 1970) and Jupp (1969, 1972).

Finally, from the important point of view of construction and validity of the Third Integral of motion, several works by Contopoulos (1963, 1965, 1966, 1968, 1970) are highly recommended since, beside careful analytic developments, they always include extensive and impressive numerical applications and verifications, a very effective process of excluding mere academic exercises, and convince the reader of the real usefulness of averaging perturbation techniques.

REFERENCES

1. American Mathematical Society Translation (1962), Series 1, vol. 5, "Stability and Dynamical Systems", Providence, R.I. (Contributors: Malkin, Almuhamedov, Bautin, Nemytskii, Elsgolz).

2. Andronov, A. A. et al, 1966, "Theory of Oscillations", Pergamon Press, New York.

3. ____________, 1967, "Theory of Bifurcations of Dynamical Systems on a Plane" (in Russian) Izdatelstvo "Nauka" Glaunaya Redaktsiya, Fiziko-Matematicheskoi Literatury, Moscow.

4. Arnol'd, V. I., 1961, "On the Generation of a Quasi-Periodic Motion from a Set of Periodic Motions", Dokl. Akad. Nauk USSR 138, 13-15.

5. ________, 1963, "Small Denominators and Problems of Stability of Motion in Classical and Celestial Mechanics", Uspeki Mat. Nauk 18, 91-192 (Russ. Math. Surv. 18, 85-191).

6. Bakai, A. S., 1966, "Resonance Phenomena in Nonlinear Systems", Diff. Eq. 2, 479-491.

7. Baker, H. F., 1916, "On Certain Linear Differential Equations of Astronomical Interest", Phil. Trans. Roy. Soc. London (A) 216, 129-186.

8. Berger, M., 1969, "On One Parameter Families of Real Solutions of Nonlinear Operator Equations", Bull. Amer. Math. Soc. 75, 456-459.

9. Bhatia, N. P. and Szegö, G. P., 1970, "Stability Theory of Dynamical Systems", Springer-Verlag, New York, Berlin.

10. Blaquière, A., 1966, "Nonlinear System Analysis", Acad. Press, New York.

11. Bogoliubov, N. N., 1945, "On Some Statistical Methods in Mathematical Physics", Published by Izd. Akad. Nauk USSR, Moscow.

12. __________ and Mitropolskii, Y. A., 1951, "Asymptotic Methods in the Theory of Nonlinear Oscillations", Gordon and Breach, New York.

13. Bohlin, K., 1887, "Uber die Bedeutung der Prinzips der lebendigen Kraft für die Frage von der Stabilität dynamischer Systeme", Acta Math. 10, 115-138.

14. Brown, E. W., 1932, "Elements of Theory of Resonance", Rice Inst. Publ. 19.

15. Cesari, L., 1940, "Sulla Stabilità delle Soluzioni dei Sistemi di Equazioni Differenziali Lineari a Coefficienti Periodici", Atti Accad. Ital. Mem. Classe Fis. Mat. e Nat. 11, 633-692.

16. ______, 1959, "Asymptotic Behavior and Stability Problems in Ordinary Differential Equations", Springer-Verlag, Berlin (2nd Ed. Acad. Press, N. Y., 1963).

17. Chen, Y., 1966, "Vibrations: Theoretical Methods", Addison-Wesley, Reading, Massachusetts.

18. Cole, J. D., 1968, "Perturbation Methods in Applied Mathematics", Ginn-Blaisdell, Waltham, Massachusetts.

19. Contopoulos, G., 1960, "A Third Integral of Motion in a Galaxy", Zeits. für Astrophys. 49, 273-291.

20. ______________, 1962, "On the Existence of a Third Integral of Motion", Astron. J. 68, 1-14.

21. ______________, 1963, "Resonance Cases and Small Divisors in a Third Integral of Motion. I", Astron. J. 68, 763-779.

22. ______________, 1965, "Resonance Cases and Small Divisors in a Third Integral of Motion. II", Astron. J. 70, 817-835.

23. ______________, 1966, "Resonance Cases and Small Divisors in a Third Integral of Motion. III", Astron. J. 71, 687-698.

24. ______________, 1968, "Resonant Periodic Orbits", Astrophys. J. 153, 83-94.

25. ______________, 1970, "Resonance Phenomena in Spiral Galaxies" in "Periodic Orbits, Stability and Resonances" (G. E. O. Giacaglia, Ed.), D. Reidel Pub. Co., Dordrecht, Holland.

26. Danby, J. M. A., 1970, "Wild Dynamical Systems" in "Periodic Orbits, Stability and Resonances" (ibidem).

27. Diliberto, S. P., 1960-1961, "Perturbation Theorems for Periodic Surfaces. I-II", Rend. Circ. Mat. Palermo 9, 265-299 and 10, 111-200.

28. Elsgolts, L. E., 1952, "An Estimate for the Number of Singular Points of a Dynamical System Defined on a Manifold", Am. Math. Soc. Transl. No. 68, Providence, R. I.

29. Garfinkel, B. et al, 1971, "A Recursive von Zeipel Algorithm for the Ideal Resonance Problem", Astron. J. 76, 157-166.

30. ________________, 1972, "Regularization of the Ideal Resonance Problem", Cel. Mech. 5, 189-203.

31. Gelfand, I. M. and Lidskii, V. B., 1958, "On the Structure of the Regions of Stability of Linear Canonical Systems of Differential Equations with Periodic Coefficients", Am. Math. Soc. Transl. (2) 8, 143-182.

32. Giacaglia, G. E. O., 1968, "Double Resonance in the Motion of a Satellite", Symp. Math. 3, 45-63 (Publ. Acad. Press, New York, 1970).

33. Giacaglia, G. E. O., 1969, "Resonance in the Restricted Problem of Three Bodies", Astron. J. 74, 1254-1260.

34. ________________, 1969, "Parametric Representation of Resonance in the Restricted Problem", Mem. Soc. Astron. Ital. 40, 499-515.

35. ________________, 1970 (Ed.), "Periodic Orbits, Stability and Resonances", D. Reidel Publ. Co., Dordrecht, Holland.

36. ________________, 1970, "Two Centers of Libration" in "Periodic Orbits, Stability and Resonances" ibidem.

37. ________________, 1971, "Characteristic Exponents at L_4 and L_5 in the Elliptic Restricted Problem of Three Bodies" Cel. Mech. 4, 468-489.

38. Gustavson, F. A., 1966, "On Constructing Formal Integrals of a Hamiltonian System Near an Equilibrium Point", Astron. J. 71, 670-686.

39. Hagedorn, P., 1971, "The Inversion of the Stability Theorems of Lagrange and Routh" (to appear in Arch. Rat. Mech. Anal.).

40. Hale, J. K., 1954, "On the Boundedness of the Solutions of Linear Differential Systems with Periodic Coefficients", Riv. Mat. Univ. Parma 5, 137-167.

41. __________, 1969, "Ordinary Differential Equations", Wiley-Interscience, N. Y.

42. __________ and LaSalle, J. P. (Eds.), 1967, "Differential Equations and Dynamical Systems", Acad. Press, New York.

43. Henon, M. and Heiles, C., 1964, "The Applicability of the Third Integral of Motion; Some Numerical Experiments", Astron. J. 69, 73-79.

44. Henrard, J., 1969, "Periodic Orbits Emanating from a Resonant Equilibrium", Boeing Sci. Res. Lab. Publ. D1-82-0874.

45. Hori, G., 1960, "The Motion of an Artificial Satellite in the Vicinity of the Critical Inclination", Astron. J. 65, 291-303.

46. _______, 1967, "Non-linear Coupling of Two Harmonic Oscillations", Publ. Astron. Soc. Japan 19, 230-241.

47. Jupp, A. H., 1969, "On the Ideal Resonance Problem", Astron. J. 74, 35-43.

48. __________, 1970, "On the Ideal Resonance Problem. II", Mon. Not. Roy. Astron. Soc. 148, 197-210.

49. __________, 1972, "A Second Order Solution of the Ideal Resonance Problem by Lie Series", Cel. Mech. 5, 8-26.

50. Kevorkian, J., 1965, "The Two Variable Expansion Procedure for the Approximate Solution of Certain Nonlinear Differential Equations", Yale Summer Institute in Dynamical Astronomy (in "Lectures in Applied Mathematics" vol 7, p. 206,

Am. Math. Soc., Providence, Rhode Island).

51. Krylov, N. and Bogoliubov, N. N., 1934, "The Application of Methods of Nonlinear Mechanics to the Theory of Stationary Oscillations", Publ. Ukrai. Acad. Sc. No. 8, Kiev.

52. Kyner, W. T., 1968, "Lectures on Nonlinear Resonance", Yale Summer Institute in Dynamical Astronomy, Purdue Univ. (42 pp.).

53. LaSalle, J. P. and Lefschetz, S., 1961, "Stability by Lyapunov's Direct Method", Acad. Press, New York.

54. Lefschetz, S., 1967, "Geometric Differential Equations" in "Differential Equations and Dynamical Systems" (J. K. Hale and J. P. LaSalle, Ed.) Acad. Press, New York.

55. Lyapunov, A. M., 1966, "Stability of Motion", Academic Press, New York.

56. Malkin, I. G., 1952, "Theory of Stability of Motion", Izdat. Gos., Moscow-Leningrad.

57. __________, 1956, "Some Problems of the Theory of Nonlinear Oscillations", Gostekhisdat, Moscow.

58. Mitropolski, J. A., 1964, "On the Construction of Quasi-periodic Solutions of Nonlinear Differential Equations by a Method Ensuring Rapid Convergence", Ukrain. Mat. J. 16, 475-501.

59. Moser, J., 1956, "The Resonance Lines for the Synchrotron", Proceed. CERN Symp., Geneva (pp. 290-292).

60. ________, 1958, "New Aspects in the Theory of Stability of Hamiltonian Systems", Comm. Pure Appl. Math. 2, 81-114.

61. ________, 1966, "On the Theory of Quasi-periodic Motions", SIAM Rev. 8, 145-172.

62. ________, 1967, "Convergent Series Expansions of Quasi-periodic Motions", Math. Ann. 169, 136-176.

63. ________, 1968, "Lectures on Hamiltonian Systems", Mem. Amer. Math. Soc. 81, 1-60.

64. Nemitskii, V. V. and Stepanov, V. V., 1960, "Qualitative Theory of Differential Equations", Princeton University Press, Princeton.

65. Nemitskii, V. V., 1965, "Some Modern Problems in the Qualitative Theory of Ordinary Differential Equations", Russ. Math. Surveys, 1-34.

66. ______________, 1967, "Topological Classification of Singular Points", Diff. Eq. 3, 359-370.

67. Peixoto, M. M., 1967, "Qualitative Theory of Differential Equations and Structural Stability" in "Differential Equations and Dynamical Systems" (J. K. Hale and J. P. LaSalle, Ed.), Academic Press, New York.

68. Poincaré, H., 1898, "Les Méthodes Nouvelles de la Mécanique Céleste" (vol. 2) Gauthier-Villars, Paris (Dover Reprint, New York, 1957).

69. Roels, J., 1969, "Orbites de Longues periodes resonantes autour des points equilateraux de Lagrange", Astron. Astrophys. 1, 380-387.

70. ________, 1971, "An Extension to Resonant Cases of Lyapunov's Theorem Concerning the Periodic Solutions Near a Hamiltonian Equilibrium", J. Diff. Eq. 9, 300-324.

71. Roseau, M., 1966, "Vibrations non linéaires et théorie de la stabilité" Springer-Verlag, Berlin, New York.

72. Sansone, G. and Conti, R., 1964, "Nonlinear Differential Equations", Pergamon Press, New York.

73. Siegel, C. L., 1956, "Varlesüngen über Himmelsmechanik", Springer-Verlag, Berlin.

74. Urabe, M., 1967, "Nonlinear Autonomous Oscillations", Academic Press, New York.

75. Whittaker, E. T., 1937, "A Treatise on the Analytical Dynamics of Particles and Rigid Bodies", Cambridge University Press, London.

76. Zeipel, H. von, 1916-1917, "Récherches sur le Mouvement des Petites Planèts", Arkiv. Astron. Mat. Fys. 11, 12, 13.

APPENDIX

REMARKS, SOME OPEN QUESTIONS AND RESEARCH TOPICS

Averaging Methods have been in use now for some time and we have accumulated a great deal both in experience and insight. They have given surprisingly good results, in general much better than expected from mathematical estimates. No matter how good an estimate may be, it is obvious that for every particular problem, a better estimate may be possible. Another point to consider is that, especially in problems where the angular frequencies can be obtained directly from observations, a much better result is obtained where such frequencies (numbers) are used in the analytical theory. The periodic oscillations around mean values, obtained by analytical developments, show remarkable agreement. This is probably one of the main reasons of success of the Hill-Brown Lunar Theory as opposed to that of Delaunay. Considerations of this sort show that the old question remains open and no general "modern" technique is available for the computation of the actual frequencies of a nonlinear system. For applications, this problem is still unsolved, since series approximations, convergent or only formal, can be computed with a finite, and generally very small, number of terms. Unless one can find the general term and sum the series. Only for trivial examples this could be achieved in the past. Thus, we look for a nonlinear method of solution. We may add that accelerated methods of convergence of the Newton type, as introduced by Kolmogorov, are a real progress toward that goal. But even then, it is a known fact that Newton's Method does not always give quadratic convergence: some conditions have to be satisfied. Besides, until now, only numerical applications of the method have been developed. In numerical analysis there also exist methods of solution of differential equations which eventually give better than quadratic convergence. They are known generally as Aitken's Methods. For example, having reduced the integration to a fixed point problem, if x_{n-1}, x_n, x_{n+1} are three successive approximations to the point, then an excellent estimate of the fixed point (or root) is

$$x_{n+2} = x_{n+1} - (x_{n+1} - x_n)^2/(x_{n+1} - 2x_n + x_{n-1}).$$

This relation estimates exactly the geometric series, in the sense that if one takes $x_{n-1} = 1$, $x_n = 1 + \epsilon$, $x_{n+1} = 1 + \epsilon + \epsilon^2$, it follows that $x_{n+2} = 1/(1-\epsilon)$. This and other interesting methods are discussed by Feagin (1972). How can one make use of such accelerated methods in an analytic theory, is also an open question.

There are other methods beside those using averaging techniques but, as far as actual applications to the solution of problems by power series in a (small) parameter, they did not become as popular as these. Averaging techniques are simple, easily visualized and understood, and, most important, systematic. That is, easily transformed into automatic methods of solutions, by iteration, in an electronic computer, with an algebraic symbol manipulator. Computer processors especially designed for averaging techniques have been built and results published by Deprit, Rom (1970-1971) and Jefferys (1970). Also J. Cherniak has developed a FORMAC based language, to handle equivalent problems, at the Smithsonian Astrophysical Observatory in Cambridge, Massachusetts.

In his long paper in 1963, Arnol'd analyses a certain number of unsolved problems, and, to our knowledge, most of them remain unsolved today. Although much progress has been made in the qualitative understanding of Dynamical Systems, still very little is known for two-dimensional systems and much less for higher dimensional cases. The first question raised by Arnol'd is that of stability of an elliptic equilibrium point in systems with more than two degrees of freedom. Of the same difficulty would be a proper generalization of Poincaré-Birkhoff Fixed Point Theorem to higher dimensions. As another matter, Arnol'd deals with metric stability of Dynamical Systems, which is related to the fact that the majority of changes in the initial conditions maintains the quasi-periodicity character of solutions. On the other hand, since the gaps between the invariant tori, in dimensions greater than 2, are connected and unbounded, a trajectory departing from in between two invariant tori does not necessarily stay close to either or is bounded: this, he calls the topological instability. The question is whether such a situation is representative of Dynamical Systems (Hamiltonian) with more than two degrees of freedom. Arnol'd mentions the fact that no rigorous proof exists for the existence of Poincaré's zones of instability (wild motions) in the neighborhood of hyperbolic points. We wish to indicate that, since then, Danby (1970) has shown the actual existence of such zones in a particular dynamical system. Although the work is numerical, it clears any doubts that such zones might not exist. Another problem is that of large perturbations. Since the original question on the existence of invariant tori for strongly perturbed systems was proposed, not much progress has been made, although from a semianalytical point of view Contopoulos (1970, 1971) has published quite interesting results, mostly related to the existence and eventual disintegration of a third integral of motion.

Next in both importance and difficulty is the proper generalization of Floquet-Lyapunov Theory for quasi-periodic systems. If y is an n-vector $(y_1, y_2, \ldots, y_n)$ and if $\dot{y} = Y(y)$ possesses a periodic solution $y = \bar{y}(\omega t)$ with period $T = 2\pi/\omega$ and $Y(\bar{y}) \neq 0$ for all t, one has the associated variational systems

$$\delta\dot{y} = \frac{\partial Y}{\partial y}(\bar{y}(\omega t))\delta y.$$

The Floquet-Lyapunov theory stipulates the existence of a system of normal coordinates in which the above periodic linear system is reduced to a linear one with constant coefficients. In other words, introducing an angle variable θ and an (n-1)-dimensional vector x, there exists a transformation

$$y = \bar{y}(\theta) + F(\theta)x,$$

where F is an $(n-1) \times n$ matrix, 2π-periodic, such that

(a) $\dot{y} = Y(y)$ is transformed into $\dot{\theta} = \omega + f(x,\theta)$, $\dot{x} = g(x,\theta)$ with $f(0,\theta) = g(0,\theta) = 0$.

(b) $\theta = \omega t + \text{const.}$, $x = 0$, is the periodic solution $y = \bar{y}(\omega t)$.

(c) $\frac{\partial g}{\partial x}(0,\theta) = \Omega$, is an $(n-1) \times (n-1)$ matrix, indendent of θ, so that the variational system takes the form $\dot{\theta} = \omega$, $\delta\dot{x} = \Omega\delta x$. The eigenvalues of Ω are

Floquet's characteristic exponents.

The question is how to generalize the above to the case when a system $\dot{y} = Y(y)$ has a quasi-periodic solution $y = \overline{y}(\omega_1 t, \omega_2 t, \omega_3 t, \ldots, \omega_m t)$, and y is an n-vector, $m \gtrless n$. More specifically we ask if there is a system of normal coordinates $\theta = (\theta_1, \theta_2, \ldots, \theta_m)$, $x = (x_1, x_2, \ldots, x_p)$ such that the equation $\dot{y} = Y(y)$ is reduced to the form

$$\dot{\theta} = F(\theta, x)$$
$$\dot{x} = G(\theta, x)$$

with F, G 2π-periodic in each θ_k and $F(\theta, 0) = \omega = (\omega_1, \omega_2, \ldots, \omega_m) = \text{const.}$ and $G(\theta, 0) = 0$. The quasi-periodic solution $\overline{y}(\omega, t)$ is transformed into $\theta_k = \omega_k t +$ const., $x = 0$, and the variational equations to

$$\dot{\theta} = F(\theta, x)$$
$$\delta\dot{x} = \frac{\partial G}{\partial x}(\theta, 0)\,\delta x$$

with $\frac{\partial G}{\partial x}(\theta, 0) = \Omega =$ constant $p \times p$ matrix. The eigenvalues of Ω would, in this case, be the generalized Floquet Exponents of the quasi-periodic solution $\overline{y}(\omega t)$. On the other hand, given a system with the above properties, that is, a priori, in normal coordinates, the problem is equivalent to define an invariant manifold $x = 0$. Such a manifold is obviously a torus of dimension m in a $p + m$ dimensional space, with angular coordinates $\theta_1, \theta_2, \ldots, \theta_m$. The eigenvalues $(\Omega_1, \Omega_2, \ldots, \Omega_p)$ of Ω and the vector components $(\omega_1, \omega_2, \ldots, \omega_m)$ of ω constitute the characteristic numbers of Moser (1967) and the ω's are, by hypothesis, rationally independent.

If such generalization exists, that would prove the reducibility of the variational system along the quasi-periodic solution. For $n = 1$ and $m \geq 2$ (quasi-periodic solution of a one-dimensional system), reducibility has been shown under certain conditions, but no other cases $(n > 1)$ are known. See Gel'man (1957) and Adrianova (1962) as cited by Arnol'd.

Also a problem of great importance, mentioned several times before, is the study of the totality of motions in the neighborhood of an equilibrium point. For non-canonical systems, reduction to normal form has been shown by Siegel (1952), but for canonical systems the necessary hypotheses cannot be satisfied, as discussed earlier. The generalized case of the study of totality of motions in the neighborhood of a periodic solution is also an open question for canonical systems. This is discussed by Siegel (1956) but left unsolved. The most general form of the present problem is the formidable task of the study of totality of motions in the neighborhood of a quasi-periodic solution. An important result on the subject was obtained by Belaga in 1962. The problem, as presented by Moser for the conservation of quasi-periodic solutions, in 1967, was discussed earlier. His results are of the same type of those obtained by Belaga, whose main theorem we give below for the benefit of the reader.

"Consider the system

$$\dot{x} = \Lambda x + f(x,y), \quad x = \text{n-vector},$$
$$\dot{y} = \omega + g(x,y), \quad y = \text{m-vector},$$

where

$$\Lambda = \text{diag}(\lambda_1, \ldots, \lambda_n)$$

and

$$f = 0(x^2), \quad g = 0(x)$$

are 2π-periodic in $y = (y_1, y_2, \ldots, y_m)$. Consider the infinitely many conditions $|(k_1\lambda_1 + \cdots + k_n\lambda_n) - \epsilon\lambda_j + \sqrt{-1}\,(\ell_1\omega_1 + \cdots + \ell_m\omega_m)| \geq K(|k_1| + \cdots + |k_n| + |\ell_1| + \cdots + |\ell_m|)^{-(m+n+1)}$ for a certain $K > 0$; $\epsilon = 0,1$; $j = 1,2,\ldots,n$ and all integers $k_1, k_2, \ldots, k_n$, $\ell_1, \ell_2, \ldots, \ell_m$, and $|k_1| + |k_2| + \cdots + |k_n| > 1 + \epsilon$. Then there exists an analytic transformation

$$X = x + \varphi(x,y)$$
$$Y = y + \psi(x,y)$$

reducing the given system to

$$\dot{X} = \Lambda X$$
$$\dot{Y} = \omega.$$

The functions φ, ψ are 2π-periodic in $y_1, y_2, \ldots, y_m$ and $\varphi = O(x^2)$, $\psi = O(x)$."

Such a theorem, however, is not applicable to canonical systems, in the sense that those which satisfy the irrationality condition above for ω, λ form a set of zero measure in the space of (ω, λ). Moser's results only show the existence and a convergent method of construction of quasi-periodic solutions. The totality is not known.

Obviously, all these questions can be generalized to Dynamical Systems with more than one independent variable, or for Functional Differential Equations. For this last subject see Hale (1971). Another problem of great interest is the question of a better understanding of the solution of systems away, toward or through conditions of resonance. When do we actually have a lock-in process and what is the preferred state of resonance of a system? For strongly perturbed systems it is a completely open question. The presence of dissipative forces is here of high importance, but might not be the decision maker in the choice of a particular resonant stable situation. An interesting example in Celestial Mechanics has been recently studied by Colombo (1965) and Kyner (1970). In fact, the occurrence of resonance among planets and satellites, both in orbital and spin motion, has been quite an issue for controversy. See, for example, Gingerich (1969), Hénon (1969), Molchanov (1969) and Roy and Ovenden (1954, 1955).

There are also several questions dealing with the study of dynamical systems in a Lie Algebra. This may or may not have connection with perturbation techniques using Lie Transforms although it is well-known that the motions described by a Hamiltonian H form the Lie Group corresponding to the Lie Algebra defined by the Poisson Parentheses $(F_i, F_j) = 0$, where F_i $(i = 1,2,\ldots,n)$ are integrals of the system generated by H, when they exist. Moser, in 1967, gives a clear picture on how to treat problems of perturbations of quasi-periodic motions from that point of view, using a Lie Algebra (of operators). By the same approach, it might be possible to describe in more precise terms the perturbation techniques introduced by Hori in 1966 and discussed in Chapter II. Essentially, the problem can be viewed as the construction of an operator algebra and the definition of its range and null-space. In the problem of formal reduction to a normal form, such technique was used by Gustavson in 1966. We are not aware of other applications, except the study of rigid motions described by Leimanis in 1965. In the field of Quantum Mechanics, however, this is quite common (Algebra of commutators). Bridges for other aspects have been established toward Celestial Mechanics by Kustaanheims and Nuotio (1965, 1966, 1970), in various papers (Algebra of Spinors). Unfortunately, these are not easily accessible to people working in Astronomy, Engineering and Physics, since they require a solid knowledge of Algebra.

The list of open questions and research topics is obviously endless and one can always think of generalizations. The difficult point is to see what generalizations might have an important role in Applied Sciences.

REFERENCES

1. Adrianova, L., Ya., 1962, "The Reducibility of Systems of n Linear Differential Equations with Quasi-periodic Coefficients", Vest. Leningrad Univ. 17, No. 7, 14-24.

2. Arnol'd, V. I., 1963, "Small Denominators and Problems of Stability of Motion in Classical and Celestial Mechanics", Russian Math. Surveys, 18, No. 6, 85-191.

3. Belaga, E. G., 1962, "The Reducibility of a System of Ordinary Differential Equations in the Neighborhood of a Quasi-periodic Motion", Dokl. Akad. Nauk USSR 143, 255-258.

4. Colombo, G., 1965, "Rotational Period of the Planet Mercury" (Letter to Editor) Nature, 208, 575.

5. Contopoulos, G., 1970; 1971, "Orbits in Highly Perturbed Dynamical Systems", I(1970); II(1970); III(1971), Astron. J. 75, 96-107; 75, 108-130; 76, 147-156.

6. Danby, J. M. A., 1970, "Wild Dynamical Systems" in "Periodic Orbits, Stability and Resonances" (G. E. O. Giacaglia, Ed.) D. Reidel Publ. Co., Dordrecht, Holland.

7. Feagin, T. W., 1972, "Numerical Solution of Two-Point Boundary Value Problems Using Chebyshev Series", Ph.D. Thesis, College of Engineering, Univ. of Texas at Austin.

8. Gel'man, A. E., 1957, "The Reducibility of a Class of Systems of Differential Equations with Quasi-periodic Coefficients", Dokl Akad. Nauk USSR 116, 535-538.

9. Gingerich, O., 1969, "Kepler and the Resonant Structure of the Solar System", Icarus 11, 111-113.

10. Gustavson, F. A., 1966, "On Constructing Formal Integrals of a Hamiltonian System Near an Equilibrium Point", Astron. J. 71, 670-686.

11. Hale, J. K., 1971, "Functional Differential Equations", Springer-Verlag, New York, Berlin.

12. Hènon, M., 1969, "A Comment on 'The Resonant Structure of the Solar System' by A. M. Molchanov" Icarus 11, 93-94.

13. Hori, G., 1966, "Theory of General Perturbations with Unspecified Canonical Variables", Publ. Astron. Soc. Japan 18, 287-296.

14. Jefferys, W. H., 1970, "A Fortran-Based List Processor for Poisson Series", Cel. Mech. 2, 474-477.

15. Kustaanheimo, P. and Nuotio, V. S., 1965, "Spinor Algebra. I". Preprint No. 1, Dept. of Appl. Math., Univ. of Helsinki.

16. ________________, 1966, "Lectures on Celestial Mechanics". Preprint No. 2, Dept. of Appl. Math., Univ. of Helsinki.

17. Kyner, W. T., 1970, "Passage Through Resonance" in "Periodic Orbits, Stability and Resonances" (G. E. O. Giacaglia, Ed.) D. Reidel Publ. Co., Dordrecht, Holland.

18. Leimanis, E., 1965, "The General Problem of Motion of Coupled Rigid Bodies about a Fixed Point", Springer-Verlag, New York, Berlin.

19. Moser, J., 1967, "Convergent Series Expansions of Quasi-Periodic Motions", Math. Ann. 169, 136-176.

20. Mulchanov, A. M., 1969, "Resonances in Complex Systems: A Reply to Critiques", Icarus 11, 95-103.

21. ________, 1969, "The Reality of Resonances in the Solar System", Icarus 11, 104-110.

22. Nuotio, V. S., 1970, "Generalized Motor Electrodynamics", Ann. Acad. Sci. Fennicae, Series A, VI. Phys. 351, 1-37.

23. Rom, A., 1970, "Mechanized Algebraic Operations", Cel. Mech. 1, 301-319.

24. ____, 1971, "Echeloned Series Processor", Cel. Mech. 1, 331-345.

25. Roy, A. and Ovenden, M., 1954-1955, "On the Occurrence of Commensurable Mean Motions in the Solar System. I; II", Mont. Not. Roy. Astron. Soc. 114, 232; 115, 296.

26. Siegel, C. L., 1952, "Über die Normalform Analytischer Differentialgleichungen in der Nähe einer Gleichgewichtslösung", Nach Akad. Wiss. Göttingen, Math. Phys. Kl. IIa, 21-30.

27. ________, 1956, "Vorlesungen über Himmelsmechanik", Springer-Verlag, Berlin.

INDEX